Collaboration in Creative Design

Panos Markopoulos • Jean-Bernard Martens
Julian Malins • Karin Coninx • Aggelos Liapis
Editors

Collaboration in Creative Design

Methods and Tools

 Springer

Editors
Panos Markopoulos
Department of Industrial Design
Eindhoven University of Technology
Eindhoven, The Netherlands

Jean-Bernard Martens
Department of Industrial Design
Eindhoven University of Technology
Eindhoven, The Netherlands

Julian Malins
Norwich University of the Arts
Norwich, UK

Karin Coninx
Hasselt University – tUL – iMinds
Expertise Centre for Digital Media
Diepenbeek, Belgium

Aggelos Liapis
Intrasoft International
Markopoulou-Peania Avenue
Athens, Greece

ISBN 978-3-319-29153-6   ISBN 978-3-319-29155-0   (eBook)
DOI 10.1007/978-3-319-29155-0

Library of Congress Control Number: 2016940098

Printed on acid-free paper

This Springer imprint is published by Springer Nature
The registered company is Springer International Publishing AG Switzerland

Contents

Part III   Designing with Stories

**Part IV   Tools for Creativity and Collaboration in Early
          Design**

Contributors

Berke Atasoy Department of Industrial Design, Eindhoven University of Technology, Eindhoven, The Netherlands

Tilde Bekker Department of Industrial Design, Eindhoven University of Technology, Eindhoven, The Netherlands

Paul Bermudez Centre for Human-Computer Interaction Design, City University London, London, UK

Carole Bouchard Laboratory of New Products Design and Innovation, Arts and Métiers ParisTech, Paris, France

Aarnout Brombacher Department of Industrial Design, Eindhoven University of Technology, Eindhoven, The Netherlands

Derya Özçelik Buskermolen Department of Industrial Design, Eindhoven University of Technology, Eindhoven, The Netherlands

Jacob Buur Mads Clausen Institute, University of Southern Denmark, Kolding, Denmark

Karin Coninx Hasselt University – tUL – iMinds, Expertise Centre for Digital Media, Diepenbeek, Belgium

Peter Dalsgaard CAVI, Aarhus University, Aarhus, Denmark

Gurjot Dhillon Industrial Design Department, Eindhoven University of Technology, Eindhoven, The Netherlands

Berry Eggen Department of Industrial Design, Eindhoven University of Technology, Eindhoven, The Netherlands

Joep (J.W.) Frens Department of Industrial Design, Designing Quality in Interaction Group, Eindhoven University of Technology, Eindhoven, The Netherlands

Mathias Funk Department of Industrial Design, Eindhoven University of Technology, Eindhoven, The Netherlands

Pelin Gultekin Department of Industrial Design, Eindhoven University of Technology, Eindhoven, The Netherlands

Mieke Haesen Hasselt University – tUL – iMinds, Expertise Centre for Digital Media, Diepenbeek, Belgium

Kim Halskov CAVI, Aarhus University, Aarhus, Denmark

Sara Jones Centre for Creativity in Professional Practice, Cass Business School, London, UK

Julia Kantorovitch VTT-Technical Research Center, Espoo, Finland

Vassilis Javed Khan Industrial Design Department, Eindhoven University of Technology, Eindhoven, The Netherlands

Jung-Joo Lee Division of Industrial Design, National University of Singapore, Singapore, Singapore

Aggelos Liapis Intrasoft International, Markopoulou-Peania Avenue, Athens, Greece

James Lockerbie Centre for Creativity in Professional Practice, City University, London, UK

Yuan Lu Department of Industrial Design, Eindhoven University of Technology, Eindhoven, The Netherlands

Andrés Lucero Mads Clausen Institute, University of Southern Denmark, Kolding, Denmark

Kris Luyten Hasselt University – tUL – iMinds, Expertise Centre for Digital Media, Diepenbeek, Belgium

Fiona Maciver Norwich University of the Arts, Norwich, UK

Neil Maiden Centre for Creativity in Professional Practice, City University, London, UK

Julian Malins Norwich University of the Arts, Norwich, UK

Panos Markopoulos Department of Industrial Design, Eindhoven University of Technology, Eindhoven, Noord-Brabant, The Netherlands

Jean-Bernard Martens Department of Industrial Design, Eindhoven University of Technology, Eindhoven, Noord-Brabant, The Netherlands

Tuuli Mattelmäki School of Arts, Design and Architecture, Aalto University, Helsinki, Finland

Jesús Muñoz-Alcántara Department of Industrial Design, Eindhoven University of Technology, Den Dolech, The Netherlands

Jean-François Omhover Laboratory of New Products Design and Innovation, Arts and Métiers ParisTech, Paris, France

Maarten Piso Industrial Design Department, Eindhoven University of Technology, Eindhoven, The Netherlands

Javier Quevedo-Fernández Department of Industrial Design, Eindhoven University of Technology, Eindhoven, Noord-Brabant, The Netherlands

Kimberly Schelle Industrial Design Department, Eindhoven University of Technology, Eindhoven, The Netherlands

Jacques Terken Department for Industrial Design, Eindhoven University of Technology, Eindhoven, The Netherlands

Davy Vanacken Hasselt University – tUL – iMinds, Expertise Centre for Digital Media, Diepenbeek, Belgium

Creativity and Collaboration in Early Design

Panos Markopoulos, Jean-Bernard Martens, Julian Malins, Karin Coninx, and Aggelos Liapis

Abstract Contemporary creative design practice draws from the fields of product design and user-centered design, as boundaries between these two traditions become blurred as design thinking is embraced by industry and academia. By early design we mean design activities taking place with the formulation of an initial design challenge all the way through to the articulation of a design concept and leveling off as designers shift their attention towards more detailed considerations of form, function and interaction, refining the design concept and making the transition to development work. Novel methods applied to the early stages of design help adopt a wider societal and business perspective transcending considerations of products or systems, helping to design for latent needs and emerging user experiences. Intense collaboration with stakeholders from different organizations throughout the design process is often required and design teams tend to be distributed across organizations and geographical locations, which reinforce the need for tools that can support collaboration during the early design process.

P. Markopoulos (✉) • J.-B. Martens
Department of Industrial Design, Eindhoven University of Technology, Eindhoven,
Noord-Brabant, The Netherlands
e-mail: P.Markopoulos@tue.nl; j.b.o.s.martens@tue.nl

J. Malins
Norwich University of the Arts, Francis House, 3-7 Redwell Street, NR2 4SN Norwich, UK
e-mail: j.malins@nua.ac.uk

K. Coninx
Hasselt University – tUL – iMinds, Expertise Centre for Digital Media, Wetenschapspark 2,
3590 Diepenbeek, Belgium
e-mail: karin.coninx@uhasselt.be

A. Liapis
Intrasoft International, Markopoulou-Peania Avenue, Athens, Greece
e-mail: aggelos.liapis@intrasoft-intl.com; agliapis@gmail.com

© Springer International Publishing Switzerland 2016
P. Markopoulos et al. (eds.), *Collaboration in Creative Design*,
DOI 10.1007/978-3-319-29155-0_1

1

Introduction

During the last two decades we have witnessed a convergence between the fields of interaction design and product design with regards to both the challenges they address as well as the methods they use to address them. Product designers are expected to design products with embedded electronics endowed with computation and communication capabilities, so interaction and dynamic behavior are as much their concern as form design. Conversely interaction designers schooled in fields such as computer science, cognitive ergonomics, communication, are concerned with tangible and embodied interaction, where physicality is key. Form, movement, interactivity, and materiality are essential aspects, as is the broader societal and business contexts in which the designed system will be used.

The corollary of such a convergence in the concerns, challenges, and solutions from product design and human-computer interaction is a sharing of methods and the emergence of new methods that borrow from disparate intellectual traditions. Typical concerns include product design methods that address idea generation, creative problem solving, communication, business relevance, manufacturing feasibility, and the realization of physical prototypes. On the other hand, user-centered design approaches provide guidance for involving users in the design process, prototyping dynamic aspects of interaction, ensuring usability and valuable user experiences. Design and engineering traditions meet and may even clash in the planning and execution of design projects, in the methodology literature and even in debates held at research venues.

These developments have given rise to an amalgam of design practices and methods with their roots in the different fields discussed above. User centeredness which is core to the field of human-computer interaction, has been reincarnated in modern approaches under the label of design thinking [Design Thinking], which advocate iterative design processes involving users, and their perspectives to develop new design concepts and iteratively improve them towards a final concept.

On the other hand traditional user-centered design approaches derived from the human computer interaction field tend to favor incremental innovation rather than radical breakthroughs (Norman and Verganti 2014). Modern user-centered interaction design adapts and appropriates methods established in the broader field of design to achieve radical innovation. Transcending considerations of user needs, modern user-centered design takes a holistic perspective that combines business, socio cultural, and aesthetic considerations, which motivates an ever-expanding collection of methods, processes, and tools to support early design.

There is already an abundance of methods that designers can apply to a design problem at hand. This book does not set out to provide a comprehensive review of such methods and design processes; rather it aims to focus on some almost universal characteristics of early design and discuss the nature of design methods that best suit this phase of the design process. Secondly, the book aims to identify some developments in the field and to discuss the role of specialized software tools that can assist designers during early design. The following chapter provides an overview of the book's contents.

Early Design

There are many different views on how a design process should be structured or staged as can be evidenced by the prolific number of models of design processes that have been discussed in the design literature (Dubberly 2004). Moreover every design project needs to address its own planning constraints and its own business and organizational context. For these two reasons, it is of limited use to put forward a generic and predetermined step-by-step specification for the design process that would prescribe designer's activities and would help plan methods and techniques to be applied at its different stages. While a design process may proceed in an orderly succession of activities, it is also characterized by iteration, serendipity, by a flexible meshing of activities that may appear opportunistic or arbitrary addressing different levels of abstraction and different concerns in an unplanned order.

Nevertheless, design processes share a temporal structure and are convergent, with design ideas crystallizing as the design process progresses. In the following chapter the term *early design* is used to describe a range of activities that designers carry out at the outset of a project when they first dive into a specific challenge and until a design concept is defined and represented in a sufficient form to trigger more detailed design considerations and product development activities. As design progresses, different possibilities are identified, different paths are explored and perhaps eliminated, and the understanding of the design challenge as such consolidates gradually until a design concept can be expressed clearly and communicated to stakeholders. This book is concerned with methods that can be applied by designers to enable this transition as apposed to the phases that follow where details are worked out and the path towards realization is laid.

Rather than review or classify them, the book focuses on how the trends described above shape novel methods and tools for early design.

- Design is iterative and the design process needs to be able to adapt to changes to the context of a project, the evolving understanding of a problem domain, and the needs of different stakeholders as they are discovered and shaped while designing. Even where an engineering oriented, rationalist approach to the design process is adopted with a distinct staging of the design process, often caricatured as a waterfall model, iteration is inherent either as a way to address changing information and requirements, or as a way to refine concepts and deliverables. Representations of design concepts should suit and serve this iterative approach, allowing the focus to remain on the essence of a concept, and allowing a speed of iteration that is commensurate with an advanced design process. Sketching design ideas and concepts becomes a key practice for communication of ideas as well as enabling the designer to externalize thoughts, experiment, wrap his/her mind around a design challenge and gain a better understanding of the design problem.
- Design transcends considerations of the physical or digital artifact to focus on how using a product, system or service is lived, experienced, and appraised by people. While the term user experience design has become associated with the

very narrow field of designing web systems and services, the term experience here is used broadly pertaining to how a design affects how life is lived, experienced, and remembered (Hassenzahl 2010). Focusing on the experience rather than the artifact/system/service provides a basis for envisioning, innovation, and even evaluation during the design process. As scholars in the field of user experience design have suggested, a core activity in describing, modifying and designing experiences (or better, designing for experience) is storytelling. Correspondingly there is an increasing interest in constructing and communicating stories as a means to support early design and creative collaboration with stakeholders.

- Design involves different modes of thinking, and especially a succession of divergent thinking which opens up options, invites, external influences, brings in new information, and convergent thinking which is more analytical, structured, reducing the options considered, and elaborating design representations. Design processes often involve a succession of these two models of thinking as is discussed in more detail in chapter "Two Heads Are Better Than One: Principles for Collaborative Design Practice", in relation to the Design Council's 'double diamond' process. Early design requires both these approaches and this is reflected in the design representations, the tools and methods that are applicable at this stage.

Discovery in Design

Designers do not work in isolation but rather they are embedded in a network of stakeholders, experts and potential users. The term 'open innovation' has been used to describe this transition from a team of designers working 'in house' for a single employer to teams of designers crossing organizational and geographical borders. Communication with diverse stakeholders is of paramount importance whether this pertains to understanding their needs, enabling them to explore solutions together, communicating design concepts, or evaluating them. As design teams often span different organizations, and have different expertise, culture and infrastructure, the methods and tools they use should account for this diversity and enable fluent collaboration. Design teams are often spread geographically as well, which increases their reliance on information and communication technologies,

- Fiona Maciver and Julian Malins in their chapter entitled 'Two heads are better than one', discuss collaboration in the design process and argue how being connected to others can enhance the overall creative effort.

Paying close attention to people's needs and behavior has always been at the heart of design and is key to the user-centered design approaches developed in the field of human computer interaction. Labels such as 'design research' and 'design ethnography' are often used to refer to approaches designers use to get to know the individuals they design for. Methods borrowed from social sciences, particularly cognitive science, ergonomics, and social anthropology, tend to bring about a lot of precision and with that some workload and rigor in the analysis of user behavior and needs, with a rich variation in the observation and survey methods used. For example, one might apply task analysis to get a detailed understanding of how people carry out tasks, what ergonomic and cognitive support they might need to carry them out better, what kind of errors plague the execution of their tasks [Dan Diaper]. Such methods have been developed and elaborated in the field of human computer interaction, but are more useful later in the design process where a direction has been given to the design and the scope of a design challenge is well understood. More suitable for early design are ethnographic approaches that help understand the context of use of the intended product, system or service. Ethnographically inspired approaches have been developed to help systems design, e.g., (Beyer and Holtzblatt 1997), and (Barab et al. 2004). These tend to be quite detailed in their consideration of context and behaviors and bring about rigor in their analysis.

Early design favors a broad-brush approach and techniques that provide a direction for a project. They can be imprecise, empathic and intuitive, but they are also vehicles for obtaining user insights and establishing a direction of a project. An approach that emphasizes the values of people, their aesthetics and latent emotional needs are probing techniques introduced to the design world by (Gaver et al. 1999, 2004). Probes are packaged materials and tasks that are sent to remote participants to collect data. Originally aimed to be deliberately ambiguous and value laden, they have been adapted to support ethnographic approaches and the information collection needs of designers.

- Tuuli Mattelmäki, Andrés Lucero and Jung-Joo Lee discuss probing. A method aimed at understanding users and inspiring design, as a process of collaborative discovery and learning, and as a tool for entering the users' contexts.
- Carole Bouchard and Jean-François Omhover introduce the Conjoint Trends Analysis method as a way to structure and operationalize information collection during the discovery or information phases of the design process, and discuss the requirements for new tools.

Generating Ideas and Concepts

Early design requires designers to be creative and able to challenge established ways of thinking, to identify opportunities and even to be able to reframe and re-conceptualize situations. Designers need to generate concepts that will deliver clear value to stakeholders; they should be able to communicate concepts, and collect suitable feedback but also communicate the value proposition of their design to different stakeholders. The design literature is rich with methods that help generate new ideas and concepts, or engage creatively with different groups. A relatively recent development is the popularity of design card decks summarizing and physicalizing design knowledge in a way that it can be applied fluently and playfully in the design process. Canvas models that emphasize value flows and the broader business context of a design process have also been gaining popularity.

- Andrés Lucero, Peter Dalsgaard, Kim Halskov and Jacob Buur discuss design cards, a low-tech, tangible, and approachable way to introduce information and sources of inspiration as part of the design process.
- Pelin Atasoy, Tilde Bekker and Yuan Lu introduce the Value Design Method. This is a canvas-based method that allows designers to gain an overview of the different expectations of different stakeholders, early in the design process.
- Vassilis-Javed Khan, Gujrot Dhillon, Maarten Piso and Kimberly Schelle discuss modern trends that rely on the collective intelligence and creativity of crowds accessed over the Internet to support early design activities.

Even in today's computer age, the most widespread tools for supporting creative thinking are pen and paper. Donald Schön (1983) was the first to provide an elaborate explanation of designing in terms of a reflective conversation with the materials that characterize a situation (such as the people, objects and places). He used sketching as a prototypical example of such a reflective conversation. In his view, designers start by shaping and presenting the situation according to their initial understanding. This initial representation "talks back" to them, in the sense that it helps them to better understand the situation and the possibilities offered. They respond by making changes, improving and correcting their ideas in successive steps. In this way, the problem definition and solution evolve simultaneously (Dorst and Cross 2001).

Design representations should allow designers to work "sketch-like" way, that is, without drawing much cognitive effort away from the design process itself. The increased interest in the design of interactive systems implies that design skills such as drawing, which are borrowed from the graphical arts, are no longer sufficient for all aspects of the design process. Therefore, new representations are proposed to complement more traditional ones, especially in order to clarify the fourth dimension in the design, i.e., time. Sketching, originally referring to fast drawing, has been extended now to refer to very lightweight, selective, changeable or even throw-away design representations (Buxton 2010) that allow designers to experiment, reflect and learn by designing.

- Frens provides a tutorial introduction to cardboard modeling seen as a low-tech approach to explore, experience and communicate interaction design concepts.

Designing with Stories

A creative product or idea should be both new and of sufficient quality, where there is some consensus (Dean et al. 2006) that quality should be characterized in three dimensions: specificity, feasibility and relevance of an idea or product. Note that these quality dimensions agree one-to-one with the "what", "how" and "why" questions that have been proposed for structuring the design of experiences

(Hassenzahl 2010). Especially arguing the relevance of a product cannot be accomplished without clarifying the context in which such a product can produce value to its users. Such values often do not show up instantly, but tend to evolve over time, so that designing in the fourth dimension (time) is becoming increasingly relevant (Buxton 2010). It has been argued independently by several authors that communicating and discussing the relevance of design ideas and concepts, especially in an environment with stakeholders from diverse disciplines, can most effectively be accomplished by exchanging stories about the envisioned use and potential value of products. The ability to capture, convey and assess details in narrative form is something that almost all humans share, and it is therefore no wonder that scenarios are already well established in the conceptual design of future systems and products. Scenarios and stories are not only relevant in conceptual design but often continue to be used throughout the entire process of design and development, sometimes even incorporating marketing and advertising concepts.

- Atasoy and Martens introduce Storiply, a method for structuring the design of user experiences by incorporating the craft of storytelling, as practiced in the movie industry.
- Haesen, Vanacken, Luyten and Coninx discuss storyboards as a means of communication in multidisciplinary design teams.
- Ozcelik and Terken in turn focus on how to involve the end user into the process of creating and assessing stories.

These approaches all aim to facilitate communication about tentative concepts in order to understand needs, attitudes, and values, well before the designer is able to articulate precisely the functionality and nature of such systems. The book discusses two methods that explicitly support the use of stories in design.

- Quevedo and Martens have conceived and implemented the tool idAnimate, which extends traditional sketching towards the sketching of animations, making explicit use of the multi touch interaction offered by interaction devices such as the iPad.
- Markopoulos discusses an alternative or complimentary approach where video prototypes are used as a way to represent design concepts within contexts or to illustrate interaction scenarios.

Tools Supporting Design

Schön's perspective on design practice discussed above emphasizes some essential characteristics of designers, that is, their ability to "know in action", "reflect in action" and "reflect on action". Such abilities become apparent in the activities that designers undertake:

- Collecting or creating inspirational materials
- Browsing through and selecting from the collected materials

- Organizing materials in a way that reflects a more coherent understanding
- Communicating the design vision and concrete design ideas
- Reflecting on the design problem and suggested solutions

Tools for supporting design should support all these activities. This section examines current practices regarding the use of collaboration technology in design, and the nature of tool support that is needed to support creativity and collaboration in early design.

- Bermudez and Jones review how tools and technologies currently used to support collaborative creativity and problem solving in design providing an overview of early stage activities and the role of tool support within them.
- Lockerbie and Maiden introduce Bright Sparks, a web-based software tool that supports the popular 'Hall of Fame' creativity technique.
- Malins and Maciver review 'Design Thinking' as an approach that can inform development processes contrasting incremental development and radical innovation, and discussing the role of end users therein.
- Funk discusses linking data and information to creative design, focusing on collaborative processes at early phases of the design with data.
- Liapis, Haesen,and Munoz present a survey of current practices and needs of designers regarding software support during early design and discuss how future software tools should support them.

Summary

Contemporary design faces new and changing challenges, which necessitates the development of new methods and tools to support designers. Not only the design activities themselves, but also the setting in which such activities take place are changing rapidly. Increasingly, design is taking place in multidisciplinary teams with team members that are only remotely in contact with each other most of the time (as they might actually be in different cities, countries or even continents and time zones). The chapters of this book capture some of the latest developments in terms of methods and tools that can assist designers with coping with such changing demands.

While many of the chapters provide a link to relevant theories, the primary objective of the book is to offer methods that are feasible in actual design practice. Inevitably, only some of the proposed tools and method have been validated extensively, while others are more speculative; nevertheless as editors we are convinced that the tools and methods being described in this book are representative of what "research for design" has been able to offer to design practitioners in recent years. We hope you enjoy getting acquainted with and trying out some of these recent design methods and that they assist you (and your team) when tackling the fuzzy process of creative designing taking place against a backdrop of increasing complexity and that they help you to produce successful design solutions.

Acknowledgements This work has been partially funded by the EC under the 7th Framework Programme, under grant agreement number FP7-ICT-2013-10 – 610725- COnCEPT COllaborative CrEative design PlaTform.

References

Barab SA, Thomas MK, Dodge T, Squire K, Newell M (2004) Critical design ethnography: designing for change. Anthropol Educ Q 35(2):254–268

Beyer H, Holtzblatt K (1997) Contextual design: defining customer-centered systems. Elsevier, Burlington

Buxton B (2010) Sketching user experiences: getting the design right and the right design. Morgan Kaufmann, Amsterdam

Dean DL, Hender JM, Rodgers TL, Santanen EL (2006) Identifying quality, novel, and creative ideas: constructs and scales for idea evaluation. J Assoc Inf Syst 7:30

Dorst K, Cross N (2001) Creativity in the design process: co-evolution of problem–solution. Des Stud 22:425–437

Dubberly H (2004) How do you design. Compend Models

Gaver B, Dunne T, Pacenti E (1999) Design: cultural probes. Interactions 6:21–29

Gaver WW, Boucher A, Pennington S, Walker B (2004) Cultural probes and the value of uncertainty. Interactions 11:53–56

Hassenzahl M (2010) Experience design: technology for all the right reasons. Synth Lect Hum-Centered Inform 3:1–95

Norman DA, Verganti R (2014) Incremental and radical innovation: design research vs. technology and meaning change. Des Issues 30:78–96

Schön DA (1983) The reflective practitioner: how professionals think in action, vol 5126. Basic books, New York

Part I
Discovery

Two Heads Are Better Than One: Principles for Collaborative Design Practice

Fiona Maciver and Julian Malins

Abstract Design is an inherently complicated activity, reliant on the input of many other disciplines, stakeholders, and users. Over recent years, product designers, clients, suppliers and customers have become even more close and connected, and working together has become paramount in the design process. This chapter looks at the notion of collaboration in design, and suggests that being connected to others can enhance the overall creative effort. As the prevalence of interdisciplinary teams and global work practices grows, this is relevant across all design disciplines. An interdisciplinary team approach provides benefits that can bring about innovation. However, teams are also idiosyncratic and serendipitous. Since design itself is equally unpredictable, there is a need to structure collaborative working. This chapter aims to provide creative practitioners and students with a set of methods by which a collaborative approach can be fostered and maintained in contemporary design practice.

Introduction: Organising for Uncertainty

Design and its concern with improving the future is marked by uncertainty. The outcome of a project rests squarely in the hands of the design team. At the outset of the process, the shape or form of a final outcome is unknown, and only results from progression through the design journey. Indeed, the design problem is in itself unpredictable or 'wicked' (Rittel and Webber 1973), meaning that there is no single, correct way of framing or solving a given problem. Therefore, the outcome depends on the experience, emotions, and subjectivity of those addressing the issue. These characteristics have made it problematic to model a framework that adequately describes and generalises the process: every design situation is unique.

The decision whether to cooperate or compete with others is a pivotal issue in management, and there are benefits to adopting either strategy (De Wit and Meyer 2005). In design, however, all projects – from products to software, from clothing to buildings to experiences – occur within commercial parameters, and require

F. Maciver (✉) • J. Malins
Norwich University of the Arts, Francis House, 3-7 Redwell Street, NR2 4SN Norwich, UK
e-mail: f.maciver@nua.ac.uk; j.malins@nua.ac.uk

© Springer International Publishing Switzerland 2016
P. Markopoulos et al. (eds.), *Collaboration in Creative Design*,
DOI 10.1007/978-3-319-29155-0_2

the input of a number of specialists. Design has always been inextricably linked to business: the 'industrial' design profession came about as businesses sought to mass-manufacture products for the open market (Woodham 1997). Moreover, the results of design are to be used by real people. However, when the factors which influence the process differ in every situation, and when the people involved affect the outcomes of the process, how can firms whose business depends on the outcome of the design process – the client, suppliers, manufacturers, creative consultancies – ensure that the final results will be adequate for all concerned?

In the modern day context, the design process has come to rest on communication: the only route forwards is working together to achieve optimal results. In this respect, design is a people-centric discipline, reliant on a dynamic social setting. These interrelationships have become even more important over the last two decades as the discipline has undergone great change. Major shifts in the technical, commercial, cultural and social landscapes have fundamentally altered how goods and services are developed, manufactured and sold, and have impacted the underlying methodologies of the profession. The need to work together – to cooperate, collaborate and communicate – with clients, users and a wide network is paramount when developing the technical, networked and sophisticated products required in today's fast-moving, dynamic marketplace (Press and Cooper 2003).

It follows that there needs to be structure around the process to support the collaborative effort. It is necessary for stakeholders to come together – face-to-face or remotely – to share approaches and grow ideas, and to work together creatively to realise a goal, particularly since the number of players and information involved in the development process is currently growing in complexity. This chapter outlines tools and methodologies which can help in supporting a design team, in allowing stakeholders to communicate throughout the project, and in managing the creative process.

The chapter looks at the emerging shape, tools and structure of the contemporary design process. It examines how designers and design studios are embracing technical, commercial, cultural and social changes, and how they are adapting to cope with an increasing need to collaborate in the design process. First, the design process is examined in depth, and it is suggested that different types of collaborative practice occur throughout the process. Since every design situation is unique, the tools comprising any collaborative strategy will always be different. Second, the latest tools that are being adopted by design studios to catalyse successful creative collaboration are assessed. Many of these are information communication technologies (ICTs) which are freely available, and these bolster more conventional methodologies. A case example suggests when and how these tools may be used. Third, these ideas are drawn together in a set of principles which are intended as a guide to assist and encourage designers and students in the creative process. Being inclusive, embracing insight and having confidence to change direction, and incubating and developing ideas, are considered to enable practitioners to foster an ethos of innovation. The authors have engaged in qualitative research activity with design consultancies in Europe and the US, including interviewing and research of an ethnographic nature, and these principles are based upon those insights, alongside their experience in teaching and working design and business students, and review of design success stories and of relevant literature.

Strategies for Collaboration During the Design Process

There are several reasons to explain the shifts that characterise modern day design practice. First, global connectivity and instant communications have enabled a diverse network inside and outside of the design process, comprising, for example, designers, clients, users, customers, engineers, researchers, manufacturers, suppliers, retailers and others. Hence, the design team can be distributed geographically across countries and even continents. Indeed, most manufacturing is now outsourced for economic reasons, and this gives rise to the need for international collaborative working (Kolarevic et al. 2000). Second, increasingly sophisticated technologies of production and manufacture have resulted in more technical and networked products. In turn, this creates a more complicated design process requiring the input of many experts (Press and Cooper 2003). Third, customers have become more informed, more powerful, and less loyal to one particular brand. To be competitive, there is a growing need to understand their needs and desires, and tailor products and services to match these precisely. Similarly, digital communications mean that trends change quickly, and so the cycle of new product development (NPD) is accelerating.

These trends drive more dynamic, fast-moving and connected modes of product development. Research by Dell'Era and Verganti (2010) shows that coordinating the input of a range of different sources in a well-balanced team tends to produce more innovative design results. A network provides benefits such as a greater breadth of specialist skills and knowledge, more expansive insight, and the opportunity to cross-pollinate and develop ideas.

A collaborative strategy can be deployed outside as well as inside the design team. User-centred design, which focuses on human-centred research, seeks to involve end users in the development process to understand specific needs and problems to be addressed. Each of these strategies results in different design outcomes, such as whether it is radically innovative, or a redesign of a previous product. Therefore, the collaborative strategy should be tailored to meet the project objectives.

Varying Forms of Collaboration

To understand the precise form and characteristics of collaboration in creative projects, it is useful to look at the phases and activities across the design process. It has been observed that each design project differs, however research by the

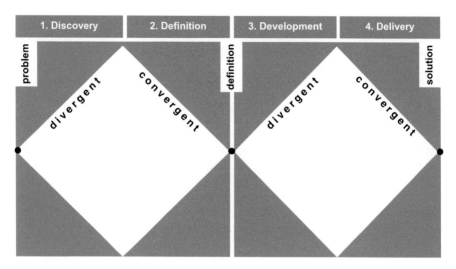

Fig. 1 The Design Council's 'double diamond' model of the design process

Design Council (2007) has uncovered four phases through which any design process passes. The 'Double Diamond' model, shown in Fig. 1, delineates four broad areas – 'discovery', 'definition', 'development', and 'delivery' – this allows design teams to explore ideas, test solutions and innovate. The model also indicates that iteration can occur during phases, and that previous phases may be revisited during the process. This proposition is useful since it acknowledges the different modes of thinking (for example, divergent thinking which is broad and externally-focused or convergent thinking which is more focused and internal-facing) throughout the process. It follows that modes of collaboration also vary depending on stage of the project.

Addressing the aims and objectives of the design process requires a different set of activities and approaches, as summarised in Table 1. For example, at the *discovery* phase, a large pool of ideas needs to be generated. A common approach may be to use brainstorming techniques internally to generate ideas. Involving stakeholders from different disciplines and areas of expertise is likely to enlarge the scope of the problem frame. Depending on the scope of the project, field research may be conducted with the target user group. This approach is an example of divergent thinking, and may yield a significant quantity of information.

Collaboration may occur at any stage of the design process, and Table 1 illustrates how different interfaces propel the different tasks required to progress through a project. In managing a successful design process, maintaining a relationship and open lines of communication with these parties is crucial (Maciver 2012). Indeed, working with stakeholders is a common approach in business, and research has suggested that strong design 'clusters' are more competitive (Verganti 2006) than those working in isolation. The banding together of people, teams, and organisations is recognised as a means to achieving greater goals than could be accomplished alone.

Table 1 Summarising the objectives of each phase of the design process

PHASE	OBJECTIVE OF PHASE	REQUIRED TASKS AND COLLABORATIONS
1. Discovery	Divergent thought and ideas - orientation characterise this phase where the design 'problem' is explored, investigated and questioned in full	*Internal project team:* Idea generation, brainstorm *Clients:* discussions around brief, aims and requirements *Users:* gaining broad understanding of current situation
2. Definition	Convergent thought, where the insights are collated, result in the definition of a clear problem space to be addressed	*All stakeholders:* requirements agreement and analysis *Users:* research specific needs and challenges *Internal project team:* collect ideas, discuss and brainstorm around these, select winning concepts
3. Development	Divergent thought patterns predominate this phase, where the fundamental aspects and details of the solution are investigated in detail	*Creative team:* design development *Stakeholders:* collaborative prototyping, including testing *Manufacturer:* production iterations *Users:* testing of prototypes with real people
4. Delivery	Convergent thought brings the 'winning' concept to reality in this phase, when production, manufacture and launch take place. Feedback loops can enable iteration and improvement	*Client:* discussions around marketing strategy *All stakeholders*: achieve consensus of strategy to bring product to market *Users:* final testing, and research on launch strategy

Source: The authors

Despite these benefits, there are many challenges to establishing and maintaining creative partnerships. The next section looks at the problems faced by designers.

Challenges to Collaborative Creative Networks

The input of many people creates a very complex ecosystem, and there has been scepticism on the part of design practitioners as to how creativity can be enabled in a group context. It is suggested that designers have found the transition to overt collaboration challenging (Sonnenwald 1996). Whilst the cross-fertilisation of ideas is common in the industry, it is normal for professionals to be protective when it comes to their intellectual property (Mun et al. 2009). Cooperation throughout the design process is typical, however may be unacknowledged, or treated with suspicion, by designers (Edmonds et al. 2005). This can, in part, be attributed to the tools that are common yet unobtrusive in design studios (for example email, teleconferencing, FTP (file transfer protocol), and instant messaging) and which provide a constant flow of communication across distributed teams, as well as more sophisticated tools (such as video conferencing, browser-supported file sharing and online whiteboard brainstorming sessions) which enable the sharing and real-time editing of project work. The use of such platforms is widespread and increasingly essential in the design industry. A survey conducted by Liapis et al. (2014) indicates a varied and comprehensive range of software used in design studios in Europe.

In terms of the collaborative process itself producing more creative and innovative products, there is also scepticism. Research by Norman and Verganti (2014) suggest that consulting users does not reap frame-breaking outcomes: for example, Apple does not conduct user research in the belief that users are unaware of the potential of technology. Conversely, it can also be argued that there is value in research for keeping the firm's offering current, innovative and relevant on the market, and for the user (Bailetti and Litva 1995; Veryzer and Borja de Mozota 2005). Such findings can enable a deep level of insight into customer aspirations, and alters how organisations innovate (Aula et al. 2005; Lojacano and Zaccai 2004). This approach tends to lead to incremental developments based on user needs rather than radical forms of innovation (Malins and Gulari 2013).

The logistics of enabling group creativity is also an issue to be considered: working with a team distributed across countries, with different languages and time zones, complicates communication, and finding consensus and agreement on all matters can be problematic. Nevertheless, it is now paramount to communicate and collaborate more intensively with international partners in order to bring new ideas to fruition. Designers are finding ways to creatively collaborate with stakeholders (Arias et al. 2000; Simoff and Maher 2000). Technology, particularly ICT, is becoming central to the business operations of design organisations. There has been a steep learning curve over the last decade, as design firms have had to find new means to communicate with a group. Moreover, where there is such a breadth of information and voices entering the design process, there is the need to manage projects adequately. The following section describes a variety of methodologies and tools that are adopted in design practice to support creative collaboration.

Tools to Support Collaborative, Creative Environments

Emergent methods and tools are fortifying the conventional techniques used in traditional design practice. New technologies, the internet and the platforms powered by it, are being adopted for collaborative creative working. Many of the ideas described in this section are practical with a tangible purpose, while others are conceptual methodologies flexible with application to different design

Fig. 2 Challenges in
collaborative creative
processes

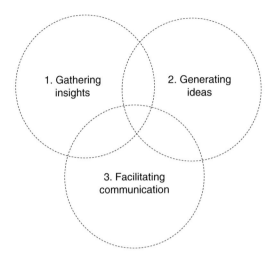

Fig. 2 Challenges in collaborative creative processes

disciplines. It is significant to note the democratic availability of modern design and productivity tools. The software used by professionals is readily available in education. Therefore, student projects can assist in preparation for real-life work situations. Indeed, the tools already in place in university faculties, such as peer-to-peer communication platforms and interactive course material, are formative for professional practice. Students can use these in innovative ways, such as to collect and share resources, to better reflect professional practice in the studio environment.

This section delineates three interlocking challenges faced in collaborative creative processes, illustrated in Fig. 2: gathering insights, generating ideas, and facilitating communication. It analyses how modern methodologies may be responsive to addressing these questions in creative collaboration:

1. Gathering insights: What methods enable the gathering of a range of insights from many stakeholders?
2. Generating ideas: How can ideas be generated collectively when dealing with vast amounts of information and insights?
3. Facilitating communication: When working with a distributed team, how can the channels of communication be opened and maintained?

A Collaborative Approach for Gathering Insights

People are the core of any design project, and having a good range of insights is crucial for a successful project result. The ideas of the omnipotent, individual designer contrasts starkly with the need to collect insight from other people. Box 1 compares the approaches adopted by the lone mavericks such as Philippe Starck and Gordon Murray, versus the conditions created at firms like Pixar and IDEO for collective creativity.

BOX 1: The Creative 'Maverick' *Versus* the 'Collective' Approach
There has been research conducted to decipher whether the best creativity results from lone mavericks or group situations. Since real people have to use the products of design, the design historian Adrian Forty (1986: 245) concludes that "no design works unless it embodies ideas that are held in common by the people for whom the object is intended". In this respect, successful designers temper personal style to balance conflicting forces. Lloyd and Snelders's (2003) study into designer Philippe Starck's creative process in arriving at the Juicy Salif lemon squeezer concludes that it does not fulfil many of the functional attributes of a squeezer: the legs bend when pressure is applied, there is no bowl to catch the juice, and the metal tarnishes on persistent contact with the acidic juice. However, it does possess a set of non-functional qualities: it is a dramatic kitchen item, it is decorative and it acts as a conversation starter. Likewise, Cross and Cross's (1996) study of the creative process Gordon Murray undertook when developing the Brabham and McLaren Formula 1 cars of the 1970s and 80s also suggests a very personalised and therefore unpredictable approach to design decisions.

Starck and Murray can be considered quite exceptional and individual in their fields (in which they are world renowned), however, in organisations there is a need to temper individual style to incorporate an egalitarian philosophy. Co-founder and president of Pixar, Ed Catmull (2008: 66), upholds that: *"filmmaking and many other kinds of complex product development, creativity involves a large number of people from different disciplines working effectively together to solve a great many problems"*. To foster such 'collective creativity', Catmull has instilled a flat culture at Pixar where there is no hierarchy and everyone can talk to everyone, where creatives are empowered to take decisions, where all in the 250-strong production team can make suggestions, and where learning happens through 'post mortems'. Likewise at IDEO, there is recognition that all members of an interdisciplinary group instil different qualities (Kelley and Littman 2006). IDEO partners suggest that acknowledging a range of different roles in the creative team is important in order to achieve an egalitarian approach which is considered to result in better designs.

Individual and highly personalised ideas and approaches are typical of a discipline as subjective as design, however the strategies deployed at IDEO and Pixar demonstrate how these may be tempered and integrated into a harmonious group approach.

Identifying Suitable Partners

Early in the process, online resources enable the creative group to be established, if it is not already complete. Online communities allow people to make connections with others who have the skills and expertise, or financial backing, required to ignite a project. Platforms that connect potential partners and stakeholders across disciplines (for example, creative with technical talent) can enable global collaboration for specific projects. For example, Kickstarter connects would-be innovators with investors to help bring an idea to fruition, and has funded a range of creative projects including movies (e.g. Veronica Mars), games (Exploding Kittens received 110,000 backers), and many technological projects (e.g. the Pebble watch, the Ouya games console, and the Form 1 three-dimensional printer). Indeed, NESTA (National Endowment for Science, Technology and the Arts), an independent UK charity that supports innovation, reports how such resources drive entrepreneurship in the economy (see: Stokes et al. 2014).

Research and Road Testing

Throughout the process and particularly at the early stage of design, insight can be gathered from many types of experts. For example, a project to design a piece of smart, wearable technology for an elderly person may begin by speaking to people who have specialist information on the issues at hand: doctor, occupational therapist, physiotherapist, carer, nano technology specialist, as well as potential wearers of the device, to fully understand the issues. This method will ensure that the ensuing product design process will investigate the right issues, and that the result will have relevance and value to the end user. Research by Leonard and Rayport (1997) reports that an *empathic* approach allows firms to innovate using techniques such as: customer observation, data gathering through visual, auditory and sensory cues, analysis of data, brainstorming, and the development of prototypes of possible solutions. As the design process moves towards the 'development' phase, testing rough prototypes throughout keeps sight of the problem at hand, and keeps the core attributes of the product on track.

Design consultancy IDEO has pioneered this mode of human-centred design. Its *Open IDEO* arm aims to *'design a better world'*, and its projects involve addressing social problems on a global scale. Via its website, the organisation seeks to mobilise users from around the world. By providing guidelines in the form of design and HCI toolkits, members can provide local and personal information allowing the designers to home in on intricate issues on which information may not normally be forthcoming.

Generating Insights by Establishing Communities

Key insights can be generated using technology from a range of different people. Online platforms and communities can again be deployed for the organising of research and the collection of first-hand data leading to new insight. Joining or creating groups online can allow the right people to be reached, as well as allowing access to a large number of opinions. Asking the right questions on general or specialist forums, can be used to garner opinion from target user groups.

A Collaborative Approach for Generating Ideas

The early stages of the design process calls for the synthesis of the insights gathered from different sources. At this stage the problem to be addressed is still uncertain. Therefore, ideas generated through brainstorming hold the key to the solution and all have potential: the design team has to assess all insights. These are later analysed and filtered when data from the rest of the group is complete. Working in such a creative group environment requires a democratic approach, where each input is considered valid.

Gathering Inspiration and Idea Generation Tools

Gathering inspiration and stimulus material is a good starting point when beginning a project, as it can spark ideas and lead to new areas previously unconsidered. The fashion designer Paul Smith's first book is entitled '*You can find inspiration in everything – and if you can't, look again*', and in it Smith (2003) asserts that many of his designs were the result of things observed in everyday life. As advocated by Smith, being open to a wide range of interdisciplinary influences can allow the development of new ideas and thought patterns. Acting like a sponge to soak up influences outside of the immediate problem frame allows concepts to be considered in new and different ways.

Online platforms can support the quest to find and sort stimulus material. Virtual, cloud-based image repositories allow users to upload their own, or collect existing inspiration images in mood boards, with the benefit of being able to access their data on any system. Such platforms can also make recommendations of similar images, and connect users to other boards with similar themes. The ability to share such stimulus material with others in their design team supports group communication and collaboration. This is a valuable time saving platform for design teams, and can also assist the team when presenting a narrative at the end of the project.

News, media and culture provides design teams with up to date information regarding design trends. Specialist design agencies work specifically in predicting and disseminating upcoming trends. However, the power of personal, niche experience is not to be underestimated. Many products have resulted from a designer or entrepreneur's own dissatisfaction with the current situation sparking a desire for improvement. For example, the OXO Goodgrips products were born when the company's president observed his arthritic wife's difficulties using regular kitchen utensils, and led to a collaboration with Smart Design which spawned a full-range of kitchen products designed with this type of user in mind. Empathic modelling is a method by which designers can 'put oneself in another's shoes' and test out the user's current situation (McDonagh and Formosa 2011). Experiencing the user's challenges allows insights to be gained based on the user's perspective.

Developing a Creative Environment

Physical setting and environment is considered to be important in catalysing creativity. Whether ideas emerge from a meeting in the boardroom or around a conversation over coffee, it is impossible to predict exactly when new ideas will appear. However, it seems that the most relaxed moments and settings bring the most creative results: Roam (2009) notes the value of jotting down ideas on napkins or the back of envelopes when caught unprepared.

The social setting is certainly important to how nascent ideas can be nurtured and developed. Some firms, like Pixar and Google, encourage staff to mingle, and recreation areas are centralised to achieve this. The rationale is that by mixing different functions in a social setting, there is a greater likelihood that ideas will be cross-pollinated. Google has been innovative in its office interior architecture, and has included features such as break out areas away from the desk, recreational activity spaces and 'chill out' zones complete with beanbags, with the expectancy that relaxation of the body will also relax the mind. Lego practices a policy of 'serious play' where groups solve problems together using Lego to construct and tell stories.[1] While it is impossible to generalise how a creative environment should be designed, it is clear that communication is crucial when developing ideas. Moreover, sharing and discussing at the early creative stages has a greater likelihood of developing the idea in new directions, and flagging different insights.

Facilitating Collaboration and Communication

The final dimension of the collaborative design project is facilitating communication, maintaining the group, harmonising contributions, and keeping sight of purpose and process. Allowing creativity to flourish when the design team is distributed is particularly challenging. Online platforms have been developed to enable sophisticated modes of creative communication. For example, online group sketching sessions are possible using dedicated online platforms. These allow users to log in and simultaneously sketch digitally on a blank page which has been designed to replicate a virtual whiteboard. Voice and video web conferencing are freely available, and surmount language barriers through use of shared screen environments and visual communication. Instant messaging services allow informal communications across the team during the working day. Since project material can take many forms – such as images, diagrams, objects, three dimensional prototypes, rough sketches, experience and journey maps, wireframes, websites and videos, amongst others – being able to share these is paramount when it comes to presenting

---

[1] For further information on Lego's methodologies for creativity, see: http://www.lego.com/en-gb/seriousplay/.

the story of the design process. Presentations and project files can be shared and organised, and updated collectively in real-time using online platforms.

It seems that the key to smooth and creative partnerships occur when teams are open and transparent. An egalitarian team approach, where information and sharing is full and complete, and where stakeholders have appropriate access to information, helps to bring about positive relationships and a smooth interchange of ideas. A project management channel for communication can help to maintain democratic access to information. These steps assist in cross-pollinating ideas, skills and insights, and foster the optimal conditions for innovative results.

A Framework for Creative Collaboration

The previous sections have examined some of the challenges of creative collaboration, and have analysed how the available tools may be able to address these issues. While it has been noted that the methods and collaborative strategy followed in a creative process cannot be generalised, it is possible to present a framework for enabling the right type of collaboration for any project. Table 2 presents a case study, suggesting and summarising the techniques and collaborative requirements that may be used in a hypothetical design project. It also indicates where potential tools may assist different activities of the design process. This lays the foundations for highlighting patterns and principles in the creative process.

Principles for Fostering Collaborative Creativity

The chemist Louis Pasteur is quoted as saying of his seemingly serendipitous discoveries that "*Chance favours the prepared mind*". Similarly, Thomas Edison proclaimed: "*Genius is one per cent inspiration, and ninety nine per cent perspiration*". These advocacies for preparation, consistency and reliability in the creative *engine room*, rather than reliance upon sudden flashes of brilliance, are equally true today. The lead author witnessed examples of protracted design processes being punctuated by creative breakthroughs during research conducted at Design Partners, a leading Irish design consultancy. The consultancy can be considered commercially successful and indicative of best practice, as it expanded during the recession of 2008–2013 to open new studios in continental Europe and North America. On one project, led by a consultant named Rob, the brief was to design an implement that both measured and scooped baby milk formula, in different quantities depending on the age of the child, for a baby-care company. Through interviews with parents recruited on a local parenting website, and with several members of the team having young children themselves, the designers understood the effort to remember to count the number of scoops when making formula for a crying baby. Since the product was to be used at any time of the day or night, it had to be as simple to operate

Table 2 Framework for collaboration across the design process

Design brief: To design and develop a budget-end desk lamp with an adjustable, moving arm, which uses a sustainable and eco-friendly light source

Phase/associated issues	Methodologies/actions	Collaborative requirements	Potential collaborative tools
1. Discovery			
What are the parameters of the problem?	Meeting with client to understand and fine-tune parameters	Meetings with stakeholders, formal and informal	A virtual meeting space; instant messaging to communicate informally
Who are the stakeholders?	Meet with project team	Find experts in lighting design	Expert forum/community membership
Whose input is required?	Instigate brainstorm session with creative team to understand who the user is	Recruit potential users to participate in research (e.g. students, homeworkers)	Virtual brainstorm environment which records train of thought
How can you communicate with these people?	Ad hoc acting out of use scenario within creative team	Brainstorm session	Image database/repository
How will you organise the process?	Initial online research, gathering of stimulus material	Gather stimulus material and create mood board	Project management platform
	Conduct preliminary user research	Assign team roles	
	Evaluate ideas within team		
2. Definition			
Who do you need to talk to?	Ranking of priorities in the design, and agreement on parameters with client team	Discussions with wider NPD team	Virtual meeting space
Where can you find the necessary expertise to understand the problem?	Conduct more in-depth user research in real-life setting (homes were light may be used). Ethnographic research plus users recording experiences digitally	Show initial sketches and share visuals	Tool to share visuals and sketches (digital or otherwise) online
How can you find the right information?	Research on: lighting, materials, light sources, engineering, production	Story board the use case scenario	AI tool which can suggest material to be added to story board
What methods will you use to brainstorm as a group?	Sort, filter and storyboard insights from research	Sort ideas from stakeholder workshop	Analysis of keywords and semantics arising from group discussions

(continued)

Table 2 (continued)

Design brief: To design and develop a budget-end desk lamp with an adjustable, moving arm, which uses a sustainable and eco-friendly light source

Phase/associated issues	Methodologies/actions	Collaborative requirements	Potential collaborative tools
	Sketching of ideas in design team, reaching consensus as to features of design	Prototype, build and model solutions in a craft workshop	Tool to allow research participants to record and send information digitally
	Encourage all stakeholders to prototype ideas in a workshop	Collaboratively build presentation for client	Cloud-based presentation software
	Discussion with NPD team to agree on features of the concept		
	Pitch idea to client team		
3. Development			
What ideas are going to be carried forth?	Detailed development prototyping in workshop alongside lighting and engineering experts	Focus group with group of end users to discuss the solutions	Space to share visuals with producers, instigate real time dialogue
Is it important that everybody in agreement?	Testing prototypes with end users	Construct a CAD file and send to production plant	
How can the prototypes be shared with the group?	Commission a production model	Liaise with producers sharing visuals to surmount language barrier	
How will the prototypes be tested?			
4. Delivery			
How can we present the information?	Meeting with client to present the story of the finished desk light	Update storyboard	Sort and timeline progress of the project
How can we get feedback on our ideas?	Analyse feedback and iterate idea	Gather original stimulus material	Presentation synthesis in the cloud
	Market research re-starts iteration loop….	Integrate client requests into CAD file	

as possible. In getting to grips with the details required to design this product, Rob consulted with experts in measurement at the School of Food and Nutritional Sciences at University College Cork, while the design team created profiles of intended users, made storyboards and constructed a mood board. Gathering together this research, the design team developed six concepts in the weeks leading up to the client presentation which were modelled by design engineers and built by in-house model makers. Many centred on the idea of a set of different sized spoons, however none emerged as a clear winner. The evening before the pitch, Rob was reading an online forum and out of the corner of his eye saw an advertisement for a washing detergent. He wasn't quite sure why the detergent measuring cup sparked a sudden inspiration to solve the niggling design dilemma. He envisaged and roughly sketched a moving mechanism inside the scoop which could be adjusted to precisely measure different quantities of formula. Texting his colleagues about the break-through, Rob and the team convened early the next morning in the studio. Everyone was enthusiastic about this new idea – there was a gut feeling that it would work, and that it could be produced within the parameters of the brief. The team sketched, storyboarded and presented the idea to the client at 11 am. It turned out to be the winning concept.

Rob's story illustrates two key ideas: first, the innate *designerly* passion for arriving at the best ideas which transcends barriers and boundaries, and second, that the winning concept resulted in the hours of due diligence that Rob, the team and their collaborators spent talking, sharing knowledge, and devising seemingly random ideas. Rob and the team conducted a range of experts, and used personal experience to gain an innate understanding of the problem. They researched, brainstormed and prototyped several ideas. They pondered and reflected on how their solutions addressed the issue. They observed and absorbed different sources of inspiration. They laid the groundwork over weeks of effort, however they were open to new ideas and had the courage to change their path and go with a new idea at the last minute.

Creativity in action has been the subject of in-depth investigation. Research by Dunbar (1997) attempted to examine the "messiness" of innovation and invention. During a year long, ethnographic study of scientists in their labs, Dunbar discovered that the majority of breakthroughs were not expected nor predicted. In fact, Dunbar discovered that the process of trying – and failing – led to the most significant discoveries in the lab. Mistakes, failures and anomalies in experiments, alongside a patchwork of lab meetings, team discussions on experiment failures and anomalies, random corridor conversations ('watercooler moments') provoked learning and thinking about discoveries in new ways. Furthermore, Dunbar concluded that the subjectivity of the scientists themselves – their choices of what's interesting and worthy of further investigation – are major precursors to new findings and innovations. Johnson (2010) likewise suggests that the sudden 'flash', 'lightbulb' or 'Eureka!' moment that individuals often report having which lead to an idea are actually the product of years worth of overlapping and often quite serendipitous experiences, conversations and failures. It seems that Pasteur and Edison's assertions hold true today.

These stories highlight the underlying premise of this chapter that creativity is rarely a solo activity, but rather burns slowly and unfolds organically in ways that can't quite be predicted. Despite this non-conformity, we can generalise some patterns and therefore learn how to promote creativity and encourage new ideas. Three broad principles are outlined below which can be deployed in every day collaborative creative practice, and which can help groups establish and maintain creative rhythm.

Principle I: Taking an Interdisciplinary Approach

Taking an interdisciplinary approach means being inclusive about who can play on your team. Creating a network to establish collaboration, embracing the differences of members rather than seeking homogeneity, and operating a philosophy where all ideas are acceptable supports the development of new ideas. Mobilising users and potentially insightful experts is vital in gaining the right information. Rob showed that openness to embrace knowledge from other disciplines, talking to different groups, and seriously processing their input enabled learning which resulted in a better result. This principle also applies inside firms where the flat structure at Pixar, and its democratic approach allows learning. Allowing others to ask questions prevents domination of one idea, and prepares a good starting discussion for innovation. On the micro level, being interdisciplinary means not restricting your own self to one particular domain, but having an open attitude. Design is a team sport and, while there may be a leader, everyone's input is valuable.

Principle II: Changing Direction and Embracing New Insight

We know that, given the uncertainty and unpredictability that characterises the design process, novel thinking is required to arrive at a great solution. Flexibility in approach to the process is a great asset. Embracing new insight involves being a 'sponge', absorbing influences and experiences from a range of contexts – professional and personal, first-hand or observed. Rob and the team were genuinely interested in parents' problems, and desired to create the best possible result. Moving the goalposts and switching course on the last day shows commitment to their client, as well as the trail of related stakeholders. By constantly being aware, observant and mindful, Rob remained attuned and was able to spark ideas until the last possible moment.

Principle III: Incubating Ideas – Persevering, Iterating, Testing, Improving

Its ambiguous foundations in discovery-driven observations, rich, context sensitive data and serendipitous circumstances mean that design work is slow burning. The ability to persevere, improve and incubate ideas determines how the team can grow and change its focus over time: Dunbar's scientists failed on many occasions before stepping towards a new discovery. As the environment at Google shows, a relaxed mind appears to reap greater creative results. Many design teams advocate stepping away from the project. For example, interdisciplinary New York design studio Sagmeister Walsh closes for one full year in every seven. Partner Stefan Sagmeister[2] notes that the majority of projects and work arising in the following seven years comes from work and ideas derived from this sabbatical. Likewise, it was in Rob's free time when the winning idea appeared. By taking time to regain perspective, new ideas have space to come to the surface. Trying different approaches, prototyping, trying again and not settling for the first answer requires tenacity and an overall creative confidence.

Conclusion

In an era of interdependency, this chapter has examined the realms of cooperation in contemporary, creative practice. Uncertainty, heterogeneity, individualism are key themes, all of which are inconsistent with the uniformity expected in commercial networks. However, it has been suggested that such a serendipitous approach – tailored to the group, the project, the unique situation – is how innovation typically unfolds. Indeed, the connected group effort is posited to enhance creative results. This is an especially important concept in the design process, where a 'design thinking' approach implies solving problems by reconciling people, technology and commerce. This raises a number of questions for the reader: does the creative process need a leader? Are all voices equal? Who takes the final decisions?

While the course of innovation may not be predictable, creating the best conditions and adopting the right mind-set can encourage a creative ethos. By examining when and how collaboration happens in the design process, it has been possible to suggest a range of methods that are equally applicable for use in many types of design situations. Again, tailoring the strategy is key.

As collaboration becomes paramount for the development of new products, services and experiences, technology is becoming a strong enabler by putting structures and frameworks around the creative process. It facilitates new types

---

[2]See 'The Power of Time' by Stefan Sagmeister available at http://www.ted.com/talks/stefan_sagmeister_the_power_of_time_off#t-945073.

of collaboration, and bolsters existing methodologies. ICT is making it easier to assemble groups, gather, sort and present insight and communicate with a diverse network, and it is predicted that digitisation will continue becoming an intrinsic component of the design process in coming years. Providing the best interfaces to support creative approaches continues to be a priority for software developers.

References

Arias E, Eden H, Fischer G, Gorman A, Scharff E (2000) Transcending the individual human mind – creating shared understanding through collaborative design. ACM Trans Comput Hum Interact (TOCHI) 7(1):84–113

Aula P, Falin P, Vehmas K, Uotila M, Rytilahti P (2005) End-user knowledge as a tool for strategic design. In: Joining forces. University of Art and Design, Helsinki

Bailetti AJ, Litva PF (1995) Integrating customer requirements into product designs. J Prod Innov Manag 12(1):3–15

Catmull E (2008) How Pixar fosters collective creativity. Harv Bus Rev 86:64–72

Cross N, Cross AC (1996) Winning by design: the methods of Gordon Murray, racing car designer. Des Stud 17(1):91–107

De Wit B, Meyer R (2005) Strategy synthesis: resolving strategy paradoxes to create competitive advantage, 2nd edn. Thomson, London

Dell'Era C, Verganti R (2010) Collaborative strategies in design-intensive industries: knowledge diversity and innovation. Long Range Plan 43(1):123–141

Design Council (2007) Eleven lessons: managing design in eleven global companies. The Design Council, London

Dunbar K (1997) How scientists think: on-line creativity and conceptual change in science. In: Conceptual structures and processes: Emergence, discovery, and change. American Psychological Association Press, Washington, DC

Edmonds EA, Weakley A, Candy L, Fell M, Knott R, Pauletto S (2005) The studio as laboratory: combining creative practice and digital technology research. Int J Hum Comput Stud 63(4):452–481

Forty A (1986) Objects of desire: designs and society 1750-1980. Thames and Hudson, London

Johnson S (2010) Where good ideas come from: the natural history of innovation. Penguin, London

Kelley T, Littman J (2006) The ten faces of innovation: IDEO's strategies for defeating the Devil's advocate and driving creativity throughout your organization. Profile Books, London

Kolarevic B, Schmitt G, Hirschberg U, Kurmann D, Johnson B (2000) An experiment in design collaboration. Autom Constr 9(1):73–81

Leonard D, Rayport JF (1997) Spark innovation through empathic design. Harv Bus Rev 11:102–113

Liapis A, Kantorovitch J, Malins J, Zafeiropoulos A, Haesen M, Gutierrez Lopez M, Funk M, Alcamtara J, Moore JP Maciver F (2014) COnCEPT: developing intelligent information systems to support colloborative working across design teams. In: Proceedings of the 9th international joint conference on software technologies, Vienna, Austria, 29–31 August

Lloyd P, Snelders D (2003) What was Philippe Starck thinking of? Des Stud 24(3):237–253

Lojacono G, Zaccai G (2004) The evolution of the design-inspired enterprise. MIT Sloan Manag Rev 45(3):75–79

Maciver F (2012) Diversity, polarity, inclusivity: balance in design leadership. Des Manage Rev 23(3):22–29

Malins J, Gulari M (2013) Effective approaches for innovation support for SMEs. Swed Des Res J 2(13):32–39

McDonagh D, Formosa D (2011) Designing for everyone, one person at a time. In: Kohlbacher F, Herstatt C (eds) The silver market phenomenon: business opportunities in an era of demographic change. Springer Verlag, Berlin, pp 91–100

Mun D, Hwang J, Han S (2009) Protection of intellectual property based on a skeleton model in product design collaboration. Comput Aided Des 41:641–648

Norman DA, Verganti R (2014) Incremental and radical innovation: design research vs. technology and meaning change. Des Issues 30(1):78–96

Press M, Cooper R (2003) The design experience: the role of design and designers in the twenty-first century. Ashgate, Aldershot

Rittel HWJ, Webber MM (1973) Dilemmas in a general theory of planning. Policy Sci 4:14

Roam D (2009) The back of the napkin: solving problems and selling ideas with pictures. Penguin, London

Simoff SJ, Maher ML (2000) Analysing participation in collaborative design environments. Des Stud 21(2):119–144

Smith P (2003) You can find inspiration in everything – and if you can't, look again. Thames & Hudson, London

Sonnenwald DH (1996) Communication roles that support collaboration during the design process. Des Stud 17(3):277–301

Stokes K, Clarence E, Anderson L, Rinne A (2014) Making sense of the UK collaborative economy. NESTA report, September. http://www.nesta.org.uk/publications/making-sense-uk-collaborative-economy. Accessed 5 Sept 2014

Verganti R (2006) Innovating through design. Harv Bus Rev 84(12):114–122

Veryzer RW, Borja de Mozota B (2005) The impact of user-oriented design on new product development: an examination of fundamental relationships. J Prod Innov Manag 22(2):128–143

Woodham JM (1997) Twentieth-century design. Oxford University Press, Oxford

Probing – Two Perspectives to Participation

Tuuli Mattelmäki, Andrés Lucero, and Jung-Joo Lee

Abstract Practitioners from different fields of design and research apply the 'Probing' method as means of getting a better understanding of their users and to inspire their designs. During the 15 years since its first appearance, the probing method has been extended for deployment in different contexts and for different uses. In this chapter we first briefly introduce what probes are about, then we look at probing from two perspectives: (a) as a process of collaborative discovery and learning, and (b) as a tool for entering the users' contexts. We illustrate these perspectives through cases in which probes have been introduced in educational and professional environments. Based on the findings, we discuss how a making process of probes can engage a design research team to the issues of concern, and present a set of problems and challenges encountered while probing professional work. Finally, we propose a set of considerations for designing probes for different purposes.

Introduction

Human computer interaction (HCI) and user-centered design (UCD) practitioners apply experimental methods such as probes as means of understanding genuine experiences of users and inspiring design. There are variations in the applications of probes but in general they are based on (a) user participation by means of self-documentation (b) for studying user's personal context and perceptions (c) by applying exploratory mindsets and materials (Mattelmäki 2006). In this chapter we

T. Mattelmäki
School of Arts, Design and Architecture, Aalto University, FI-31000 Helsinki, Finland

A. Lucero (✉)
Mads Clausen Institute, University of Southern Denmark, DK-6000 Kolding, Denmark
e-mail: lucero@acm.org

J.-J. Lee
Division of Industrial Design, National University of Singapore, 4 Architecture Drive, 117566 Singapore, Singapore

© Springer International Publishing Switzerland 2016
P. Markopoulos et al. (eds.), *Collaboration in Creative Design*,
DOI 10.1007/978-3-319-29155-0_3

will first give an introduction to probing, and secondly, deepen the understanding of the approach by considering it as a process of collaborative exploration and learning, and as a tool to enter the users' world. Finally we will end by listing key considerations for the application.

Gaver et al. (1999) first introduced Cultural Probes as a form of exploratory and design-oriented self-documentation method. Cultural probes are collections of evocative tasks meant to elicit inspirational responses from people – not comprehensive information about them, but fragmentary clues about their lives and thoughts (Gaver et al. 2004). Cultural probes were purposefully against scientism, open-ended, and designer-centered when approaching users: *"these packages of maps, postcards, and other materials were designed to provoke inspirational responses from elderly people in diverse communities. Like astronomic or surgical probes, we left them behind where we had gone and waited for them to return fragmentary data over time."* (Gaver et al. 1999, p. 22). An aesthetically well-designed probe kit is given to volunteers, who then complete the assignments and send them back to the researchers. The contents of the probe kit differ from one design or research project to another, but the assignments and materials typically are purposefully ambiguous, trying to stimulate the mind of the participants and capture their experiences.

Since the original probes, the development has been active as researchers and practitioners in the design community have extended probes for different contexts and uses, including Technology probes (Hutchinson et al. 2003), Mobile probes (Hulkko et al. 2004), Empathy probes (Mattelmäki and Battarbee 2002), Urban probes (Paulos and Jenkins 2005), and Design probes (Mattelmäki 2006), just to name a few.

Since the Cultural Probes were introduced 15 years ago the probes method has become a phenomenon, as stated by Wallace et al. (2013). It has been studied and discussed in research literature widely. In their excellent review on probes uses in HCI, Boehner et al. (2007) already counted 90 papers citing the use of probes in the ACM guide to computing literature. Despite the fact that Bill Gaver and his colleagues have been critical on the misinterpreted applications of the experimental method, both researchers and practitioners continue probing in various ways and for a number of reasons. One might say that it is not a specific method per se but rather a family of approaches that have been influenced by the Cultural Probes. Some of the probes approaches have a closer relationship to the original one, however, as identified also by Boehner et al. (2007), probes have quite often been reported as a data collection method similar to questionnaires.

Gaver et al.'s (1999) Cultural probes was done for inspiration and information. Based on empirical data and literature Mattelmäki (2005) later identified four reasons for using design probes in product development and concept design context: (1) the probes data and the whole process of probing can fuel design inspiration, (2) at best probes also provide useful information about users' context and needs, (3) they allow participation by including tools to reflect and express participants' needs and ideas, and to participate in design, and (4) they foster empathy and dialogues between participants and researchers/designers, and moreover within design teams.

Later, to stress the collaborative and exploratory nature of the probing process, she elaborated the reasons for probing as follows (referring to Brandt's (2006) work with Exploratory Design Games):

- *"To support creative thinking, to explore novel or unconventional perspectives and to inspire designers and other stakeholders;*
- *To engage and empower various participants in an exploratory design process, to reflect and create new ideas based on their experiences and insights;*
- *To ease the social collaboration in multidisciplinary teams and with users;*
- *To involve collaborative people and organizations in human-centered design dialogues. These dialogues are part of developing the understanding of the users, making sense of the design space and its opportunities and supporting the exchange of information and learning in collaborative teams;*
- *To enter the individual zones of the people that are studied. Probes aim to foster subjective and empathic insights into the other participants as well, be they designers or other collaborative experts"* (Mattelmäki 2008; 67).

Following the same mode of thinking Wallace et al. (2013) give a rather open-ended definition for probes as tools for design and understanding to be used as empathic engagement with participants in the search of what is personally meaningful.

What Is Probing?

Although the probing approach escapes from one definition or an agreed procedure, we may still talk about the *how of probing* based on an essential structure and practices commonly found in precedents, in order to provide an entry point for beginners.

Probing is typically applied in the early phases of the design process where the questions and the design directions are explored. Mattelmäki (2006) has identified five steps in the probing process:

1. Tuning in to the topic, i.e., designers and researchers collaboratively explore the experiential elements of the topic, and plan and design the probes kits and the assignments.
2. Probing by users, i.e., self-documentation and reflection of the experiences at the users' context.
3. First interpretations by designers and researchers, i.e., the returned probes are studied for further questions.
4. Deepening together by users and designers, i.e., follow-up of the probe materials in an interview with the users.
5. Interpretations and outcomes, i.e., the researchers and designers collaboratively make sense and create interpretations of the probes. The process results in empathic understanding and descriptions of the users, their contexts and the phenomena that are explored, design ideas or clarified directions, and further questions.

The original Cultural probes process consisted of three steps, namely 1, 2 and 5, as it aimed at design imaginations based on the returned probes, rather than valid understanding of user needs (Gaver et al. 1999). For user-centered design projects whose aim is mostly to reach a valid understanding of users' world, the authors recommend considering the 5-step process.

A probes kit can be a combination of a number of assignments that allow written and visual expressions. The assignments both document the current situation, e.g., in diaries, with open questions and by taking photographs, and trigger thinking of potential future experiences, e.g., by open questions, visual collage or mapping tasks or even some form of early design ideas.

It is worth noting that the probing process and kits are in principle meant to be designed and redesigned specifically for each project. The open-ended nature of the probes supports that and the design of the probes per se is an important part of the probing process. For this reason, probes purposefully avoid to pin down a unique procedure. The authors of Cultural Probes and several followers underline ambiguity and open-ended interpretation, rather than rigid guidelines in the application.

Building on these notions, in this chapter we look at probing from two perspectives in order to further elaborate the exploratory, reflective and collaborative process of probing.

First, we consider probing as a process of collaborative discovery and learning. This perspective mainly looks at probing from the designers' or researchers' point of view. It also focuses on benefits in the process of method-making, i.e., the collaborative exploration and empathic design process that starts already when the design team starts to consider what are the probing instruments and what are the questions to ask (Lee 2014). Mattelmäki (2008) calls this phase *"tuning in for co-exploring"*. This part also addresses the materiality of the probes engagement including the character and design of the probes tasks (see also Wallace et al. 2013). To illustrate this perspective we present an example case in which design students reflect their experiences while probing.

Second, we consider probes as a tool for entering users' contexts. This perspective focuses on the participants' view and emphasizes in line with Wallace et al. (2013; 3) that *"probes need to work hard to facilitate a participant's reflection, deploying a range of multi-angled methods"*. The success of the probing requires that the participants invest their time and thoughts when working with the probes. Furthermore, users' active role in the design process has been stressed by, for example, Liz Sanders (2001) who envisions that designers should be engaged in building scaffoldings that support everyday people's generative design thinking. Probes as a tool can offer such scaffolding.

However, we have identified problems of probing in professional contexts and based on examples provide a set of considerations for designing probes that are at the same time pleasurable and easy (e.g., Lucero and Mattelmäki 2007). To illustrate and provide a context for the discussion on the proposed considerations, we present how probes were applied in professional environments, including a study with industrial designers and other studies in which probes have been applied. Our findings suggest that the main challenges researchers will face when designing probes, especially for professional contexts, include aspects related to (1) reducing the demands placed on participants, (2) encouraging a fluent and playful process for participants to avoid 'obligation', (3) being sensitive to the special nature of the work that is being studied, (4) supporting different strategies for using the materials, and (5) motivating the participants. In the following, we will introduce these two perspectives with examples and end the chapter with an outline of considerations on how to make probes work in practice.

Probing as a Process of Collaborative Discovery and Learning

Like many other user research methods in design, the design of probes requires careful considerations to *make the probes work* (Lucero et al. 2007). Probes, as a self-documentation method, needs to communicate to, inspire and engage the user participants during the probing process almost on its own, going beyond designers' or researchers' control. As the probes often aim to trigger participants' reflection and imagination, the triggering mechanism should be carefully designed. Probing tasks often involve designerly activities, such as drawing, visual collaging, photographing, or low-fidelity modeling, to help users reflect on their own experiences and express them through various means.

Because of these reasons, the design of probes tasks and packages often involve designers' hands-on making. In principle, this making needs to be done in each application case in order to fit in the idiosyncratic context of the project. This nature of the probes is precisely what Bill Gaver and his colleagues have aimed to highlight:

> Just as machine-addressed letters seem more pushy than friendly, however, so might a generic approach to the probes produce materials that seem insincere, like official forms with a veneer of marketing. The real strength of the method was that we had designed and produced the materials specifically for this project, for those people, and for their environments (Gaver et al. 2004, p.29).

Some may view the making-phase of the probes time- and resource-consuming, or as *extra effort*, which could have been minimized by pinning it down to the standardized form of the probes. However, in terms of the essential aim of the probes, i.e. allowing open interpretation in users' own context and enabling personal dialogues and empathic engagement between designers and users, the idea of pinning down the process could conflict with what the probes is actually for and can do. In fact, the making phase of probes brings benefits to the design team, going beyond being more relevant when users fill it in. Recent studies have highlighted that the design team could build empathic engagement to the user context and sensitivity to the users already when making the probes (c.f. Lee 2014; Mattelmäki 2008; Wallace et al. 2013). They address the design team's collaborative learning from the making of the probes in following aspects:

- First, the design team's exploration into communication methods and materials of the probes sensitize the team with the topic and helps them build empathic mindset to users.
- Second, visual and tangible construction of the probes allow externalization of designers' inner hypothesis, thus leading them to early exploration of possible design space.
- Third, the design team's collaborative discussions and decision-making during the probes-making allow the team to have a shared understanding of the probes aim and the users.

Lee (2014) has discussed the above-mentioned effects of the probes-making from the students' case. She analyzed 50 students' learning diaries written during the one project course from master's program of Industrial and Strategic Design in Aalto University for 2 years (25 students each year). In that course called *User-Inspired Design*, the students learned empathic design approaches for collaborating with users and exploring future design opportunities, beyond the scope of traditional user-centered design.

During the 9-week course, the students worked in a group and went through a comprehensive concept design process by using various empathic and collaborative design methods. One of the often-used methods was probes. Each week during the course period, the individual students wrote the learning diary, reporting and reflecting on the challenges, activities and accomplishments during the project. The diaries contained lively stories about challenges that the students encountered, how they organized their actions to the challenges, and how they tried to make the probes work – those are *behind-scene stories* which we seldom find from academic papers or handbooks of the method. Thus the diaries exhibited what kind of situated work the probes-making entailed and what learning was going on during that work.

Stepping into Users' Shoes for Making the Probes Work

Designing the relevant tasks for probing was the students' major concern. The *aesthetics* (i.e., look and feel) and *usability* of the probes required huge efforts. The aesthetics and usability of the probes were important criteria for the students to motivate the users' participation in the probes.

> We designed buttons that they can attach to the bag [a bag for the probes package]. It might not be related to our research directly, but we made it for motivating teenagers [with a] jolly-looking kit. We had such heated debates within our team to decide the colors, too. The teenage girls would like vivid colors but boys would not, and so on. It was interesting to hold such debates, imagining the teenagers' feelings and preferences while doing our probes. (A quote from one student for the project on designing for teenagers' peer-to-peer activities)

The students' diary stories showed that the work for probes-making, for example, holding discussions on what colors the teenagers would like and making bags and badges as the probes components, kept the team discussion oriented towards topics of what the teenagers would prefer and what they would be like. The students also discussed what time of the day the teenagers would keep the probe diary, how they would carry the probe kits with them and so on. This kind of practical work for making the probes gradually engaged the students in the users' situations by talking about the users and simulating user experiences, for example, simulating what the users would feel when they touch the probes materials and answer the probes questions.

> First of all, I realized how important it is to concern our target users over the whole process of user research. Of course it sounds so self-evident, but it also means that we should carefully consider them when we make the materials, such as diary or social map, for the probes. Which font size is enough for our users to read? What kind of language is more understandable for them? We should really consider characteristics of our users to get the right results. (A quote from the student for the project on enhancing social interactions of elderly people)

Considering font sizes or colors might be a peripheral issue. Yet, by orienting the design team's actions towards such peripheral, physical details, the design team could become more and more sensitive to the users and their contexts, and build emotional engagement with them. This observation is in line with Hemmings et al. (2002) observations on Gaver's team when designing the Domestic Probes. They discovered that in the early phases of the project talking played a central role. Through discussions the team shared their knowledge of design issues, reached an understanding of the probes' qualities, and, as reported by Hemmings et al. (2002; 45), "spent a lot of time arguing and joking, made up stories, made sketches, kept notes, and talked over previous and possible scenarios".

Knowing the Designer's Own Assumptions Through Probe-Making

The probes-making process could enable the design team to realize their own assumptions to the user groups and preoccupations to the topics by making them externalized. In another example from the students' diaries, one student team aimed to design a service in the outskirts of Helsinki that could support elderly people to be more active and visible in the local society. The students wanted to apply the probes in order to understand elderly people's past memories, emotional experiences, daily activities and wishes. Initially this student team had the idea of a daily probe tasks, which would be delivered to the elderly people on a daily basis. A different probing task delivered each day was the students' tactic to make the whole process exciting and fun for the elderly participants.

The student team visited one community facility where the elders spend their time together, and tried to recruit the participants for their probes study. Soon the students realized that their daily probes idea would not work out. Different from the students' expectations, the participants' daily schedules were too busy to meet the students everyday.

Fig. 1 Re-designed probes for elders: This student group included separate envelopes inside the folder to be opened on daily basis (*Left*). This group personalized each probes kits by placing each participant's names on the package. They also placed their design team's identity (3P as a group name) in the probes packages so that it can create feeling of dialogic, personal communication (*Photo courtesy: Sam Dunne, Jari-Pekka Kola, Joanne Lin, Otto Miettinen, Milla Toukkari*)

> *In our own study, we had already thought a lot about the probes tasks before we met our users for the first time . . . It became obvious that we needed to adjust the tasks we had planned for the probe kit to better suit their [the participants'] preferences. In particular, the elderly ladies were afraid of having to use [too] much of their time for the probes. Contradicting to our stereotypical thinking, they [the elderly participants] were extremely busy!* (A quote from the student' diary)

In this story, their realization on 'busy elderly people' not only led them to redesign their probe package (they re-designed the probe package that contained the daily tasks in different sealed envelopes so that the elders could open one each day) (Fig. 1), but also to reframe the whole direction of the design opportunities. After noticing the elderly people's busy schedules, the student team reframed their project aim, from 'how to activate the elderly people's life' to 'how to cultivate on this active elderly group to spread their spirit to the society' (Fig. 2).

In this story, the making process of the probes enabled the student team to see a truer picture of the users. The sequence of actions in the making of the probes made the students' own assumptions and intentions more tangible so that the students themselves could recognize them. Just as ethnographers conduct auto-ethnography for externalizing their own assumptions for writing about others (Ellis 2004), the process of the visual and tangible creation of the probes could help designers understand their own assumptions to the users before interpreting the probes results and generating design ideas.

The students' cases above imply that the process of making the probes, including iterative hands-on making of the probes package, group discussions and decision-makings on the choices for the probes components and so on, improved not only the relevance and efficacy of the probes itself, but also the students' understanding of what actually matters to users. The local sensitivity and contextual knowledge

Fig. 2 Student's drawing on their re-conceptualization of elders after the probe-making. They reframed the characteristics of their target group from passive elders to active elders who enjoy their life, *Granny Ludens*, inspired by Homo Ludens (*Photo courtesy: Sam Dunne, Jari-Pekka Kola, Joanne Lin, Otto Miettinen, Milla Toukkari*)

developed through making the probes led the team to identify meaningful design opportunities, as illustrated in the case of the students who changed their design aim from activating passive elderly people to facilitating active elders to influence their community.

These observations lead us to consider the making process of the probes as the externalization and manifestation of designers' tentative hypothesis of users and future design opportunities. This notion is in line with what Wallace et al. (2013) talked about the probe designs as "forms of tentative hypothesis towards empathic understanding and also future design ideas that are informed by aspects of particular contexts we have hunches about." In this sense, probes-making can be understood as a form of articulated introspection into what the designer already knows, through iterative externalization of what the designer wants to know in relation to an instrumental goal. In itself, the making process of the probes carries values and benefits for design, enabling the design to understand users and speculate possible design solutions.

Probes as a Tool for Entering the User's Context

Most published probes studies have been carried out in domestic contexts. There are, however, cases in which probes have also been experimented to study work contexts such as nurses and clinical collaboration at hospitals (Jääskö and Mattelmäki 2003), e-work (in which the domestic and the working context become blurred), and ageing workers' well being (Mattelmäki 2006). Our experience while applying probes both in domestic and professional environments indicates that applying the probes approach at work has special characteristics that have not been formally addressed. For example, introducing probes in the work place can have a negative effect due to interruptions to the work of the participants. Answering questions on a diary can be a significant distraction from the participant's main task. Participants can be reluctant to take part in these studies (Carter and Mankoff 2005).

With professional probes, Lucero and Mattelmäki (2007) looked at the use of probing in professional contexts. Using a case study with industrial designers as a basis for the discussion, they also drew on other projects to further illustrate their findings. The 'Augmenting Mood Boards' case (Lucero and Martens 2006; Lucero 2009) was a project that studied the impact of augmented reality systems in work practice. The project tried to assess whether professional users would change their current work practices favoring the use of augmented reality tools that provide support for their work. Probes were applied to open a dialog with professional users (i.e., industrial designers) and find opportunities for augmented reality interaction techniques to support their work.

Seventeen practicing industrial designers were recruited for this study. They all initially agreed to participate in the study although ultimately only ten worked on the probes and sent them back. The participants varied in their education (university/academy), in age (between 24 and 50), and in gender (six women, four men). A wide variety of contexts were obtained, ranging from an office in a large company, to freelance work performed at home. Participants worked with the probes for seven consecutive days in their design studios and were free to choose the day of the week in which they would start. To increase motivation, all participants were given the probe kit during a personal meeting. All participants signed a consent form in which their anonymity was guaranteed.

The materials included in the kits probed different aspects of the life and design practice of an industrial designer. We describe the probe kit (Fig. 3) using Mattelmäki's properties of probe objects (Mattelmäki 2006). First, the kit included

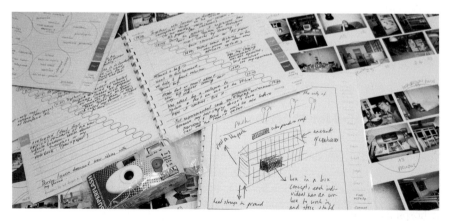

Fig. 3 The industrial designers' probe kit including a diary, a disposable camera, and some of the 200 pictures that participants made during the study

a 'Design Studio' diary that allowed probing several aspects. The diary included (1) a 'Timeline' to probe the daily thoughts and activities of the participants, (2) closed questions covering different aspects of routines, collaboration, and use of technology, (3) open questions to make people tell stories and express their opinions, (4) a map to allow self-expression, and (5) an 'Ideal Design Studio' drawing exercise to probe the dreams and aspirations of industrial designers. Second, the kit included a disposable camera to probe the environment and take pictures to visually support some of the experiences they had while working on the probes. Instead of suggesting pictures on the camera itself by re-packaging it, a 'Picture Record Table' was included in the diary where participants kept track of their pictures. Some suggestions for shots were made but half of the pictures were intentionally left unassigned so they could share different aspects of their environment or activities. In total, participants made over 200 pictures with the disposable cameras. Half of the participants personally returned the probe while the other half sent their probes by mail in the self-addressed and stamped envelopes included in the kit.

The findings from the probe study concerning the designers' way of working are reported elsewhere (Lucero 2009). We will now present the main findings in relation to the challenges of applying professional probes. To guide the discussion we use the previously described probes study with industrial designers. We also provide some illustrative examples from other projects, some of which the authors have directly been involved in.

High Demands on the Participants

Several participants dropped the study after they had initially agreed to participate. There were different reasons for not completing the study although lack of time was often mentioned. However, the energy and time demanded from participants to fill in the diaries proved to be a major problem. One participant summarizes the main difficulties participants encountered with the diaries:

- *"I must say it is a BIG job, much more than I thought. Keeping your diary has a big impact on the way I work, so I wonder if this probe is actually useful."*

Participants indicated the diary should be less time-consuming and should involve less writing. High demands on participants' efforts to complete diaries in the work environment have also been reported by Carter and Mankoff (2005).

The challenge of filling in the diaries in work context has been addressed also in a study about mobile work where participants used camera phones (i.e., mobile probes) as a way to report their experiences (Hulkko et al. 2004). In this case the participants were sent SMSes during the study with tasks for messaging and taking pictures. In another study in which camera phones were used for probing at hospitals (Mattelmäki 2006) the phones had a special probes application. In that study the participants were asked to check daily the tasks from the probes application whenever it was best suited for them. This was thought to be less intrusive for their work than SMSes arriving at inappropriate moments. However, the feedback from the participants indicates that even opening the probes application for self-reporting required a lot of activeness and remembering. Some of the nurses would have preferred receiving tasks in messages instead. Hence, the balance of activating, interfering and remembering in midst of working is delicate.

Professional probes should aim at low time-consuming activities that reduce the demands on the participants. Alternatives to diaries should be considered. In the study with designers, participants reported taking pictures with the disposable cameras as easier than writing down text. Carter and Mankoff (2005) propose a hybrid between photo and audio capture for studies in which detail is important. Pictures are most appropriate for easy capture and later recognition, while audio is better suited for annotation.

Probing as an Obligation

Several participants from the study with designers reported that filling-in the diary at times felt like an obligation, something they 'had' to do. This created a negative effect making participants often forget about working on the diary:

- *"I think I would be able to give a clearer impression (of my work) in a simpler way if I could use this study as a pleasurable extra, more like a break. The writing gave me the feeling of something that required extra attention."*

When probes become an 'obligation' participants can lose motivation and perceive working with the probes as a cumbersome task (Lucero et al. 2004).

Practical design of the probe kits and diaries can support motivation and reporting at work. The use of stickers and easy-to-access illustrations make diary keeping more playful for users (Mattelmäki 2003). The use of hints such as graphical elements, words and pictures to stimulate associations is recommended. A brief note made on the spot can later trigger deeper reflections in interviews.

Professional probes should encourage a fluent and playful process while documenting the participants' work. The materials should be easily approachable and should avoid the feeling of being an 'obligation'. One of the aims of probing is to sensitize and activate participants to reflect on everyday experiences with fresh perspectives. Thus, the probes should give motivational clues so participants can pay attention to their experiences, and have perhaps even a funny character, a pleasurable extra for work.

Understanding the Specific Work Domain

When planning the probes the nature and context of the work should be considered. In the industrial design study, the placement of the probes was closely looked into.

Fig. 4 The nurses' probes including cards and a diary

Fig. 5 The ageing workers' probe kit

To create less mess on the sometimes-cluttered desks of designers, most probe materials were concentrated into one booklet. In the study where nurses were involved (Jääskö and Mattelmäki 2003) the diaries were designed to be small and plastic covered to fit the pockets (Fig. 4). In the ageing workers' study (Mattelmäki 2006) the participants mostly worked at schools. Thus, the diaries were in form of school agendas folded into plastic pockets with clips to hold in their clothes or cleaning trolleys (Fig. 5).

The planning of probe tasks for work contexts should consider organizational and management concerns as well. In the industrial designer study, a few participants were concerned about confidentiality issues in relation to their work. This problem was addressed by first reassuring designers, indicating to them that the consent form included in the diaries explicitly considered this aspect. Participants were also free to choose the week in which they would work on the probes if they felt one project was more confidential than another. In the nurses study (Hulkko et al. 2004) the subjective character of the probes approach and its playfulness, openness and inspirational quality raised management concerns at one of the contacted hospitals. The hospital administrators were thoughtful if the patients' ethical rights were respected and if the self-reporting at work time would risk the quality of the patient care. The grounds for these concerns were legitimate, because during the study we learned that nurses were not able to complete many words in diaries and for taking pictures they had had to make special arrangements such as covering the faces of the patients. However, the probes even when partly completed did spark reflections during work that were later documented or discussed in the interviews.

Another aspect is to carefully consider topics that may be sensible in certain work environments. To study clinical collaboration the participants considered the question "describe a panic situation at work" as highly unprofessional. Panic is not a word to be used in hospital context and in patient care. Thus a provocative wording can influence strong opinions, which are sometimes aimed at, but also negative attitudes to filling-in the probes.

Professional probes should be tuned in to the special nature of the work that is being studied. Aspects of the (1) placement of the probe, (2) management concerns, or (3) the use of provocative wording should be closely looked into to allow the probes to successfully enter the environment they were sent to study.

Different Strategies to Use the Materials

Professional probes can be applied for various purposes. The probe kits, questions and the tasks often vary in each case. The reason why probes are used, the focus and the objective of the study affect how the participants should be supported in using the materials. If the aim is to focus on a specific experience, procedure or activity then the probes should be there reporting on the spot. If one is more interested in participants' characteristics, feelings and considerations, and values, then filling-in a diary is appropriate whenever it feels meaningful to the participant. In the study with designers, participants displayed a rich variety of strategies while working with the diary. Participants either filled-in the diary: (1) as they worked, incorporating the diary as a new task in their normal work, (2) at the end of each task, (3) whenever they would remember, or (4) at the end of the day. Supporting these different strategies had not been initially considered.

Similar aspects have been reported in relation to participants using photos. To document experiences, photos should be taken when these experiences occur to represent the real situation. However, if they are taken later, they can have a hidden story about the lived experience that should be traced in a following interview. As an example, in the nurses' study an anesthesia nurse took a photo of an anesthesia desk to describe 'hurry at an operation.' The researcher was confused with the photo of a piece of furniture trying to interpret the hurry in it. Later in the following interview, the nurse described that the desk illustrated a dramatic story about a situation about the operation and pointed out some clues invisible for the researcher. The notes in the diary, in which this event was described, what had happened, and how the nurses felt after completed this story. The photo, when explained afterwards, included details that to a nurse represented hurry, which were not evident to a researcher.

Professional probes should be flexible enough to allow and encourage the use of different strategies for participants to work with them.

Participants' Motivation

In the study with designers, a considerable amount of work and resources was destined to create an inspiring probe kit. The booklet itself was designed in a way that designers would hopefully appreciate that it was handcrafted and especially made for them. Upon receiving the materials, designers had very positive comments and reactions. One participant said, *"This is so nice. It really looks and feels like a diary."* The booklet was designed to visually stimulate writing. A handwriting-like font was used to communicate directly to our participants' heart and to trigger an intimate sharing of their experiences while filling-in the diary. A blue color was used for the text to further elicit that it was handwritten with a ballpoint pen. We were successful in conveying this aspect to designers as two participants asked us, *"Did you write this down manually?"* The effort put in designing the probes was rewarded by the participants' dedication to work on the probes. Similar positive comments about the aesthetics and personal touch of the material and their effect on the participants' motivation have been reported also in other studies (Lucero et al. 2004).

In the nurses' study some participants enjoyed that they were asked to study their work from many perspectives. This holistic view was very different from the way company developers normally approach their work. Usually they are asked to evaluate the technology or usability and focus on specific tasks or practices. Some participants said the probe study was valuable because they felt they also learned something new themselves. It is worthwhile indicating however, that in probe studies some participants have been confused and uncertain of the value of the subjective focus, openness and exploring character of probes. This way of approaching research was contrasting with the natural science research methods they were familiar with. For this reason in a study considering clinical collaboration (Mattelmäki 2006), tasks with professional content were added to motivate the participants from the operation theatres. Again, both positive and negative comments on the tasks were heard.

The nurses' probe tasks had visual elements in them and included collage-making assignments. Although there are individual differences how these kind of generative tasks are considered (Mattelmäki 2005), some of the participants had clearly been motivated by them. One of the nurses commented that making the visual assignments made him think in a new, more visually oriented way, which he appreciated. Two nurses later said that they kept reflecting on the probes' tasks even when the study was completed.

The nurses' probe kits (Fig. 4) included a set of illustrated cards with open questions. One task from the cards was found surprisingly inspiring and successful. The illustration had five characters: Marilyn Monroe, Florence Nightingale, an athlete showing off his muscles, Doctor Ross from an American TV soap opera, and a Finnish male pig cartoon character with individual but creative personality. The nurses were asked, "Do these characters work at your work place?" All of the participants were able to identify their co-workers and created humorous answers describing the social atmosphere at work.

Professional probes should aim at motivating participants by providing inspiring probe materials that are made especially for the study that is being undertaken and by tailoring its contents to the specific work domain. Participants will pay more attention when they feel that the questions and messages included in the probe materials are tailored for them (Fogg 2003). Using the professional jargon of the participants can support creating empathy both for participants and designers. Designing probe kit materials as handmade documents especially prepared for each study has an important effect in supporting the credibility of the material (Mattelmäki and Battarbee 2002).

Considerations for Making the Probes Work

The two probing perspectives to participation presented above help us expand our understanding of what it is like to use the probes in the design process. The first case on students' stories illustrates how the students' work on making the probes could already help them gain empathic mindset to users and sensitivity to user context. The second case on the professional probes shows what considerations should be taken in order to engage the users in the probing process. As the two cases clearly imply, the benefits of the probes for design team's collaborative learning and entering the users' world cannot be simply achieved without the design team's careful considerations and sensitivity. To summarize we suggest the following considerations on *how to make probes work*.

- **Probes should be tuned into the special nature of the participants' context:** Having a casual meeting with the participants before the probe making can greatly improve the contextual fitness of the probes' design. Especially when applying the probes in a specific work context (e.g., professional probes), the probes design should aim at low time-consuming activities that reduce the demands on the participants. Photo and audio capturing should be considered as alternatives to diaries.
- **The design team should pay their attention to new discoveries and group discussions during the making process of the probes:** Designers can gain context knowledge and build tentative design hypothesis through the making

process of the probes, including a casual meeting with the participants, group discussions and material explorations for the probes design. Be aware that this discovery can lead to essential knowledge about users and new design opportunities, even before collecting the probes returns.

- **The probes design can be a reflection of their pre-assumptions and tentative design hypothesis:** The probes questions, types of tasks and material designs are the results of the designers' pre-understanding of the topics and tentative design focus. If the designers do not reflect on how their assumptions directed them to the probes design, they might lose a chance to identify new design opportunities. In other words, making the probes is an opportunity for the designers to realize and reflect on their own assumptions and tentative hypothesis.
- **Probes can provide a 'pleasurable extra' for user's everyday routines, which can trigger their motivations and inspirations:** Probes should encourage a fluent and playful process for participants while documenting. The materials should be easily approachable perhaps even have a funny character to be perceived by participants as a pleasurable extra for their everyday routines or work.
- **Probes should be flexible enough to encourage the use of different strategies for filling in:** The probes design should allow participants to work in ways that are meaningful and relevant at various situations. For example, in professional context, the participants or their management have concerns on the ethics and confidentiality as well as time resources. Flexibility on reporting time and means should be provided to avoid these problems. The means of reporting should be open enough that the participants can frame their answers without risking their privacy. We have suggested that one solution is to allow quick and dirty filling-in strategies, i.e., to note meaningful insights or experiences during the work and to be able to deepen the information and reflections afterwards.
- **The customized design of probes can enhance participants' motivation and commitment:** The customized, handmade probe materials can create impressions to the participants that the probes were made especially for the participants, valuing their participation.

One of the aims with probes is to sensitize participants to the design topic, as well as the experiences and practices that might be relevant for the design. In that way, participants are invited to participate in a co-design process. To facilitate this process, participants can be provided with clues, "things to think with" (Papert 1980) to enable 'designerly' change oriented thinking, to be able to express their needs and dreams with regards to future experiences.

For professional contexts, supporting dialogues to enable design empathy is important too. This can be facilitated with self-made, personal probes kits. When effort is put into the customized research material we expect to motivate the participants to go beyond the official professional roles and to express their personality and their subjective experiences in relation to their work.

Key References for Further Reading

Boehner K, Vertesi J, Sengers P, Dourish P (2007) How HCI interprets the probes. In *Proc. of CHI '07*, ACM Press, 1077–1086.
Gaver W, Dunne T, Pacenti E (1999) Cultural probes. Interactions 6(1), January, ACM, 21–29.
Gaver W, Boucher A, Pennington S, Walker B (2004) Cultural probes and the value of uncertainty. Interactions *11*(5), September, ACM, 53–56.
Lucero A, Lashina T, Diederiks E, Mattelmäki T (2007) How probes inform and influence the design process. In *Proc. DPPI '07*, ACM Press, 377–391.
Mattelmäki T (2006) *Design probes*. Dissertation. University of Art and Design Helsinki, Finland.
Wallace J, McCarthy J, Wright PC, Olivier P (2013) Making design probes work. In *Proc. of CHI'13*, ACM Press, 3441–3450.

References

Boehner K, Vertesi J, Sengers P, Dourish P (2007) How HCI interprets the probes. In: Proceedings of CHI '07, ACM Press, 1077–1086
Brandt E (2006) Designing exploratory design games: a framework for participation in Participatory Design? In: Proceedings of PDC '06, ACM Press, 57–66
Carter S, Mankoff J (2005) When participants do the capturing: the role of media in diary studies. In: Proceedings of CHI '05, ACM Press, 899–908
Ellis C (2004) The ethnographic I: a methodological novel about auto ethnography. AltaMira Press, Walnut Creek
Fogg BJ (2003) Persuasive technology: using computers to change what we think and do. Morgan Kaufmann, Boston
Gaver W, Dunne T, Pacenti E (1999) Cultural probes. Interactions 6(1):21–29, ACM
Gaver W, Boucher A, Pennington S, Walker B (2004) Cultural probes and the value of uncertainty. Interactions 11(5):53–56, ACM
Hemmings T, Crabtree A, Rodden T, Clark K, Rouncefiled M (2002) Probing the probes. In: Proceedings of PDC '02, pp 42–50
Hulkko S, Mattelmäki T, Virtanen K, Keinonen T (2004) Mobile probes. In: Proceedings of NordiCHI '04, ACM Press, pp 43–51
Hutchinson H, Mackay W, Westerlund B, Bederson BB, Druin A, Plaisant C, Beau-douin-Lafon M, Conversy S, Evans H, Hansen H, Roussel N, Eiderbäck B (2003) Technology probes: inspiring design for and with families. In: Proceedings of CHI '03, ACM Press, pp 17–24
Jääskö V, Mattelmäki T (2003) Methods for empathic design: observing and probing. In: Proceedings of DPPI '03, ACM Press, pp 126–131
Lee J (2014) The true benefits of designing design methods. Artifacts 3(2):1–12
Lucero A (2009) Co-designing interactive spaces for and with designers: supporting mood-board making. Doctoral dissertation, Eindhoven University of Technology, The Netherlands
Lucero A, Martens J-B (2006) Supporting the creation of mood boards: industrial design in mixed reality. In: Proceedings of TableTop 2006, IEEE
Lucero A, Mattelmäki T (2007) Professional probes: a pleasurable little extra for the participant's work. In: Proceedings of IASTED-HCI 2007, ACTA Press, pp 170–176
Lucero A, Lashina T, Diederiks EMA (2004) From imagination to experience: the role of feasibility studies in gathering requirements for ambient intelligent products. In: Proceedings of EUSAI 2004. Springer, Berlin/Heidelberg, pp 92–99
Lucero A, Lashina T, Diederiks E, Mattelmäki T (2007) How probes inform and influence the design process. In: Proceedings DPPI '07, ACM Press, pp 377–391

Mattelmäki T (2003) Probes – studying experiences for design empathy. Empathic design – user experience in product design. IT Press, Helsinki, 119–130

Mattelmäki T (2005) Applying probes – from inspirational notes to collaborative insights. CoDesign 1(2):83–102, Taylor & Francis, London

Mattelmäki T (2006) Design probes. Dissertation, University of Art and Design Helsinki, Finland

Mattelmäki T (2008) Probing for co-exploring. CoDesign 4(1):65–78

Mattelmäki T, Battarbee K (2002) Empathy probes. In: Proceedings of PDC 2002, CPSR, pp 266–271

Papert S (1980) Mindstorms – children, computers and powerful ideas. Basic Books, New York

Paulos E, Jenkins T (2005) Urban probes: encountering our emerging urban atmospheres. In: Proceedings of CHI '05, ACM Press, pp 341–350

Sanders EB (2001) Virtuosos of the experience domain. In: Proceedings of the 2001 IDSA education conference

Wallace J, McCarthy J, Wright PC, Olivier P (2013) Making design probes work. In: Proceedings of CHI'13, ACM Press, pp 3441–3450

Supporting Early Design Through Conjoint Trends Analysis Methods and the TRENDS System

Carole Bouchard and Jean-François Omhover

Abstract This paper introduces methods and tool support to enrich the early design activity, especially in what we call the informational phase, where inspiration sources are of crucial importance. The Conjoint Trends Analysis method is put forward as a way to structure and operationalize the informational phase of early design; a model of the design process underpinning the method is introduced and guidelines are provided for its application in an industrial context. Based upon studies of designer's activity during early design formal models of this process have been created that have been used for a partial digitalization of the informational phase of design in the TRENDS system.

Introduction

Modeling the Early Design Process

Researchers in design science tend to model the design process in order to optimize it. The design process is mostly shown into linear sequential steps (Pahl and Beitz 1984; Andreasen and Hein 1987; Jones 1992; Hubka and Eder 1996; Ullman 1997; Baxter 1995; Ulrich 2000; Cross 2000; Dorst and Cross 2001; Howard et al. 2008). Some other models describe it as the succession of elementary design cycles (Lebahar 1993; Gero and Kannengiesser 2004; Boehm 1988; Blessing 1994; Roozenburg and Eckels 1995), which may also represent divergence and convergence (Van Der Lugt 2003; Design Council 2007), or the transition from an abstract space to a concrete one (Suh 1999; Tichkiewitch et al. 1995).

As early design, we consider both phases of Planning and Concepts Development (Ulrich 2000). These phases are mostly implicit until the concepts are fixed. Indeed, there are not or not systematically some explicit representations which are produced by the designers at this stage. Only in a very formalized process, for instance in

C. Bouchard (✉) • J.-F. Omhover (✉)
Laboratory of New Products Design and Innovation, Arts and Métiers ParisTech,
151 Bd de l'Hôpital, 75013 Paris, France
e-mail: carole.bouchard@ensam.eu; jean-francois.omhover@ensam.eu

© Springer International Publishing Switzerland 2016
P. Markopoulos et al. (eds.), *Collaboration in Creative Design*,
DOI 10.1007/978-3-319-29155-0_4

car design, designers elaborate some moodboards. Despite its implicit character, the early design process is strategic for future product development and for the optimization of the design process (Cross 2000; Zeiler et al. 2007). It is the place where major conceptual orientations are fixed with still high divergence capability for low costs. In order to optimize the digital chain (Computer Aided Design, Product Data Management, Product Life Cycle Management), the goal is currently to develop new computer support tools for the phase of early design. Among the various models of design, few of them propose a description of this specific phase. This description is by nature difficult because early design information is rather implicit, vague and ill-defined (Eastman 1969; Simon 1973). Information here encompasses a big diversity of heterogeneous data encompassing complex notions such as semantics, emotions and feelings. Besides, the first representations reflect poorly the cognitive and affective processes involved. Studying early design activity has enabled us to explicate skills, knowledge and expert rules that are come into play during early design. These studies showed the importance of the semantic and emotional dimensions into the design process (Bouchard et al. 2007a; Kim et al. 2008).

Modeling the early design process is necessary to improve the whole CAD-CAM chain. We propose here an information processing model of the early design activity which was translated in a methodology and some design tools that have been experienced and validated in industry. This model is structured into four phases of information, generation, evaluation and materialization (see Fig. 1). According to the model, the design process is a succession of micro-cycles including the phases of information, generation, evaluation and materialization, which enable a progressive concretization. Indeed the design process can be seen as the succession of iterative cycles, each following the four stages of information, generation, evaluation and materialization.

The information phase corresponds to the integration of inspiration sources and data by the designers. The generation phase consists in ideas and concepts generation through applied creativity. The best solutions are chosen according to the brief's criteria in an analytical way. The evaluation phase is based on the

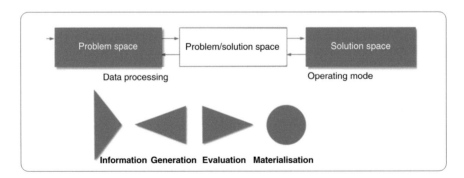

Fig. 1 Model of information processing in early design, From problem space to solution space (*top*) through a succession of divergence, convergence, and materialization phases (*bottom*)

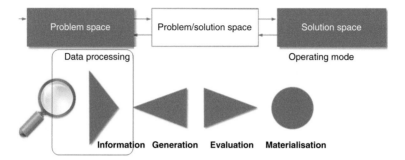

Fig. 2 Formalization of the information phase

measure of the user experience (perceptive, cognitive, affective aspects, usability and usefulness) through various levels of materialization with digital or physical prototypes and representations. If so far the phases of generation and evaluation have been well studied, in particular sketching semantic evaluation of intermediate representations, the information phase was less studied because of its implicit character. On the other hand, if creative process is well modeled in detail design, for instance with the Triz approach, it is not the case yet in early design. The link between the informative and generative phases is neither yet well established. As the materialization phase is well covered in many scientific fields and engineering, the research presented here is more about the three first ones which are less modeled yet (Fig. 2).

Computer-Aided Design Tools in Early Design

The concepts development phase is modeled in order to develop new computer aided design tools (Tovey 1992, 1994a, b, 1997; Tovey and Owen 2000; Tovey et al. 2003; Scrivener and Clark 1994a). Modeling this phase also helps to support or to improve individual and collective creativity in design (Koestler 1964; Lewis 1988; Lewis 1995; Vangundy 1992; De Bono 1995; Isaksen et al. 2000; Bonnardel 2000; Syrett and Lammiman 2002; Alberti et al. 2007; Buzan 2009). Sketches have been considered as the main media for concepts elaboration and evaluation (Schön 1983; Tovey et al. 2003; Van Der Lugt, 2000, 2001, 2003, 2005). The field of sketching and shape generation is still strongly studied, both in architecture and product design (Schön 1992; Scrivener and Clark 1994b; Do et al. 2000; Van der Lugt 2005; Goldschmidt 1994; Goldschmidt and Tatsa 2005; Goldschmidt and Smolkov 2006; Suwa and Tversky 1997; Purcell and Gero 1998; Bilda and Gero 2007). Besides, another field of research has been about concepts evaluation, with semantic analysis (Osgood et al. 1957; Osgood 1979), later sensory analysis, and more recently emotional and experience evaluation (Smets and Overbeeke 1995; Norman 2004; Green and Jordan 2001; Desmet 2002, 2008; Desmet and Schifferstein 2010). User centered design has been mostly popularized by (Don Norman 1998) in the field of products and services. Even if the research related to the generative phase of design is well modeled and disseminated, this is not yet the case of the inspirational phase which, despite its implicit character (Eckert and Stacey 1998, 2000; Büscher et al. 2000; McDonagh and Denton 2005), plays a crucial role in the creative process. Computer aided design and manufacturing tools are well utilized by industry during the detail design (Hsiao and Liu 2002). The progressive digitization of the design process increasingly compresses the transition from design towards product development. Further gains in shortening the length of the design process and the time to market needs now to address early design phases. Early design modeling is harder because of its subjective, and also affective, implicit and fuzzy character. Kansei engineering has been shown to provide a scientific framework for the elaboration of new digital tools, with the integration of semantics and emotions. It has develop tools and algorithms for design generation and evaluation (Nagamachi 2002; Berthouze and Hayashi 2002; Hayashi and Hagiwara 1997; Hsiao and Liu 2002; Ishihara et al. 1995; Schütte 2005; Schütte et al. 2006). Tool support can speed up changes and decrease design time by allowing designers to shift attention from less creative and laborious tasks to the most creative ones (Resnick 2007) in Kim et al. (2009).

Formalization of the Information Phase

The Information Phase in Early Design

The information phase covers the inspirational process, the watch actions, and all the explorations carried out to complete the design brief. This phase stimulates the emergence of new ideas and concepts. Inspiration search is performed through specific sectors of influence via various media such as the web, magazines, exhibitions, etc. The selection of relevant images is depending on color harmonies, style, object type, emotional impact, semantics and values. Designers operate a watch process going from a simple observation in their everyday life to the regular consultation of specialized data (Bouchard 1997). Even if modeling the designers' cognitive activity is of growing interest, the information phase has been less studied so far (Eckert and Stacey 2000). This phase enables to inform the generation of new solutions by referring to precedents and other sources (Lloyd and Snelders 2003). It follows a voluntary action of more or less focused research which is necessary for problem understanding (Cross 2000), goals definition (Wallas 1926; Schneiderman 2000; Amabile 1983) and functional adequacy (Osborn 1963; Schneiderman 2000). This phase involves analogical reasoning, with the extraction of design elements from adjacent sectors which are then transferred into the reference sector during concepts generation. Knowing that creative thinkers use more environmental data (Ansburg and Hill 2003), the degree of novelty of solutions is highly linked to this phase. Inspiration sources play a major role in contextual definition (Eckert and Stacey 2000). They stimulate creativity and support mental representation. They may be more or less far from the reference sector and related to functional, structural or affective dimensions (Bonnardel and Marmeche 2005; Lim et al. 2006).

Design Information and Expert Rules

Design information may be categorized in high-level (semantic, emotional, sensory descriptors, sociological values), middle-level (concepts, sector names) or low-level information (shape, color, texture). The use of abstraction levels opens the way to apply formalisms originating from the fields of artificial intelligence (Black et al. 2004; Bouchard et al. 2009) and of marketing (Valette-Florence and Rapacchi 1990; Valette-Florence 1993a, b,1994). It also reflects the expert rules implemented by the designers, who have this very particular ability to link reciprocally high-levels of abstraction of information with very concrete object attributes. In the field of marketing, the method of cognitive chaining of means-ends (Valette-Florence 1994; Bouchard et al. 2007b) aims at formalizing the relation between high-level and low-level information with a chain of values-functions-attributes. We investigated the value-function-attributes chain in the framework of car design projects. For instance, to improve a technological innovation on the dashboard, design orientations were the following: family, cohesion, family unit, cocooning, peace, smoothness.

Designers' categorization processes are poorly referenced in literature (Büscher et al. 2000; Bianchi-Berthouze and Hayashi 2002). However designers use intensively grouping strategies according to particular harmonies of color, texture, shape, semantics. The difficulty to understand this cognitive operation comes from its subjectivity, multi-dimensionality, and also from the holistic nature of visual information. Designers build more or less implicitly categories or sub-categories of visual, lexical or multi-sensory information. The categorization task requires a special expertise. Categories are specific. Their consistency is due to the coherence of the different Kansei dimensions (shape, colors, textures, semantics, values), but also the homogeneity between semantics and low-level features, and the presence of harmonies. Designers build some ambiences from these categories and sub-categories, from which they will extract relevant harmonies that they reuse in the futures solutions. Images are first groups by color harmonies, then shape and texture. Going from words to images, or vice-versa, is very fluid and dynamic. The categories are then annotated with a name and some keywords. Harmony rules apply to colors, textures or shapes, and their relation to semantics (Bouchard et al. 2009).

In the example shown Fig. 3, we recognize a common color harmony revealed by the presence of very saturated purples, pinks, oranges, in the set of images. Shapes similarities are also particular with rounded forms, like distorted plastics . . . Specific keywords come into our mind when watching these images: pop, inspired by seventies, plastics … There may be sub-categories. For instance there is chic and luxurious version of pop characterized by shiny textures. Specific name and keywords are then given to each category by annotating sticky notes. Sometimes

Fig. 3 Category Pop

they come directly from a magazine. The passage from words to images and vice-versa is iterative and very dynamic. The proposed names combine semantic adjectives, object attributes and sociological values. Only the strongest categories in terms of emotional impact, aesthetic and coherence are selected.

The Conjoint Trends Analysis Method

After studying designer's cognitive activity in early design when they look for inspiration sources and other information, we model the information phase into a procedure namely the Conjoint Trends Analysis method (CTA) (Bouchard et al. 1999, 2008). This method, based on the design rules and routines we identified, aims at anticipating trends in industrial design. The Conjoint Trends Analysis (CTA) enables to make explicit new intermediate representations before sketching activity: semantic mapping and mood-boards. It starts firstly by investigating the sectors of inspiration of the designers. These sectors are then used for analogical reasoning. Secondly, images and keywords are synthetized in mood-boards. The value of the CTA method, from theoretical point of view, is its support for formalizing coherent relations between high-level descriptors (values, semantics) and low-level attributes (shape, color, texture), while increasing the number of processed data. These relations characterize designer's expertise.

The Conjoint Trends Analysis follows three main phases. Phase 1 is about identifying the sectors of influence, Phase 2 is about trends identification, Phase 3 is about trends integration.

Phase 1: Identification of Trends Sectors of Influence

The diffusion of trends corresponds to the transfer of formal, functional or technical attributes coming from other sectors of analogy. The identification of the potential **sectors of influence** is a crucial step in the CTA method application. A sector of influence is any sector used as the source sector (for instance animals) which will enable to extract and transfer some formal or functional attributes in the sector of reference (for instance car design). The first phase of the CTA method is so to identify the sectors of influence from which the inspiration sources can be directly extracted. These sources are usually words, images or any other multi-sensory and experience stimuli. The example of biomorphism and of car design is a well known example. Biomorphism consists in taking shape expressions in the field of animals in order to apply them in the design of new artifacts. A quick and easy way of doing the identification of the **sectors of influence** is the elaboration of a **semantic mapping**.

The **semantic mapping** is a 2D visual representation built with many product images coming from the **sectors of influence** by the design team, including designers, engineers, ergonomists, and people from marketing. This kind of representation allows in a second time to highlight the **sectors of influence**. The **sectors of influence** are those used by designers to apply analogical reasoning and finally stimulate the generation of new concepts. The interpretation of the **semantic mapping** by designers leads to the visual identification of **sectors of influence** which may have influenced the design of the various products displayed. Visualizing the **semantic mapping** also offers a clear and quite exhaustive view of the product competitors. This way, it supports decision making for positioning and leading future products (Figs. 4 and 5).

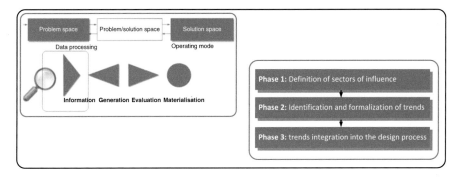

Fig. 4 Formalization of the information phase

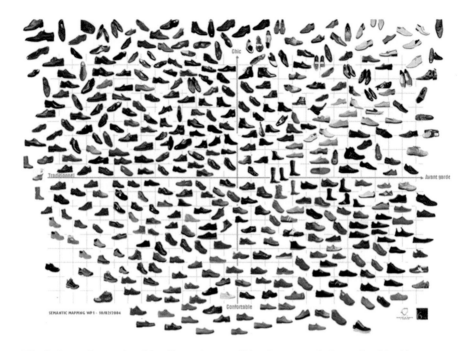

Fig. 5 Semantic mapping, Men Shoes, Axes: traditional-avant garde & comfortable-chic

Building the **semantic mapping** goes through the following steps:

(a) Gathering product images in the sector of reference (about several hundreds).
(b) Semantic characterization of the products gathered by annotation.
(c) Positioning of the images according to the semantic axes processed intuitively (semantic axes considered as the strongest or the most discriminating).
(d) Extraction of the sectors of influence by observation: for instance « this product has been designed by taking inspiration in an insect ». This step needs the presence and skill of a designer at least.
(e) Summarizing the sectors in a table or in a list.

Phase 2: Moodboards Elaboration

The goal in this second phase is to realize digital or physical **moodboards**. It firstly consists in making explicit the atmosphere or **ambiences** which integrate both consumer values and the transverse attributes such as: shape, color, usability principles, texture, semantics, values, and to extract the related pallets of attributes (see Fig. 6).

This phase is based on an iterative search of words and images realized from the content analysis of images and related texts, and then linking the values,

Fig. 6 Explicit representations elaborated through CTA approach: Semantic mapping (on the *left*) and mood-boards (on the *right*)

functions and attributes between the various collected sources. Gathering images is realized in the sectors of influence identified during the first phase. It is guided both by data related to the design brief (style, semantics, values) and the words and images included in the various inspiration sources (mainly from magazines or internet). All this information is then categorized in a coherent way in **homogeneous ambiences**, homogeneous in terms of style, values and semantics. **Moodboards**, representing these atmospheres, correspond to the compositions of images and keywords expressed by the design team. Again, the skills of the designer are crucial in this phase for guarantying this coherency and the use of harmonies rules when establishing the specific atmosphere of the **moodboard**. The information so categorized from the various sectors of influence reveals the trends which are transverse by nature.

Building the **moodboards** goes through the following steps:

1. Gathering images and keywords in the various sectors of influence identified in the first phase, in complement to the lists of specifications included in the brief (style, semantics, values, targeted population).
2. Categorization of the images by keywords (semantic descriptors and values of the targeted population) and formal attributes (colors, shapes, textures).
3. Composition of the atmosphere or ambience with coherent and homogeneous sets of keywords and images. The coherence obeys to harmony rules, for the color (rules of contrasts or complementary colors, . . .), as well as the shape, the texture, the semantics and the usability.
4. Description of atmosphere. This step consists in a textual description of the atmosphere by using and defining keywords that characterize it (rules of relations between abstract values, semantics and formal attributes).

5. Elaboration of pallets of shape, color, usage, texture. From the atmospheres constituted in step 3, the major harmonies are extracted and shown in the pallets, with strong and significant elements for the related atmosphere, such as relevant colors and textures, and eventually usage and interaction principles.
6. Description of the pallets. The pallets are described in text.

Phase 3: Trends Integration in the New Design Process of New Product Design

The third phase of integration into the design process consists in the transfer in the new design solutions of some harmonies selected in the **pallets** of **shape**, **color**, **usage**, **textures**. This integration takes place during the concept's generation. It relates firstly to the shape, then to usage principles, through concepts sketches, and finally to colors and textures. The elements extracted from the **pallets** are integrated into the concepts, by following **harmony rules**. These harmonies are compositions in which combinations of **shapes**, **colors** and **textures** offer an aesthetic set, functional, coherent and balanced.

The integration of **shape** inspiration may be done by shape interpolation or morphing. For instance, this is the case when car designers provide an aggressive character to a sport vehicle, by designing the silhouette of the front face in a way that it will remind directly of shape elements transferred from other sectors such as that of animals. Another example is aerodynamic shape given to some pens. Shape elements integration is led at two different levels: 2D sketches and 3D digital or physical models. The design team has to be aware about the possible loose of information between these two steps (from the sketch to the volume). Indeed, during the progressive concepts materialization (prototype, pre-serie, serie), the choices realized may change the nature of the positioning.

Regarding **color**, harmonies of complementary colors, contrasts or opposition are selected in the pallets and then extracted and re-interpreted in the design context. The application of colors and textures can be simulated on quick sketches or 3D digital models. The integration of usage principles, as soon as it refers to digital interfaces, is more complex and needs to be worked out in global scenarios.

Moodboards are often used as inspiration sources during the first sessions of brainstorming during concepts generation. The selection of the trends to be developed for the concepts generation has to be done according to the marketing strategy of the company. This strategy may be a leading strategy, in this case the trends to develop should be cutting-edge, or it may be a follower strategy, which will orient towards more mature trends.

The **integration of the trends in the design process** goes through the following steps:

1. Selection of harmonies in the pallets realized during the preceding step (shape, color, texture, usage).
2. Application of the integration of shape and usage, interpolations, morphings, during concepts generation, etc.

3. Application of the integration of color and texture during the concepts generation.
4. Use of keywords in moodboards to establish an argumentation related to the proposed concepts.
5. Following of design and development phases in order to be in adequacy with original design orientations and in coherency according to the selected trends.

Limits of CTA Method

From 1998 until today, The Conjoint Trends Analysis (CTA) has been disseminated in industry in marketing and design and innovation departments. It was used in different sectors of application and proved to be useful and relevant for professional designers. Applying this method has enabled to refine the theoretical model in the information phase and to highlight some shortcomings of this method. These shortcomings relate to the difficulty and effort of seeking, gathering and processing information. For this reason, a tool that partially digitizes the CTA method has been developed based on information available on the web. This tool called TRENDS is described in the remainder of this chapter.

The TRENDS Tool: Digitizing the Informational Phase of Early Design

Towards a Computer Support Tool for the Information Phase

Designers tend to build digital databases which are more and more important in their activity (Restrepo et al. 2004; Büscher et al. 2000; Restrepo and Christiaans 2005; Stappers and Sanders 2005), but which do not offer relevant search engines adapted to their needs. Indeed, these information systems are rather based on concepts-based search using the objective objects' character, more than semantic and emotions based search. Resaerchers have argued that future information systems used for creative purpose should enable navigating into the representations of knowledge, visualizing, for sharing and developing new representations in a playful way (Schneiderman 2000; Bonnardel 2000; Keller 2005; Nakakoji 2006). Besides, image search should involve a certain amount of serendipity, with the possibility to use abstract values as well as semantic descriptors. Considering all of these objectives, we set out to design a new system that would be in synergy with the cognitive processes of the designers. To this end, it was first necessary to describe their expertise in a way that could be implemented by algorithms, and following the CTA structure. This tool should provide in real time quite exhaustive results by enlarging the corpus of available images in the various sectors of influence. It would support the creative phase of concepts generation by providing some inspirational material. Data gathering by designers was achieved through fictive design scenarios and through the data extraction from previous projects where CTA was applied.

Table 1 Sectors of influence of car designers (70 % of similar sources)

Year	1997	2006
Designers	*40 (10 professional, 30 students)*	*30 professional*
Nationality	*French, English, German*	*Italian, German, Britishn French*
Sectors	1 Car design & automotive	1 Car design & automotive
	2 Aircrafts, aeronautics	2 Architecture
	3 Architecture	3 Interior design & furniture
	4 Interior design & furniture	4 Fashion
	5 Hi-Fi	5 Boat
	6 Product design	6 Aircraft
	7 Fashion	7 Sport goods
	8 Animals	8 Product design
	9 Plants	9 Cinema & commercials
	10 Science Fiction	10 Nature &urban ambiances
	11 Virtual reality	11 Transportation (moto, trucks)
	12 Fine arts	12 Music
	13 Cinema	13 Fine arts
	14 Music	14 Luxury brands
	15 Travels	15 Animals
	16 Food	16 Packaging & advertising

Kongprasert et al. (2010), Bouchard (1997); Bouchard et al. 2007a, 2008 and Mougenot et al. (2007)

The first result concerned the list of the sectors of influence used by designers (see Table 1). Table 1 shows the sectors of influence identified in 1997 and 2006 through a series of interviews held with car designers. It is interesting to observe that these sectors were more or less the same in both occasions. This means they are relatively stable for designers, suggesting that accessing related information should be supported digitally (Bouchard et al. 2008, 2009, [TRENDS D2.3]).

The second result was a table of sets of data which show the correspondence between the different categories of information processed by the designers and classified by levels of abstraction (see Table 2). This structure has been used in order to establish design ontology (Fig. 7).

Towards a Computer Support Tool for the Information Phase

This section introduces the TRENDS system an experimental tool to the application of the CTA method by using material from the web (see Fig. 8).[1] It provides

[1] The tool has been developed by a European project called TRENDS (Trends Research ENabler for Design Specifications). This project has been a collaboration between several scientific or industrial partners : INRIA Rocquencourt, PERTIMM, CARDIFF University, LEEDS University and ROBOTIKER. The end users were FIAT RESEARCH and STILE BERTONE.

Table 2 Abstraction levels and design information

Level	Category	Code	Description	Examples
High level (H)	Values	Hv	These words represent final or behavioral values	Security, well-being
	Semantic words	Hs	Often adjectives related to colour, form, or texture but also impressive words in the field of KE	Playful, romantic, aggressive
	Analogy	Ha	Objects in other sectors with features to integrate in the reference sector	Rabbit→Speed
	Style	Hy	Characterization of all levels together sketch a specific style	Edge design, classic
Middle level (M)	Sector name	Ms	Object names describing one sector or sub sector being representative for expressing a particular trend	Sports
	Context	Mc	User social context	Leisure with family
	Function	Mf	Function, usage, component, operation	Modularity
Low level (L)	Colour	Lc	Chromatic properties using qualitative or quantitative	Yellow, light blue
	Form	Lf	Overall shape or component shape, size	Square, Wavy
	Texture	Lt	Patterns (abstract or figurative) and texture	Plastic, metallic

Bouchard et al. (2007a, 2009) and Kim et al. (2009)

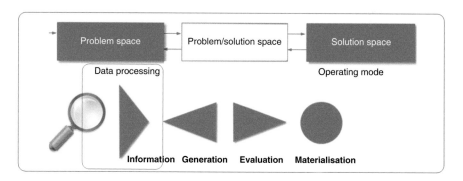

Fig. 7 Computation of the information phase

inspirational words and images to the designers in real time. This tool is dedicated to various users' profiles such as design, marketing, ergonomics. It enables image retrieval from semantic descriptors, or from image example. It also applies image processing facilities such as semi-automatic categorization and pallet generation, and statistics on words and images. The system provides values in its capacity

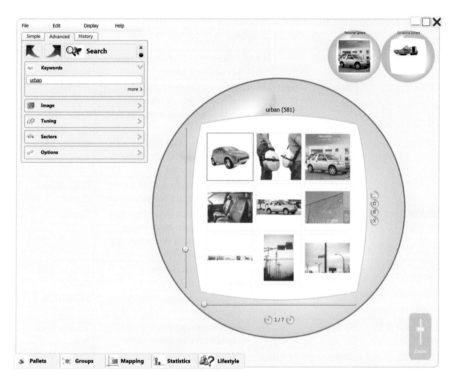

Fig. 8 Prototype TRENDS. We implemented the following functions: random image search, semantic text search, content-based search (by the similarity to an image), or mixed text/image search

to link different levels of abstraction of information and to propose new types of processing for this information. The digital transition enables automatic image grouping by color harmonies and the automatic generation of pallets of color harmonies with clustering algorithms using these harmony rules as filters. Applying Principal Components Analysis on the adjectives which are associated to the images on the websites allows a semi-automatic semantic mapping. A statistic function examines in real time the level of representativeness of a word or an image in the inspiration sectors by using similarity rules or the presence of words. The content and structure of design information processed by TRENDS system is represented Table 2. This information has been grouped by decreasing abstraction levels, going from low-level (shape, color, texture) to high-level (semantics, values). Harmony rules may apply to one of these levels (color harmonies) or to the link between different levels (semantic search).

The prototype gives access to these functions via a graphical user interface. It relies on a search engine connected to a database of approximately 2.000.000 images. This database is structured in terms of sectors of influence, the very sectors of influence that have been identified using the CTA method. The data have been grabbed using a list of source web sites within these sectors. The architecture of the system is quite open. It supports the communication of the various components (text search engine, image search engine, interface, database, etc. Some demonstration videos are available at www.trendsproject.org or contact caroll.bouchard@gmail.com).

Conclusion

Research in design science is oriented towards industry needs: increasing the variety of design solutions in less time and better answering to user needs. We modeled the information process in early design, where much is still implicit even in industry. We divided each micro-cycle in four phases: information, generation, and evaluation oriented towards decision and materialization. The information phase has been modeled in the framework of the elaboration and experimentation of the Conjoint Trends Analysis method. The progressive formalization of these different phases helped identify where computers may bring value by supporting or even automating routine tasks. However it is essential to preserve the hedonic aspects of a design activity, by providing user friendly tools, playful and attractive for the designers in some tedious or limited phases. Our research was then oriented towards digitizing the informational phase of early design with the TRENDS software. This phase is by nature ill-defined, imprecise, subjective and affective, but also individual and collective, and multi-disciplinary. This makes difficult the formalization and the computation of routines and rules in this phase. We have addressed this challenge in terms of a semi-automatic tool for supporting the information phase.

Perspectives

The formalization and the digitization of informational processes for elaborating models and contributing to the development of computational systems for the information phase is currently being enlarged to the other phases of generation and evaluation. Initial models are progressively enriched with the integration of methods and tools coming from Kansei engineering, artificial intelligence, cognitive ergonomics and applied creativity to concurrent engineering. Kansei engineering models helps us to reflect on cognitive and affective relations as well as computational rules that may exist between evaluation and generation phases. The scientific goal of is to define new computer aided design systems for enriching and improving the design process. These models need to adapt more and more to concurrent engineering, also the reduction of times and costs, and more variability in design response. Modeling the cognitive and affective processes in early design remains a field of interest, as well as the elaboration of design-oriented formalisms and the identification and selection of algorithms to enable digitization of early design processes. Particularly, the formalization of the relation between the deep needs of users and designers, while scanning the new technologies taking shape, will enable to develop new tools that can improve the interaction from the point of view of Kansei. This calls for keeping an eye on changes in design science, cognitive psychology and artificial intelligence. Recent evaluation methods, based on knowledge and tracks analysis, and also, new algorithms may be implemented to enrich design activity. The creation of new models will be based not only on the integration of implicit data, but also on the association of models in meta-models. Design information will also be enriched thanks to the advances in new disciplines such as sociology and neurosciences. The increasingly refined analysis of the interaction and of the physiological response in front of multi-sensory and multimodal stimuli will still enable us to model what is currently left implicit. Finally, a reflection should be conducted about the progressive translation towards a meta-model taking into account every phase of information, generation, evaluation and materialization in a whole unique digital device, through augmented and immersive technologies.

Acknowledgements We thank TRENDS consortium partners for their contribution to this work (see at www.trendsproject.org FIAT RESEARCH, STILE BERTONE, PERTIMM, INRIA, CARDIFF UNIVERSITY, UNIVERSITY OF LEEDS) as well as the European Commission for supporting the project.

References

Alberti P, Dejean PH, Cayol A (2007) How to assist and capitalise on a creativity approach: a creativity model. CoDesign Int J CoCreation Design Arts 3 (Suppl 1), Affective Communications in Design Challenges for Researchers, pp 35–44. ISSN: 1571–0882

Amabile T (1983) The social psychology of creativity. Springer Verlag, New York

Andreasen MM, Hein L (1987) Integrated product development. IFS (Publications) Ltd., Bedford

Ansburg PI, Hill K (2003) Creative and analytic thinkers differ in their use of attentional resources. Personal Individ Differ 34(7):1141–1152, http://doi.org/10.1016/S0191-8869(02)00104-6

Baxter MR (1995) Product design, a practical guide to systematic methods of new product development. Chapman & Hall, London/New York, p 308p

Berthouze N, Hayashi T (2002) Mining multimedia and complex data. In: Zaiane O, Simoff SJ, Djeraba C (eds) Subjective interpretation of complex data: requirements for supporting Kansei mining process, 2797th edn. Lecture notes in computer science series. Springer, Berlin, pp 1–17. ISBN: 3-540-20305-2

Bianchi-Berthouze N, Hayashi T (2002) Mining multimedia and complex data: KDD workshop MDM/KDD 2002. In: PAKDD workshop KDMCD 2002. Revised Papers. In: Zaïane OR, Simoff SJ, Djeraba C (eds), Springer, Berlin/Heidelberg, pp 1–17. http://doi.org/10.1007/978-3-540-39666-6_1

Bilda Z, Gero JS (2007) The impact of working memory limitations on the design process during conceptualization. Des Stud 28(4):343–367

Black JA, Kahol Kanav JR, Priyamvada T, Kuchi P, Panchanathan S (2004) Stereoscopic displays and virtual reality systems XI. In: Woods AJ, Merritt JO, Benton SA, Bolas MT (eds) Proceedings of the SPIE, vol. 5292. SPIE, Bellingham, pp 363–375

Blessing LTM (1994) A process-based approach to computer-supported engineering design. Universiteit Twente, Enschede

Boehm BW (1988) A spiral model of software development and enhancement. IEEE Comput 21(5):61–72, http://doi.org/10.1109/2.59

Bonnardel N (2000) Towards understanding and supporting creativity in design: analogies in a constrained cognitive environment. Knowl-Based Syst 13(7–8):505–513

Bonnardel N, Marmeche E (2005) Towards supporting evocation processes in creative design: a cognitive approach. Int J Hum-Comput Stud 63(4–5):422–435

Bouchard C (1997) Modelling the car design process. Design watch adapted to the design of car components. ENSAM thesis

Bouchard C, Christofol H, Roussel B, Aoussat A (1999) Identification and integration of product design trends into industrial design. In: ICED'99, 12th international conference on engineering design, Munich, 24–26 August, vol 2, pp 1147–1151

Bouchard C, Mougenot C, Omhover JF, Aoussat A (2007a) A Kansei based information retrieval system based on the Conjoint Trends Analysis method. International Association of Societies of Design Research, IASDR 2007, Design Research Society, Hon-Kong, 11–15 November 2007

Bouchard C, Mantelet F, Ziakovic D, Setchi R, Tang Q, Aoussat A (2007b) Building a design ontology based on the Conjoint Trends Analysis, I*Prom Virtual Conference, July 2007

Bouchard C, Omhover JF, Mougenot C, Aoussat A, Westerman S (2008) Trends: a content-based Information retrieval system for designers. In: Third international conference on design computing and cognition (Dcc'08), Georgia Institute Of Technology, Atlanta, 23–25 June 2008

Bouchard C, Kim J, Aoussat A (2009) Kansei information processing in design.,In: Proceedings of IASDR 2009, IASDR, Brisbane

Büscher M, Friedlaender V, Hodgson E, Rank S, Shapiro D (2000) Designs on objects: imaginative practice, aesthetic categorisation, and the design of multimedia archiving support. Digit Creat 11(3):161–172, http://doi.org/10.1076/digc.11.3.161.8870

Buzan T (2009) The mind map book. Dutton, New York. Paperback Edition

Cross N (2000) Engineering design methods strategies for product design. Wiley, Chichester

de Bono E (1995) Mind power. Dorling Kindersley, New York

Design Council (2007) Eleven lessons: managing design in Eleven Global Companies – Desk Research Report [online]. Available from: http://www.designcouncil.org.uk/Documents/Documents/Publications/Eleven%20Lessons/ElevenLessons_DeskResearchReport.pdf

Desmet PMA (2002) Designing Emotions. Delft University of Technology, Delft. ISBN: 90-9015877-4

Desmet PMA (2008) Product emotion. In: Hekkert P, Schifferstein HNJ (eds) Product experience. Elsevier, Amsterdam, pp 379–397

Desmet PMA, Schifferstein HNJ (2010). Experience driven design techniques. Den Haag: Lemma (in print)

Do EY, Gross M, Neiman B, Zimring C (2000) Intentions in and relations among design drawings. Des Stud 5:483–503

Dorst K, Cross N (2001) Creativity in the design process: co-evolution of problem–solution. Des Stud 22(5):425–437

Eastman CM (1969) Cognitive process and ill defined problems: a case study from design. In: Prooceedings of the first joint international conference on I.A., Washington, DC, cité in Garrigou A, 1995

Eckert C, Stacey M (1998) Fortune favours only the prepared mind: why sources of inspiration are essential for continuing creativity. Creat Innov Manag 7(1):9–16

Eckert C, Stacey M (2000) Sources of inspiration: a language of design. Des Stud 21(5):523–538

Gero JS (2002) Computational models of creative designing based on situated cognition. In: Proceedings of the fourth conference on creativity & cognition – C&C '02. ACM Press, New York, pp 3–10, http://doi.org/10.1145/581710.581712

Gero JS, Kannengiesser U (2004) The situated function-behaviour-structure framework. Des Stud 25(4):373–391

Goldschmidt G (1994) On visual design thinking: the vis kids of architecture. Des Stud 15(2):158–174

Goldschmidt G, Smolkov M (2006) Variances in the impact of visual stimuli on design problem solving performance. Des Stud 27(5):549–569

Goldschmidt G, Tatsa D (2005) How good are good ideas? Correlates of design creativity. Des Stud 26(6):593–611

Green WS, Jordan PW (2001) In: Green WS (ed) Pleasure with products, beyond usability. Taylor & Francis, London. ISBN ISBN 0415237041

Hayashi T, Hagiwara M (1997) An image retrieval system to estimate impression words from images using a neural network. In: 1997 IEEE International conference on systems, man, and cybernetics. Computational cybernetics and simulation, vol 1. IEEE, Orlando, pp 150–155. http://doi.org/10.1109/ICSMC.1997.625740

Howard TJ, Culley SJ, Dekoninck E (2008) Describing the creative design process by the integration of engineering design and cognitive psychology literature. Des Stud 29(2):160–180

Hsiao SW, Liu MC (2002) A morphing method for shape generation and image prediction in product design. Des Stud 23(6):533–556

Hubka V, Eder E (1996) Design science. Springer Verlag, London (also in German)

Isaksen SG, Dorval KB, Treffinger DJ (2000) Creative approaches to problem solving: a framework for change. CPSB, Buffalo. ISBN , ISBN 0-7872-7145-4

Ishihara S, Ishihara K, Nagamachi M, Matsubara Y (1995) An automatic builder for a Kansei Engineering expert system using self-organizing neural networks. Int J Ind Ergon 15(1):13–24, http://doi.org/10.1016/0169-8141(94)15053-8

Jones C (1992) Design methods, Second edition. Edition John Wiley & Sons, Inc.

Keller AI (2005) For inspiration only – designer interaction with informal collections of visual material. Ph.D. thesis, Delft University of Technology, The Netherlands

Kim JE, Bouchard C, Omhover JF, Aoussat A (2008) State of the art on designers' cognitive activities and computational support with emphasis on information categorisation. In: Yoo S-D (ed) EKC2008 proceedings of the EU-Korea conference on science and technology, Springer proceedings in physics, vol. 124, pp 355–363

Kim JE, Bouchard C, Omhover JF, Aoussat A, Moscardoni L, Chevalier A, Tijus C, Buron F (2009) A study on designer's mental process of information categorization in the early stages of design, IASDR Conference, Seoul, South Korea, October 2009

Koestler A (1964) The act of creation. Pan Books, London. ISBN ISBN 0330731165

Kongprasert N, Brissaud D, Bouchard C, Aoussat A, Butdee S (2010) Contribution to the mapping of customer's requirements and process parameters, 2–4 March, Kansei Engineering and Emotion Research Conference KEER 2010 Conference, Paris, France

Lebahar JC (1993) Aspects cognitifs du travail du designer industriel, Design Recherche n°3, février

Lewis JR (1995) IBM computer usability satisfaction questionnaires: psychometric evaluation and instructions for use. Int J Hum Comput Interact 7(1):57–78

Lewis KL (1988) Creative problem solving workshops for secondary gifted programming. Unpublished masters project, State University of New York College at Buffalo; Center for Studies in Creativity, Buffalo

Lim D, Bouchard C, Aoussat A (2006) Iterative process of design and evaluation of icons for menu structure of interactive TV series. Behav Inform Technol vol. 25, N°6 (8220). ISSN 0144-929X, Taylor & Francis, pp 511–519, December 2006

Lloyd P, Snelders D (2003) What was Philippe Starck thinking of? Des Stud 24(3):237–253

McDonagh D, Denton H (2005) Exploring the degree to which individual students share a common perception of specific trend boards: observations relating to teaching, learning and team-based design. Des Stud 26:35–53

Mougenot C, Bouchard C, Aoussat A (2007) A study of designers cognitive activity in design informational phase, ICED 2007. In: 16th international conference on engineering design, Paris, August 28th–31st 2007

Nagamachi M (2002) Kansei Engineering in consumer product design. Ergon Des 10(2):5–9, http://doi.org/10.1177/106480460201000203

Nakakoji K (2006) Meanings of tools, support, and uses for creative design processes. In: International design research symposium '06, 156–165, Seoul, November, 2006

Norman DA (1988) The psychology of everyday things. Basic Books, New York. [Reprinted MIT Press, 1998]

Norman DA (2004) Emotional design: why we love (or hate) everyday things. Basic Books, New York

Osborn AF (1963) Applied imagination: principles and procedures of creative problem solving. Charles Scribner's Sons, New York

Osgood CE (1979) Focus on meaning: explorations in semantic space. Mouton Publishers, The Hague

Osgood CE, Suci G, Tannenbaum P (1957) The measurement of meaning. University of Illinois Press, Urbana. ISBN ISBN 0-252-74539-6

Pahl G, Beitz W (1984) Engineering design. Springer, London

Purcell AT, Gero JS (1998) Drawings and the design process. Des Stud 19:389–430

Resnick M (2007) Sowing the seeds for a more creative society. Learn Leading Technol 35:18–22

Restrepo J (2004) Information processing in design. Delft University Press, Delft

Restrepo J, Christiaans H (2005) From function to context to form. In: Proceedings of the 5th conference on creativity & cognition – C&C '05. ACM Press, New York, pp 195–204, http://doi.org/10.1145/1056224.1056252

Roozenburg NFM, Eckels J (1995) Product design: fundamentals and methods. Wiley, Chichester

Schneiderman B (2000) Creating creativity: user interfaces for supporting innovation. ACM Trans Comput Hum Interact 7(1):114–138

Schön DA (1983) The reflective practitioner: how professionals think in action. Basic Books, New York, (Reprinted in 1995)

Schön DA (1992) Designing as reflective conversation with the materials of a design situation. Knowl-Based Syst 5(1):3–14

Schütte S (2005) Engineering emotional values in product design, Thesis

Schütte S, Alikalfa E, Schütte R, Eklund J (2006) Developing software tools for Kansei engineering processes: Kansei Engineering Software (KESo) and a design support system based on genetic algorithm. In: Proceedings of the QMOD conference, Liverpool, UK, August 2006

Scrivener SAR, Clark SM (1994a) Chapter 1: Introducing computer-supported cooperative work. In: Scrivener SAR (ed) Computer-supported cooperative work, Avebury Technical, Ashgate Publishing Ltd, Farnham, pp 51–66

Scrivener SAR, Clark SM (1994b) Sketching in collaborative design. In: Interacting with virtual environments. Wiley Professional Computing, Chichester

Simon HA (1973) The structure of ill structured problems. Artif Intell 4:181–201, cité in Garrigou A., 1995

Smets GJF, Overbeeke CJ (1995) Expressing tastes in packages. Des Stud 16(3):349–369

Stappers PJ, Sanders, EB-N (2005) Tools for designers, products for users? The role of creative design techniques in a squeezed-in design process. In: Hsu F (ed) Proceedings of the international conference on planning and design, NCKU, Taiwan

Suh NP (1999) Applications of axiomatic design. Integration of process Knowledge into Design Support, ISBN 0-7923-5655-1, Kluwer Academic Publishers, Dordrecht

Suwa M, Tversky B (1997) What do architects and students perceive in their design sketches? A protocol analysis. Des Stud 18(4):385–403

Syrett M, Lammiman J (2002) Creativity. Capstone, Oxford

Tichkiewitch S, Chapa Kasusky E, Belloy P (1995) Un modèle produit multivues pour la conception intégrée, Congrès international de Génie Industriel de Montréal – La productivité dans un monde sans frontières, vol. 3, pp 1989–1998

Tovey M (1992) Intuitive and objective processes in automotive design. Des Stud 15(1), PP23 à 41

Tovey M (1994a) Computer aided vehicle styling: form creation techniques for automotive CAD. Des Stud 1

Tovey M (1994b) Form creation techniques for automotive CAD. Des Stud 13(1)

Tovey M (1997) Styling and design: intuition and analysis in industrial design. Des Stud 18(1):5–31

Tovey M, Owen J (2000) Sketching and direct CAD modelling in automotive design. Des Stud 21(6):569–588

Tovey M, Porter S, Newman R (2003) Sketching, concept development and automotive design. Des Stud 24(2):135–153

TRENDS Consortium, TRENDS SCIENTIFIC REPORT D2.3 – Procedure for statistics realization

Ullman D (1997) The mechanical design process, 2nd edn. McGraw-Hill, New York

Ulrich KT (2000) In: Steven D (ed) Product design and development, 2nd edn. Eppinger, Denkendorf

Valette-Florence P (1993a) Les démarches des styles de vie concepts champs d'investigation et problèmes actuels, Recherche et applications en marketing, N°1

Valette-Florence P (1993b) L'univers psycho-sociologique des études de styles de vie apports limites et prolongements, Revue Française du marketing, n°141

Valette-Florence P (1994) Introduction à l'analyse des chaînages cognitifs, Recherche et Application en marketing, vol. 9, n°1, pp 93–118

Valette-Florence P, Rapacchi B (1990) Application et extension de la théorie des graphes à l'analyse des chaînages cognitifs: une illustration pour l'achat de parfums et eaux de toilette, Papier de recherché

Van Der Lugt R (2000) Developing a graphic tool for creative problem solving in design groups. Des Stud 21(5):505–522

Van Der Lugt R (2001) Developing brainsketching, a graphic tool for generating ideas. Idea Safari, 7th European conference

Van Der Lugt R (2003) Relating the quality of the idea generation process to the quality of the resulting design ideas. ICED '03. In: 14th international conference on engineering design. Stockholm, Sweden

Van Der Lugt R (2005) How sketching can affect the idea generation process in design group meetings. Des Stud 26(2):101–122

Vangundy AB (1992) Idea power. Amacom, a Division of American Management Association, New York

Wallas G (1926) The art of thought. Harcourt, Brace & World, New York

Zeiler W, Savanovic P, Quanjel E (2007) Design decision support for the conceptual phase of the design process. In: IASDR'07, conference by the international association of societies of design research, Hong-Kong, November 2007

Part II
Generating Ideas & Concepts

Designing with Cards

Andrés Lucero, Peter Dalsgaard, Kim Halskov, and Jacob Buur

Abstract In this chapter, we focus on design techniques that employ a particular form of design materials, namely design cards. Design cards can support different phases of a design process, from initial ideation through ongoing concept development towards evaluation of design concepts. We present three different techniques, namely PLEX Cards, Inspiration Card Workshops and the Video Card Game, and how they are used. Once we have illustrated the three techniques, we discuss general characteristics of design cards that make them great tools in collaborative design (i.e., tangible idea containers, triggers of combinatorial creativity, and collaboration enablers).

Introduction

In this chapter, we focus on design techniques that employ a particular form of design materials, namely design cards. Such techniques are good at bringing multiple participants together in making sense of observations and creating new exciting ideas.

Design cards are a low-tech, tangible, and approachable way to introduce information and sources of inspiration as part of the design process, and they have characteristics that set them apart from other media. Cards are instantly recognizable to most participants, meaning that they can serve as shared objects between diverse groups of participants. The tangible and manifest nature of design cards furthermore enable them to function as props that encourage and support design moves in a manner visible to all participants, and they are open to ongoing reconfiguration and manipulation in a very straightforward manner. Design cards can support different phases of a design process, from initial ideation through ongoing concept development towards evaluation of design concepts. Cards can

A. Lucero (✉) • J. Buur
Mads Clausen Institute, University of Southern Denmark, DK-6000 Kolding, Denmark
e-mail: lucero@acm.org

P. Dalsgaard • K. Halskov
CAVI, Aarhus University, Aarhus, Denmark

© Springer International Publishing Switzerland 2016
P. Markopoulos et al. (eds.), *Collaboration in Creative Design*,
DOI 10.1007/978-3-319-29155-0_5

be used with different sets of rules, depending on the design situation. Wölfel and Merritt (2013) provide a brief overview of design card sets by highlighting their key characteristics and differences. We will show three different techniques, namely *PLEX Cards*, *Inspiration Card Workshops* and the *Video Card Game*. Once we have illustrated the three techniques we will discuss more generally why such cards work.

Three Design Card Types

PLEX Cards

Playfulness is a state of mind whereby people approach everyday activities with a frivolous, purposeless and frisky attitude. Playfulness can be designed into (interactive) products and services to elicit more meaningful user experiences (Lucero et al. 2014). The Playful Experiences (PLEX) Cards (Lucero and Arrasvuori 2010, 2013) (Fig. 1) assist designers and other stakeholders in thinking about playfulness when designing and evaluating interactive products or services.

A deck of PLEX Cards consists of 22 cards, each describing a different playful experience framework category (Fig. 2). The top half of each card depicts different human emotions in an abstract way, with pictures of faces in black and white to help those using the cards focus on the emotion. The bottom half shows concrete examples from everyday life, with color pictures of hands suggesting possible interactions. The 22 cards cover different aspects of playfulness along *positive-negative*, *individual-social*, and *momentary-long term* dimensions (Lucero et al. 2014). Designers, researchers, practitioners and students alike have successfully used the PLEX Cards in their projects.

Fig. 1 A PLEX Cards Workshop

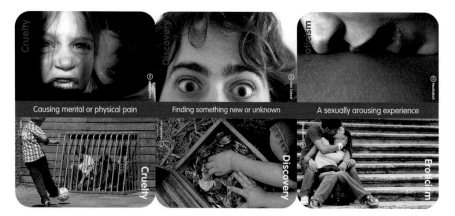

Fig. 2 Three (out of 22) PLEX Cards covering negative (i.e., Cruelty), individual (i.e., Discovery), and social (i.e., Eroticism) aspects of playfulness

Printing the Cards

Three main activities must be performed before using the PLEX Cards. The first one in order of importance is to clearly identify a design problem. The more specific the problem description and context of use are, the easier it will be to use the PLEX Cards. By combining different playfulness categories, the cards are a powerful tool to help the design team diverge and explore different aspects of the design problem. However if the design problem is too open, the PLEX Cards might simply bring about more alternatives, which might generate confusion and frustration, especially for students.

The second and third activities happen online. We originally printed and freely distributed 200 decks of PLEX Cards across different universities and research institutions around the world, but we have since run out of physical card decks. Therefore, a digital version of the PLEX Cards is now available at www.funkydesignspaces.com/plex/where people can freely download a high-resolution PDF version. The cards must be printed preferably on a color laser printer and then manually cut to form a deck of 22 cards. There are also Spanish, German, French and Polish versions of the cards available for convenience.

Using the Cards

When the PLEX cards were first created, participants would typically use the cards individually, in pairs, or in small groups of three to seven people to generate ideas. The cards would be drawn from the deck randomly, discussing one category until people felt they had to clear the table and take a new card, as they could no longer

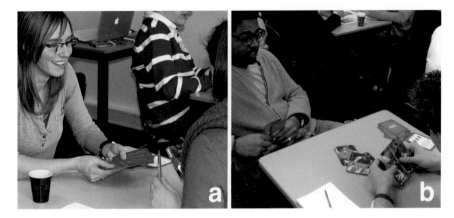

Fig. 3 The PLEX Brainstorming technique, (**a**) The first player on the right randomly picks a card from the deck (i.e., Sympathy), the seed card, which she will put face up on the table, (**b**) Players discuss an idea after they have each placed one card on top of the seed card

come up with new ideas. Two associated idea-generation techniques – namely PLEX Brainstorming and PLEX Scenario – were devised to guide and provide structure when using the PLEX Cards.

PLEX Brainstorming

The first technique is PLEX Brainstorming, which aims at rapidly generating a large amount of ideas. Participants of the idea generation session (from now on called players) are split into pairs. Each pair is handed a deck with 22 PLEX Cards. The first player randomly picks one card from the deck and places it face up on the table so that both players can see the card (Fig. 3a). This card becomes the seed card. Both players draw three extra cards from the remaining 21 PLEX Cards available in the deck. Players look at their own cards, but not at the other's. Players can now start co-constructing ideas.

The first player begins explaining the idea on basis of the seed card. The second player listens and considers the categories in their own cards. When the second player feels that they can elaborate further on the idea, they take one card from their hand, put it down on the table, and explain how it changes the initial idea. When the first player thinks they can continue with the idea based on the cards in their hand, they pick another card and place it on the table. After three cards have been dealt on the table players can freely discuss the idea (Fig. 3b). Based on the three cards available on the table, both players agree on what the idea is about and write a description of it. Once all cards have been put back in the deck and the deck has been shuffled, then the players can start a new round of idea generation. One round takes between five and ten minutes, thus three to six rounds can be completed in half an

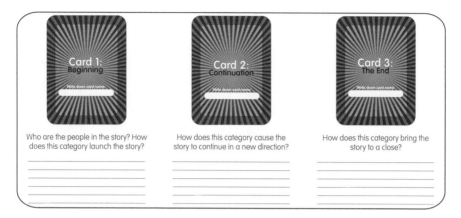

Fig. 4 The PLEX Scenario template where questions guide the scenario creation

hour. If there is an uneven total number of people, form one group of three players and allow the first player to place a fourth card on the table before discussing the idea. Having three random cards in their hands at the start of PLEX Brainstorming gives players some choice over which card they place on the table and use to extend the idea originating from the seed card.

PLEX Scenario

The second technique, PLEX Scenario, generates more complete idea descriptions in a slightly longer period of time (both compared to PLEX Brainstorming), focusing on the quality and full-roundedness of the created ideas. Similarly to PLEX Brainstorming, players are split into pairs. Each pair randomly selects three PLEX Cards from the deck of 22 cards and puts them face up on the table. Using an A3 template that can be found online (Fig. 4), players co-create a scenario using the three cards.

The scenario (or use story) is first triggered by an action related to the first card, then it is developed further by steering the story in a new direction with the second card, and it is brought to a close with the third and final card. Players are allowed to change the order in which the cards were initially drawn, until they find a combination that helps them build a scenario. The scenario is documented on the template either as text or sketched as a three-frame cartoon strip (Fig. 5a). One round of PLEX Scenario takes 10–15 min to complete, thus two or three scenarios can be created in half an hour.

In a variation of the technique, players first randomly pick seven cards and put them face up on the table (Fig. 5b). The players then create the scenario by selecting three of these available cards and place them in the order they choose. Again, you can form one group of three players with uneven numbers of people.

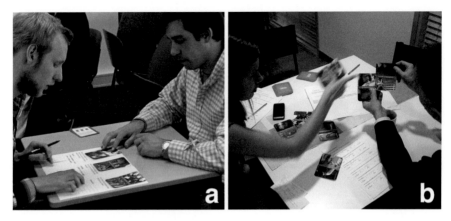

Fig. 5 The PLEX Scenario technique, (**a**) Players documenting their scenario based on three cards, (**b**) Players starting with seven random cards open on the table

After Using the Cards

One common challenge with idea generation in general is that of documentation. In the case of PLEX Brainstorming the idea can dramatically change when a new card is laid on the table. However the resulting idea is only documented when it is considered complete, i.e., after the last card has been revealed. Unless the entire session has been video recorded, interesting aspects stated in the beginning of the session may be left out of the documentation. PLEX Scenario solves this as documentation is embedded in the technique by asking players to use the A3 template to write down their ideas.

As was stated earlier, the PLEX Cards cover different aspects of playfulness, some of which one might not immediately associate with a playful state of mind. Categories that might be considered strong, controversial or difficult – such as Cruelty, Subversion, Suffering or Eroticism – can help some people think in unconventional ways, but can block others and lead them to discard some cards. When using the PLEX Cards for the first time, the facilitator should encourage people to try those potentially difficult cards as they sometimes can lead to radically new ideas.

The PLEX Cards were originally created to support people who wish to design for playfulness, and thus were meant to be used during the early stages of the design process. However, the cards have been used throughout the design process. We have seen teams keep their cards pinned to the wall as a reminder of the original idea. The cards have also operated as a checklist and a guide when evaluating the resulting product (Lucero et al. 2013).

The PLEX Cards Setup

Work space for 2–20 people:
- a laptop and projector to introduce the design problem
- enough small tables for all to fit

PLEX Brainstorming:
- a deck of PLEX cards per pair
- pens
- paper or sticky notes for documentation
- 5–10 min per round, 3–6 rounds in half an hour

PLEX Scenario:
- a deck of PLEX cards per pair
- pens
- A3 PLEX Scenario templates, 1–2 per pair
- 10–15 min per round, 2–3 rounds in half an hour

Picking which PLEX technique:
- PLEX Brainstorming better to start exploring ideas
- PLEX Scenario can then help round off ideas

Inspiration Card Workshops

An Inspiration Card Workshop (Halskov and Dalsgaard 2006, 2007) (Fig. 6) is a collaborative design event involving professional designers and participants with knowledge of the design domain in which domain and technology insight is combined to create design concepts. This method is often employed at an early stage in design projects in which designers have not yet settled on potential solutions to the design problem at hand. Alternatively, it can be employed in design projects where participants find themselves stuck or fixated on a solution they are not satisfied with and seek novel solutions.

Inspiration Card Workshops are primarily used in the early stages of a design process, during which designers and their collaborators narrow down potential future designs. The participants in an Inspiration Card Workshop are typically a combination of designers and domain experts, and the goal of the workshop is to develop design concepts from two types of inspiration cards: Technology Cards and Domain Cards. The workshop has four steps: preparation, introduction, combination and co-creation, and presentation.

Fig. 6 An Inspiration Card Workshop

Fig. 7 Inspiration Card Game cards, (**a**) Technology Card (the text translates as 'Dripping text') (**b**) Domain Card (the sign translates as 'Today's special offer')

Preparing the Cards

The main preparation activity consists of selecting and generating the two types of inspiration cards. These are index card-sized cards with a picture, a title and optionally a short text snippet.

Technology Cards, which are typically generated by the designers who participate in the event, represent technologies that may directly or indirectly be part of the design concepts. A Technology Card can represent specific technologies or interactive installations with a prominent technological component. As an example, the card in Fig. 7a, Dripping Text, is a Technology Card representing a specific application of a thermal camera tracking technology for an installation in which the silhouette of a user is tracked, allowing the user to interact with virtual text dripping down from the top of a display. To support the selection and generation of Technology Cards, we have designed a website, www.digitalexperience.dk, where inspiring interactive systems and installations are curated. Each post on the website consists of a short presentation of an innovative technology or application, and designers can create their own Technology Cards collection for subsequent printing.

In contrast to Technology Cards, which can often be reused across projects, Domain Cards represent information about the specific domain for which novel concepts are being designed. Domain cards may pertain to situations, people, settings, or themes from the domain. Domain Cards are typically generated on the basis of studies of the domain or knowledge from domain experts. While the designers who facilitate an Inspiration Card Workshop will often produce the cards, our experience shows that it is very fruitful to involve domain experts in the generation of the cards. Figure 7b is an example of a Domain Card from a department store, and it represents a prominent sales area from the store that had been identified by domain experts as particularly important to address.

Both types of inspiration cards are developed before the main workshop event. It can take several hours to select and prepare the cards; this is highly dependent on the status of the design project. In some projects, it has already been determined that certain domain aspects or types of technologies are crucial, and the selection of cards is thus more straightforward; in other projects, these aspects have not yet been decided upon, in which case the selection and production of cards can take more time.

During the Workshop

The Introduction

The workshop itself begins with a presentation of the Technology and Domain Cards selected. Each card is presented in turn, often with the help of images or video clips, to ensure a shared understanding. In general, this takes one to three minutes per card. Designers usually present the Technology Cards, while the domain participants introduce the Domain Cards. Typically, a facilitator with experience in using the method is appointed to keep the workshop on track.

Combination and Co-creation

For the subsequent combination and co-creation step the group of participants are split into teams of 4–6 people. In this step, one or more teams of participants collaboratively combine the cards and place them on posters in order to generate and document design concepts (see Fig. 8). Based on our experience, we recommend around 10–12 domain cards and 10–12 technology cards. Too few cards can constrain and limit the creative output; too many cards can lead participants to lose the overview of the options at hand and result in much time spent searching

Fig. 8 Combination and co-creation of design concepts using Inspiration Cards

through piles of cards. We advocate for combination and co-creation taking place in groups of 4–8 participants. If more people participate, we suggest splitting into several groups after the introduction. The groups can meet to present and discuss the concepts in the final presentation step.

The combination and co-creation step is often initiated by a discussion in which the participants establish a shared understanding of the cards. There are no set rules for turn taking, and cards may be combined in the way the participants deem most productive. Participants can start by selecting themes or situations from the domain that they wish to support, or transform and then select Technology Cards as a means to this end. Alternatively, they may select intriguing technologies as their starting point, and then look for situations to which they may be applied. In addition to the two types of inspiration cards, we also suggest having blank cards that participants can fill out themselves if they want to bring a specific type of inspiration into play in the workshop.

The workshop format is intentionally very open with respect to the structure of combination, emphasizing that participants are free to pursue whichever form of amalgamation to form interesting concepts. Any number of cards may thus be mingled to create a design concept. The cards are affixed to poster-sized pieces of cardboard (Fig. 9), and participants are encouraged to write descriptions and brief scenarios on the posters in order to further sketch out and articulate the concept.

In this phase, the facilitator can play an important role if ideation does not progress as intended. The facilitator may guide the discussion and ask questions to get everyone involved, in case some participants are hesitant to engage. The facilitator may also keep an eye on the types of concepts that are being created and suggest to look in new directions if participants become fixated on a particular domain or technology card. Finally, the facilitator may help ensure that all of the concepts are adequately described. A common pitfall is to rapidly develop a concept and move on in order to keep up the pace and develop as many concepts as possible,

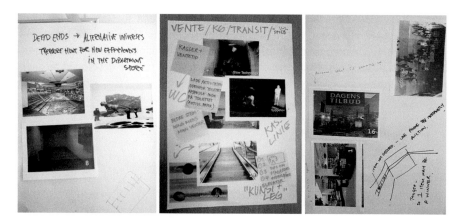

Fig. 9 Posters with cards combined to generate and capture a design concept

leading to concepts that can be hard to understand in the subsequent phases of a design process. Here, the facilitator can step in to prompt participants to add more content to the concept posters.

Presentation

After the combination and co-creation step, the participants take a short break to step back and reflect on the resulting design concepts. In the case of a single team of participants, each poster is discussed in plenum. In the case of several teams concurrently combining and creating posters, each group presents its design concepts. The object of this step is to ensure a common understanding of the concepts, rather than to evaluate them in terms of whether they are appropriate or realistic.

After the Workshop

The Inspiration Cards Workshop does not specify exact activities to be undertaken after the workshop. However, since the objective of the workshop is to develop design concepts, assembling a collection of concepts from the workshop typically follows it. Inspiration Card Workshops will usually result in 6–10 concepts per group involved in a one-hour combination and co-creation step, and in our experience these need to be documented and assembled so that participants can subsequently revisit and evaluate the concepts. In the projects we partake in, we assemble the concepts in catalogues that are shared among all participants. We will then meet up with the other participants at a later date to discuss the viability and potential of the concepts, normally leading to a selection of a subset of concepts.

Often, these post-workshop sessions can themselves lead to further refinement and concept development, e.g., when multiple concepts from the workshop are combined to form more refined concepts that are then brought on to the next phases of a design process, in which they may for instance be further developed through scenarios, mock-ups, prototypes, etc.

The Inspiration Card Setup

Work space for 3–15 people:
- a computer and a projector for viewing video
- one large table plus other for each team of 4–6 people
- a wall for attaching theme posters

Domain and technology cards:
- two copies of each domain and technology card per team
- blank cards for creating cards during the workshop

Other materials:
- pens, glue
- Sticky notes and A3 poster size paper

Video clips:
- one short video for each technology card

Time estimate:
- Preparing the cards: 2–6 h
- Introduction: 30 min
- Combination and Co-creation: 60–90 min
- Presentation: 20 min

Video Card Game

The Video Card Game (Fig. 10) provides a playful way for design teams to make sense of video recordings from user research. It was developed in industry to enhance collaboration between user-centered designers and engineering development teams and to encourage the development team to take ownership of user problems with their products or prototypes (Buur and Søndergaard 2000). The designers or researchers, who made the video recordings select many short clips from their material and produce a picture card to represent each clip. This allows a larger group of participants to each pick a random number of cards to study. Participants can form groups of cards on the table to suggest themes, and they can negotiate which cards belong to which theme. A Video Card Game session can typically cover 20–150 video clips with 4–16 participants and will take 3–5 h.

Fig. 10 A Video Card Game

In the early phases of a design project (e.g., field study and interview video), the Video Card Game helps the team make sense of recordings and form early ideas. The game typically results in often surprising perspectives on the material: the themes will describe issues worth exploring further and design opportunities that may be investigated. In later project phases, when prototypes exist (workshop and usability evaluation video), the focus will be on identifying problems, prioritizing them and finding solutions. The game encourages a focused understanding of which problems need attention.

Preparing Video Material and Cards

The Video Card Game works best with video material that contains visual activities, i.e., communicates on a non-verbal level (field observations and usability evaluation videos). Video recordings that are dominantly verbal, such as interview and discussion recordings, may better be interpreted with verbal methods, such as affinity diagramming (Lucero 2015).

In preparing the video clips and the cards (Fig. 11a) the designers or researchers who made the recordings browse their material and select clips that show the most significant actions. The clips are typically thirty seconds to two minutes long and preferably contain one closed event rather than many. There is no particular principle for selecting clips. Designers will go by their professional interests, i.e., they can pick what they find puzzling, surprising, characteristic and otherwise relevant to the project in focus. In this step they will not be expected to explain their choice of clips. The video clips will inevitably trigger observations beyond what the researchers can imagine; hence the selection of video will not steer the discussion in a very specific direction. Rather, the videos delimit the field of exploration: one cannot expect participants to talk about what they cannot see.

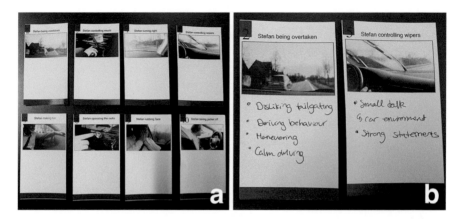

Fig. 11 Video Card Game, (**a**) Video cards representing 8 selected video recordings from user research, (**b**) Two video cards annotated with observations after watching each clip

The number of clips will vary depending on the material and on how many participants there are in the game. The card game usually works best with 30 to 100 sequences, and each participant can handle 10 to 20 cards in a reasonable time for making observations. The video clips should be available as separate digital files so they can be watched in an arbitrary order; any computer editing software will do. To strengthen the link between the clip and its card, they need to be named consistently.

The naming of cards and clips is significant as it influences the flow of the discussion. Using the name of the person(s) depicted encourages empathy (i.e., it makes a difference to talk about 'Lars' rather than 'this person'), and the activity description should be neutral and brief – to avoid suggesting a particular interpretation.

Numbering the clips makes it faster to refer to a particular clip in the heat of discussion. When more than one person prepares the clips and cards in parallel this means deciding on a numbering system upfront.

Setting Up the Game Table

The way the room is arranged for the video card game has an influence on the dynamics of the design discussion. We have learned that the players will not employ video during the discussion if the spatial barrier to grab the card and play it is too big, or if they have to stand up in front of the group whenever they want to make a point. The players need to be seated within easy reach of both the cards and the monitor.

In addition to organizing the space, the way participants are invited into the game as they enter the session affects how the game unfolds. It is important to establish a playful, yet goal-oriented, atmosphere from the start. We give the researchers time to talk about the people they have met and explain how the videos were recorded. As participants only get to see snippets of the full video material, it is important to provide some broader context. This can be in the form of portrait posters (these are the people we have studied) and brief stories from the field by the researchers, who made the video recordings. Artifacts collected in the field also help establish context.

With novice participants we prefer to start with a small interaction analysis exercise with an example card to sharpen the attention on visual content, to demonstrate how different people observe differently (and that this is beneficial), and to point out the difference between observation and interpretation (which is discouraged).

Observations are things we can actually see in the video frame: they do not need inference about what people think, or about what happened before or after. For example, an observation from a video clip from a kitchen project could be, *"The woman hands the girl a plate in the kitchen."* It is something that no one can doubt when seeing it. An interpretation of the same clip could be: *"The daughter needs her mother's help in setting the table"* – but we cannot see that she will be laying the table, or that she indeed needs help. Interpretations tend to be too speculative for the sense making process and are best left to the last step of the game.

Playing the Game

The Video Card Game runs through four steps inspired by the 'Happy Families' card game for children where participants take turns to ask each other for particular cards with the aim to collect complete families of four picture cards. The game is similar to *Quartett* in Germany or *Firkort* in Denmark.

Step 1: Dealing the Cards (30 min)
The cards are dealt randomly between. A random selection helps the players focus on the contents of each individual clip.

Step 2: Reading the Cards (60 min)
The players then split up to watch the video clips from the cards that they hold (Fig. 12a). They are encouraged to watch the clips once or twice only and make quick notes that describe observations made directly on the card (Figs. 11b, 12b). By annotating each card in their own handwriting the players come to "own" the card, which is important in the later stages. If players work in pairs, each card will encourage them to formulate observations together.

Step 3: Arranging Your Hand (30 min)
When players return to the game table they are asked to group their cards openly in front of them on the table (Fig. 13a). This encourages the players to start making sense about what might be important to them in the clips. We will refer to the groups

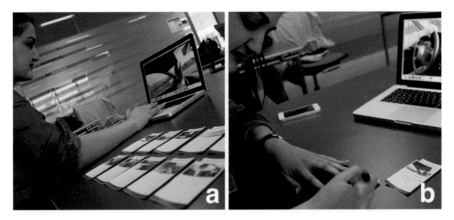

Fig. 12 Reading the Cards, (**a**) Players watching video clips in pairs from a laptop computer, (**b**) One player writes down notes with the pair's observations from that particular video clip

Fig. 13 Creating themes, (**a**) One pair arranging their hand by grouping their cards on the table, (**b**) The resulting groups of cards and their corresponding titles

as 'families.' Each player (or pair of players) around the table briefly presents their card families. There are no restrictions on how players group their cards as long as it makes sense in terms of the design activity (e.g., user activities, design problems).

Step 4: Collecting Card Families (60 min)
Each player (or pair) is then asked to choose their favorite family of cards. One after another the players describe the theme of the family they have chosen as precisely as they can. This invites the other players to contribute with cards that seem to fit into the same family. Before moving from one family to the next, the facilitator mounts the cards belonging to the family on a separate poster (Fig. 13b). If a card seems to belong in two families, the players simply make a copy. Collecting the card

families continues until all (or most) cards have found a place. The grouping of cards encourages discussion on finding the exact wording of the family heading: it needs to be precise enough to define which cards belong and which do not. By selecting their favorite family, the players also take responsibility for a theme including the labeled poster with cards.

After the Game

To gain an overview of the themes, the card family posters are pinned to a wallboard. This provides the opportunity to reflect on the immediate outcome of the game. The participants are then asked to arrange and prioritize the themes: which one do we need to discuss first? Which themes seem most important to the design project? Each 'family owner' is encouraged to lead the discussion and add notes to the poster. Since none of the players have seen all the clips, it can be advantageous to return to the video at this point. Typically each player will show and explain 'their' clips to the others, and argue how these clips are able to increase understanding about a theme.

Mock-ups, prototypes, and artifacts collected in the field have proven to be good facilitators of the discussion when they are readily available on the table to point at and think about. They help guide the discussion towards design ideas and hence help to construct a relevant focus for designing. The Video Card Game can lead beyond sense making of the material to decisions on how to move forward and what to do next. The video cards also serve as "tangible arguments" that can increase participants' confidence when they present and argue for their new ideas.

At the end of the video card game, the immediate results – the posters with video card themes and notes – are copied and circulated amongst the participants. Often this simple documentation is sufficient for participants to be able to prioritize activities and divide tasks among themselves for the next design move: who should further investigate what, or which design problems need attention.

What might go wrong? If the players choose categories of that are too general (e.g. *"Here is something about the product, and here's something about activities they do . . . "*) then the discussion will stay on a shallow level. This is an important role for the facilitator – to encourage the players to unfold their observations. Sometimes it helps to ask for a 'poetic' heading, rather than a descriptive (and boring) one!

The Video Card Game Setup

Work space for 4–16 people:
- a screen or projector for viewing video
- a table large enough for all to fit
- a wall for attaching theme posters

(continued)

Equipment for parallel viewing:
• computers for individual or paired viewing of video clips

Video cards:
• one card per each video clip

Video clips:
• 10–15 clips per participant
• duration of each clip 30 s– 2 min.

Examples of good combinations:
• 4 players with 10 cards each (40 video clips)
• 6 players in pairs with 20 cards each pair (60 video clips)
• 10 players in pairs with 15 cards each pair (75 video clips).

Time estimate:
• 3–5 h depending on number of participants and video cards

Why Do Design Cards Work?

We have presented three particular design cards and how they are used. But why
do the *PLEX Cards, Inspiration Cards*, and *Video Cards* work? There are general
characteristics with design cards that make them great tools in collaborative design
(i.e., tangible idea containers, triggers of combinatorial creativity, and collaboration
enablers). Once we understand these characteristics, we will also be able to develop
further methods with cards.

Cards Are Tangible Idea Containers

Cards act as physical carriers of ideas. Different parts of a lengthy creative exchange
between two participants can more easily be retrieved with the help of design
cards who serve as physical markers around which discussions and arguments are
anchored. During idea generation, the PLEX Cards have been described as useful at
bookmarking thoughts and ideas (Lucero and Arrasvuori 2010). Likewise, the main
feature of Inspiration Cards is exactly that they are containers of specific sources of
inspiration for ideation (Biskjaer et al. 2010).

Video cards turn video clips that are otherwise intangible into objects participants
can manipulate, point to, move around. One may understand the card as *design
material*, i.e., something designers can use to build understanding and proposals
with, rather than data to be analyzed (Ylirisku and Buur 2007). *Sense making* is just
as much a negotiation of opinions in the team, as it is finding any 'right' analysis

result. Similarly, the PLEX Cards turned a complex and difficult to communicate theoretical framework into an approachable and physical material.

The process of 'reading' the video cards banks on the *ambiguous nature* of video. The world recorded on video is so complex that different people will inevitably notice different things. More eyes see more. Merging such different observations has potential in finding the new. The cards help each participant prepare before presenting to the others.

Cards Trigger Combinatorial Creativity

Design cards support what scholars of creativity refer to as *combinatorial creativity*. Researchers point out that the new combination of existing concepts is central to creativity. In 'The Act of Creation', Arthur Koestler (1964) proposed that so-called *bisociation* of matrices, in which two concepts from different domains are brought meaningfully together to form a novel concept, is central to creativity across a range of domains. Another influential creativity scholar, Margaret Boden, has pursued this line of understanding in her study of combinatorial creativity (2004). The use of design cards makes it very concrete and easy to put concepts together in a combinatorial approach.

In Inspiration Card Workshops, for instance, cards are selected exactly so that their combination can lead to novel concepts through bisociation of matrices. People often praise the PLEX Cards for their ability to produce surprising and interesting results, ideas that they normally would not come up with. The PLEX Cards have worked particularly well to solve design problems where playfulness may not be the first natural topic to consider, i.e., elderly and falls, retrieving notifications while crossing a street (Lucero and Vetek 2014). In such disparate situations, the cards work as random input that leads to bisociation of matrices.

Cards Enable Collaboration

Despite the pervasiveness of new technologies, paper remains a critical component in many collaborative work practices. Luff et al. (2004) discuss affordances of paper that seem critical to human conduct, most of which are also applicable to design cards. A card is *mobile* as it can easily be relocated and juxtaposed with other artifacts, and *micro-mobile* as it can be positioned in delicate ways to support mutual access and collaboration. It is *persistent*, retaining its form and the character of the artwork produced on its surface. Cards can furthermore be *annotated* in ad hoc ways, allowing participants to track the development of the annotations and recognize who has done what. In the Video Card Game, annotation of the cards helps transfer *ownership* of the material – even if the video was shot by someone else, we often hear the participants talk about 'my cards' The A3 template of the

PLEX Brainstorming encourages people to make notes, write down card names, and make simple sketches as part of documenting an idea. A card also allows people to simultaneously see its contents from different angles, and it can become the focus of gestures and remarks (Luff and Heath 1998).

Cards can support and emphasize *turn taking*, as in most card games with several players. This is very prominent in Inspiration Card Workshops, in which participants will often pick up a card and present it to others, sometimes by placing it at the center of the table, as a way to emphasize the desire to add to the conversation; also, we have observed that participants in an Inspiration Card Workshop will take turns presenting a card they find particularly interesting, even if the method does not prescribe a particular order of activities. Cards support collaboration by being *shared objects for discussion* among participants. Co-creation events can be somewhat intimidating, especially to participants not accustomed to design and ideation, and a shared object can mitigate this by turning attention away from the individual participants and towards a joint marker for discussion. By lowering the participation threshold, design cards can make co-creation events more accessible to everyday people.

Conclusion

We have presented three types of design cards, namely *PLEX Cards*, *Inspiration Card Workshops* and the *Video Card Game*. We have discussed how they are used and why we believe they work in collaborative design, namely because cards act as tangible idea containers, support combinatorial creativity, and enable collaboration. The three design cards discussed in this chapter could further be combined and used in a complementary way at the start of the design process. For instance, Video Card Game cards could be prepared and be used as domain cards in the Inspiration Card Workshops. Similarly, PLEX Cards could act as domain (or experience) cards (e.g., defining a target experience without a specific context) or as technology (or emotion) cards (e.g., making people think about emotions without a specific technology in mind) in the Inspiration Card Workshops.

Key References for Further Reading

Buur J, Soendergaard A (2000) Video card game: an augmented environment for user centred design discussions. In: Proceedings of the DARE'00, ACM Press, New York, pp 63–69
Halskov K, Dalsgaard P (2007) The emergence of ideas: the interplay between sources of inspiration and emerging design concepts. CoDesign 3(4), London, 185–211
Lucero A, Arrasvuori J (2013) The PLEX Cards and its techniques as sources of inspiration when designing for playfulness. IJART 6(1) Inderscience, Geneva, 22–43
Wölfel C, Merritt T (2013) Method card design dimensions: a survey of card-based design tools. In: Human-computer interaction–INTERACT 2013. Springer, Berlin/Heidelberg, pp 479–486

References

Biskjær M, Dalsgaard P, Halskov K (2010) Creativity methods in interaction design. In: Proceedings of DESIRE 2010: creativity and innovation in design, Aarhus, Denmark

Boden MA (2004) The creative mind: myths and mechanisms. Psychology Press, New York

Buur J, Soendergaard A (2000) Video card game: an augmented environment for user centred design discussions. In: Proceedings of the DARE'00, ACM Press, New York, pp 63–69

Halskov K, Dalsgaard P (2006) Inspiration card workshops. In: Proceedings of the DIS'06, ACM Press, New York, pp 2–11

Halskov K, Dalsgaard P (2007) The emergence of ideas: the interplay between sources of inspiration and emerging design concepts. Co Design 3(4), London, 185–211

Koestler A (1964) The act of creation, University of California Press, Berkeley and Los Angeles.

Lucero A (2015) Using affinity diagrams to evaluate interactive prototypes. In: Proceedings of the INTERACT'15, Springer International Publishing

Lucero A, Arrasvuori J (2010) PLEX cards: a source of inspiration when designing for playfulness. In: Proceedings of the fun and games'10, ACM Press, New York, pp 28–37

Lucero A, Arrasvuori J (2013) The PLEX cards and its techniques as sources of inspiration when designing for playfulness. IJART 6(1) Inderscience, Geneva, 22–43

Lucero A, Vetek A (2014) NotifEye: using interactive glasses to deal with notifications while walking in public. In: Proceedings of the ACE'14 ACM Press, New York, Article 17, 10 pages

Lucero A, Holopainen J, Ollila E, Suomela R, Karapanos E (2013) The playful experiences (PLEX) framework as a guide for expert evaluation. In: Proceedings of the DPPI'13, ACM Press, New York, pp 221–230

Lucero A, Karapanos E, Arrasvuori J, Korhonen H (2014) Playful or gameful?: creating delightful user experiences. Interactions 21(3), New York, 34–39

Luff P, Heath C (1998) Mobility in collaboration. In: Proceedings of the CSCW'98, ACM Press, New York, pp 305–314

Luff P, Heath C, Norrie M, Signer B, Herdman P (2004) Only touching the surface: creating affinities between digital content and paper. In: Proceedings of the CSCW'04, ACM Press, New York, pp 523–532

Wölfel C, Merritt T (2013) Method card design dimensions: a survey of card-based design tools. In: Human-computer interaction–INTERACT 2013. Springer, Berlin/Heidelberg, pp 479–486

Ylirisku S, Buur J (2007) Designing with video. Focusing the user-centred design process. Springer, London

Combining User Needs and Stakeholder Requirements: The Value Design Method

Pelin Gultekin, Tilde Bekker, Yuan Lu, Aarnout Brombacher, and Berry Eggen

Abstract In the emerging design landscape, knowledge integration and collaboration with external partners are being valued in the design process due to the increasing scale and complexity of the design problems. It becomes important for designers to be in close contact with stakeholders, such as the people, communities and organizations who are affecting, or being affected by, the problem or the solution from the early stages of the design process. The majority of the methods that are utilized in design practice have until now been user-focused, aiming at understanding the users and designing for the user experience. Stakeholder involvement in the design process is a new topic of study in the design field. Approaches and methods that guide the designers in developing design solutions by considering diverse stakeholder perspectives are limited.

With the purpose of assisting the designers in considering the stakeholder perspectives in the design process, we present the Value Design Method that aims to integrate the user insights, business insights, and stakeholder expectations and roles at the early stages of the design process. We introduce the method alongside the Value Design Canvas. The Value Design Canvas is a visual probe that can be applied in collaborative multi-stakeholder design workshops. We provide advice on how to apply the method and on aspects that should be attended to while organizing multi-stakeholder workshops.

Introduction

Design is a creative activity. It is also a part of an interdisciplinary process of product development, which requires people/organizations with different skills and knowledge to collaborate through stages typically composed of concept generation, design, prototype development, testing, production and market introduction (Ulrich and Eppinger 2004). When a single company carries out the product development process, the decisions are usually taken within that company and collaboration

P. Gultekin (✉) • T. Bekker • Y. Lu • A. Brombacher • B. Eggen
Department of Industrial Design, Eindhoven University of Technology,
Eindhoven, The Netherlands
e-mail: p.atasoy@tue.nl; pelin@pelinatasoy.com

© Springer International Publishing Switzerland 2016   97
P. Markopoulos et al. (eds.), *Collaboration in Creative Design*,
DOI 10.1007/978-3-319-29155-0_6

occurs between the different departments of the same organization throughout the process. However, the emerging technological and economic changes, challenge this more or less closed and linear process, and require a more collaborative and flexible approach to innovation (Gardien et al. 2014).

Four economic stages have been identified from the beginning of the industrial revolution until today, to set the context for design practice; as industrial economy, experience economy, knowledge economy and the currently emerging transformational economy (Brand and Rocchi 2011; Gardien et al. 2014; based on Pine and Gilmore 1999; Drucker 1981). These stages differ in terms of what constitutes value, who is involved in value creation and how economic value is created and distributed (Brand and Rocchi 2011). Furthermore, the character and processes of design have been continuously evolved in parallel with these stages. The initial focus of designing products to fulfill functional needs, first evolved into designing for experiences, then into designing for knowledge and services, followed by an emerging focus on designing for social change (Sanders and Stappers 2008) and co-production (Drucker 1981). In this regard, it is suggested that different stages require different design processes, methods, tools and different design skills and competencies. Although companies may use the approaches and methods from earlier stages, adapting to a new stage will allow companies to extract more value through innovation (Gardien et al. 2014).

The main drivers of the emergent knowledge and transformational economy are, to a large extent, information exchange and collaboration in the product development and service delivery processes (Drucker 1981). For instance as products are being integrated with services, transactions between many service providers become an integral part of sustained product development (Basole and Rouse 2008). Also, societal issues which need joint intervention from diverse organizations and user groups receive attention and provide underexplored market opportunities. Many solutions involve consumers and consumer groups in the product development process as joint problem solvers, and value co-creation is receiving attention as an innovation approach. Consequently businesses increasingly leave the single company perspective and choose to engage in many inter-organizational relationships rather than handling all the aspects of the new product development process (Prahalad and Ramaswamy 2004; Binder et al. 2008). They adopt a more open approach to innovation, in which they share and gather information with/from external partners (i.e. other companies, competitors, non-profit organizations and users/user communities) through innovation networks. Products and services are developed and delivered to their users via complex processes, exchanges and relationships (Basole and Rouse 2008; Gardien et al. 2014).

In networked innovation, the value is created for the users through direct and indirect relationships with many partners at the network level. The design proposal and how to realize the solution are defined in relation with the input of the stakeholders based on their knowledge, resources and expectations (Basole and Rouse 2008; den Ouden and Valkenburg 2011; Tomico et al. 2010). Therefore defining the complementary knowledge and resources to generate value (how), and bringing the right collaborators together (with whom) becomes as equally important as determining the solution (what) (Brand and Rocchi 2011).

In this context designers are faced with some challenges, which require new approaches in practice. Firstly, the complexity of design problems requires designers to consider a broader technological and social context in the design process. Secondly, designing in this new domain requires consideration and inclusion of stakeholders – i.e. the people, communities and organizations who are affecting or getting affected by the problem or the solution- in the design process (Gardien et al. 2014). The focus of the design research methods has been to understand the user and the use situation (Sanders and Stappers 2014). However dealing with the challenges in the new design context requires a design perspective beyond the user centered focus. Design methods that assist the designers in gathering and integrating external knowledge into the design solution and in considering stakeholder expectations and roles while dealing with the design problems are needed.

The emerging research focus both within and outside the design discipline is on inclusive and integrative approaches in line with this need. Table 1 demonstrates how the foci of the design research and business and stakeholder management fields are evolving to support the value co-creation and networked innovation practices.

Table 1 Overview of the trends within the fields of design research, and business and strategic management fields to support the value co-creation and networked innovation practices

Field	Focus of study and methods	
	From . . .	Towards . . .
Design research	Aim for developing deep understanding of users and context variables in a holistic way	Aim for exploring design requirements and the nature of the problem by direct involvement of users and stakeholders
	Users as subjects of study	Users as partners and experts of their own experiences
	Designers and researchers have an expert mindset and they design for the users	Designers and researchers have participatory mindset, they design with the users
	Methods e.g. Ethnographic research, contextual research (Ireland 2003), personas (Pruitt and Adlin 2010), scenarios and stories (Carroll 1995; van der Bijl-Brouwer and van der Voort 2013)	Methods e.g. Cultural probes (Gaver et al. 1999), generative toolkits (Sanders 2000; Sanders and Stappers 2012), concept mapping (Visser et al. 2005)
		Emerging approaches: participatory innovation and co-design
Business and strategic management	Focus on company transactions	Focus on value exchange in innovation networks
	Analysis of business activities based on company transactions (Osterwalder and Pigneur 2010)	Integrative approach through design thinking (Brown 2009)
	Stakeholder management from company perspective (Bryson 2004)	Stakeholder management in innovation networks (Roloff 2008)

In the design research field, complex design problems for hard to empathize user groups require a more explorative approach at the early stages of the design process to understand the nature of the problem. This requires direct involvement of users in the design process, therefore a shift from an understanding of users *as a subject of study* towards an understanding of users *as experts of their own experience* is taking place (Sanders and Stappers 2014). The emerging approaches within the design field such as participatory innovation and co-design are looking for ways to support active user involvement in the design process, through collaborative ways (Buur and Matthews 2008; Mattelmäki and Visser 2011).

The business and strategic management fields are similarly calling for approaches to support networked innovation processes. While most research have been focused at the company transactions (Osterwalder and Pigneur 2010), the recent studies suggest taking a network perspective (Mason and Spring 2011). Networked innovation approach requires the design and business aspects of a design proposal to be considered in an integrated manner. Therefore exploring alternative solutions with a design-led approach is suggested to the design of strategies and business models, for instance by applying design thinking or prototyping (Osterwalder and Pigneur 2013).

Integrating business insights in the concept development stage and involving the stakeholders early in the design process are useful approaches for design problems with many stakeholders. There is a growing research direction towards more participative, integrative and design-led approaches to support the value co-creation and networked innovation practices, and methods that aid designers for these purposes are needed. To support the designers in enriching a design concept by considering stakeholder roles in the proposal, we developed the Value Design Method. The following section presents the method and an example case to explain method application.

Value Design Method

Value Design Method is developed to support making design proposals with the consideration of stakeholder expectations and relations. It aims at identifying, on the one hand, the factors that influence the design proposal to suggest the involvement of stakeholders, on the other hand, those that motivate the stakeholder participation. These insights are then utilized to enrich the design proposal and identify business aspects of the solution. The method is suitable to be applied at the early stages of design, when there is an initial design concept and a need to integrate knowledge from the experts and related stakeholders.

The method supports stakeholders to iteratively develop a design proposal by conducting pairwise comparisons between (1) design considerations (such as user and use characteristics), (2) stakeholder considerations (such as what their motivations are and what they contribute to the design proposal) and (3) business

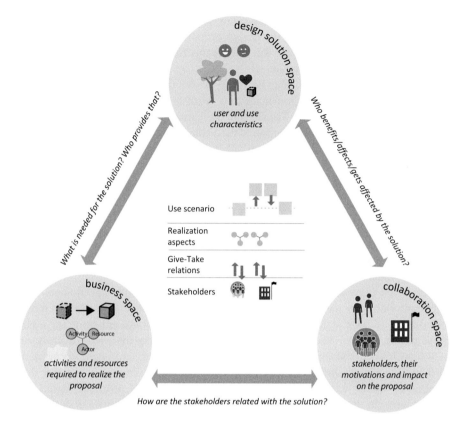

Fig. 1 The Value Design Method supports pairwise comparison between the design considerations, stakeholder considerations and business considerations based on evolving use scenarios, as shown in the diagram

considerations (such as what is needed to realize the proposal) (see Fig. 1). These considerations are brought together through use scenarios. The method utilizes scenarios as a dynamic thinking tool rather than to communicate a finalized design proposal: they evolve and become more detailed in the process with evaluations from different perspectives.

The output of the process is a refined concept consisting of: a. a user experience concept with use scenarios, b. identification of stakeholder perspectives and the conditions for their involvement in the proposal, c. insights on business considerations to realize the design solution, regarding the stakeholder roles.

The method consists of four stages:

1. Briefing & Analysing
2. Identifying values
3. Synthesizing
4. Consolidating & Evaluating

The core stage of the method is the synthesizing stage, where the three types of considerations are integrated in the design proposal. The first two stages, similar to other design processes, prepare the participants for this stage and the final stage describes and optimizes the output.

We followed a research through design approach (Gultekin-Atasoy et al. 2015) to apply this method in stakeholder ideation workshops and to gather insights for iterative method improvement. The steps of the Value Design Method are presented visually using a paper-based probe called the Value Design Canvas (see Fig. 4). This paper-based probe can also help to record the discussion. A facilitator can make notes using post-its during the discussion and place them on the relevant fields on the probe. The related spaces and considerations are placed on the probe next to each other to support the discussion. The probe layout can be adjusted according to the session requirements, i.e. level of detail of the scenarios or duration of the session.

The Value Design Method proposed here is very much related to two existing methods, namely the Value Flow Model (Den Ouden and Brankaert 2013) and the Business Model Canvas (Osterwalder and Pigneur 2010).

The Value Flow Model is a method to identify the relevant stakeholders and important values to them. It helps to create positive balance between the input and output of each stakeholder in the collaboration and commitment for their participation. It is a visualization tool that demonstrates the value exchanges between different stakeholders. The Value Design Method here is closely related to this method, yet different. The Value Design method iteratively support the creation of shared values for different stakeholders through proposals based on user insights. In this way it paves a path for designers to jump between designing user experiences and co-creating shared values with the stakeholders. It is a process approach that is particularly suitable for wicked problems in which the user insights and stakeholder insights cannot be known completely upfront. Value Flow Model can be used in combination with the Value Design Method to visualize the resulted value exchanges among stakeholders along the value design process.

The Business Model Canvas has been widely recognized as a useful tool to describe and design business models. Its strength lies in its simplicity and ease of use. Using 9 elements: customer segment, distribution channel, customer relation, value proposition, key partners, key resources, key activities, cost and revenue, it describes how a value proposition can be created and delivered to end users and how financial benefits can be created. The Value Design Method makes use of some of these components throughout the process. It creates a process towards new business models for network collaboration with specific target user groups in mind. Different elements of business model canvas can be therefore found at different moments in the approach. Consequently, the business model elements can be used as inputs to describe the resulted business model using business model canvas.

In other words, Value Design Method was designed partly based on existing innovation approaches and the aim is to create a process for designing user experience proposals with many stakeholders.

Fig. 2 Lusio prototype (Hooft van Huysduynen 2014)

The Value Design Method has been used in various contexts with different combinations of participants, including design (research) projects with professional designers (see e.g. Gultekin-Atasoy et al. 2013, 2014). To illustrate how it can be used by interaction design students, the case used in this chapter presents how a design student utilized the method in her design process to develop her design concept with the feedback from experts.

Value Design Method Application

The Value Design Method was applied to improve a design concept developed by a graduate student designer during her final Master's project. The design concept, Lusio, is an interactive decentralized platform consisting of multiple objects (Hooft van Huysduynen 2014, see Fig. 2). It is designed for primary school children with the aim to support social physical play in settings like schoolyards or gym classes. The goal of the design was to support the open ended play by using different modalities of feedback within play. Each object has the same set of fixed rules and communicates with the other objects in the set. A user can influence the color of the lights on the objects through different movements like tilting, rolling or shaking, making it possible for children to engage in social play.

The designer followed a research through design approach during her design process. She developed an interactive prototype through several iterations and conducted user observation studies which took place both at the school and outside the school context (see Fig. 3).

These iterations provided the designer with insights on how the play behavior can be shaped by the interactive properties of the play platform through different movements. However, the designer's considerations mainly focused on how children interact with the design. Other factors, such as how other stakeholders would

Fig. 3 Screenshots from the user observation studies in a gym class context, through diverse movements of tilting, shaking or rolling (Hooft van Huysduynen 2014)

benefit from the design and their roles were not considered. For instance, in a school context, the actual "users" of the product will be the children themselves, but the teachers may decide on the timing and content of play, or the school may decide on whether or not to invest in the product. How these stakeholders utilize and benefit from the solution defines a wider context for the product, and may hence affect the proposed experience and eventually the market success.

The designer wanted to evaluate the product concept with experts to improve the concept. A Value Design Workshop was organized to enrich the product concept with business and stakeholder perspectives. The participants of the session were the designer herself, two employees of a company which develops play solutions to children, one being an expert on physical education and the other being an industrial designer who specializes in play systems. A facilitator who was responsible for organizing and recording the discussion was also present during the session. The session took 4 h. In the following part, we introduce the method stages, first by explaining how to apply the method, then by providing examples from the case to clarify the discussion content and the insights gathered. Figure 4 presents the layout prompt used during the session. The layout is tailored for the needs of a specific session, in this case by limiting the number of post-its and linking the related comments at separate stages with each in order to inspire the discussion, for a session of 4 h. We present the layout along the related stages.

Briefing and Analysing

The process started with the design brief, in which the design problem and concept were introduced. This information consisted of a description of the user, the design context, the design challenge, the description of the concept developed so far and whether there are any stakeholders involved in the solution. It should be noted that the information provided does not have to be complete and the method can be applied to have a deeper understanding of the various aspects of the problem. Visual materials and/or models are useful to include at this stage.

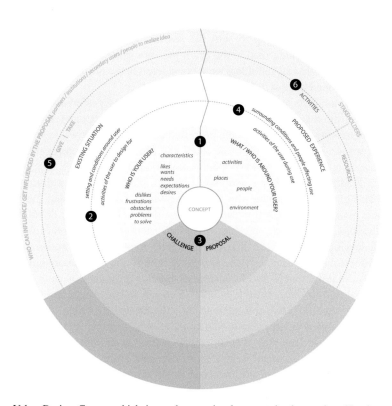

Fig. 4 The Value Design Canvas which is used as a visual prompt in the session. Numbers correspond to the discussion topics along the analysis and synthesis stages of the session, *1*: Analysis of user and use context, *2*: Analysis of the activities in the use context, *3*: Design challenges and proposals in response to the challenges in each stage, *4*: Basic use scenario, *5*: Stakeholder give and take relations, *6*: Stakeholder roles in the realization aspects

This is followed by the Analysing step. At this stage, the design problem is decomposed using a user-centered approach. The decomposition is made by defining the user characteristics, user context with surrounding conditions, available solutions on the market and people and organizations around the users as possible stakeholders (Fig. 4, area 1). In this step, a basic use context was defined, typical user activities in the use context are identified (Fig. 4, area 2), and the use scenario is structured. The challenges (Fig. 4, area 3), such as problem areas, unmet needs or conflicting interests between stakeholders are also identified. These challenges provided starting points to look for design opportunities.

In the design case, the session started with an introduction given by the facilitator about the purpose of the session and the stages of the method. Then the designer introduced the design motivations and the design concept to the other participants, using the interactive model and drawings. She gave a brief description of her design process, highlights from the user research and the design challenges that she was faced with, to make her design considerations clear. Giving information about these aspects helped the group to start the session with a shared understanding. Two user groups were identified in the brief: primary school children aged 4–6 and 10–12 years. No other stakeholders were considered in the proposal, so this information was mentioned but not detailed further at this stage.

In the analysis stage the group identified the play preferences of the two user groups from the earlier research and their own experiences and defined the context characteristics that can affect the proposed user experience: while younger children prefer individual play with basic rules, the older children prefer social play with more complex rules around cooperation or competition.

Although the designer addressed various play preferences during the design cycles to a certain extent, the experts provided a more detailed evaluation from their own expertise (in this case the cognitive point of view) about possible use situations. For instance, since the sensorial systems of the 4–6 years old children are not developed at the same time, providing different feedback modes (sound, visual, tactile) can be important. Also supporting both social and motor skill development can be an important focus of consideration while designing for play.

Analysis on the context dynamics also led to an insight: the schoolyard and gym class settings have contrasting characteristics which brings a design challenge. The schoolyard is an outside environment with a larger space and the children are involved in free play with no supervision. On the other hand, the gym class is an indoor environment with a more compact space, the pedagogical aspect is considered and children are involved in structured play with supervision. These discussions led to identification of design challenges: how to support the play preferences for different age groups? How to support supervised and unsupervised play? Another design challenge was discovered while considering the activities in the gym classes and schoolyards. The duration of play differs for two age groups: while a play session in a schoolyard is 1.30 h for younger children, it is limited to 15 min for the older group. The spotted challenge was: how to facilitate an open-ended play in a short amount of time? (Fig. 5)

Identifying Values

The identifying values stage is a sub-stage that links the analysing and synthesing stages. The most important challenges that can be solved with the design proposal are selected and transformed into the initial description of values at different levels defined by the Value Framework (Den Ouden 2012) namely: Value for the User (why the design is meaningful for the users), Value for the Market (why the design is better compared to the existing solutions), and Value for the Stakeholders (why the design is attractive to the stakeholders). These values are used as evaluation criteria for the use scenario. The A1 size layout prompt can be used (Fig. 6) to

Fig. 5 The session configuration, with the prototypes and the layout

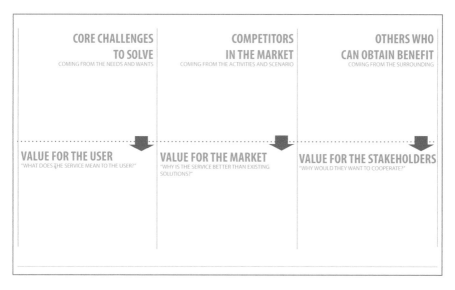

Fig. 6 A1 size template for the value identification stage, taking the user, the market and the stakeholder values into account

specify the insights gathered from the analyzing stage, at different levels. As the session proceeds, new values are identified along the process.

In the design case, at this stage, the participants discussed which design challenge was more crucial to solve and what the added value would be. Table 2 gives an overview of the values that were uncovered in the session. For instance, the group identified that meeting the varied requirements of different age groups is valuable in the gym class context. The value for the market was then identified as supporting the development of both motor and social skills of two different

Table 2 Example values identified during the session, on three different levels: value for the user, value for the market, value for the stakeholders

Value for the user (why the design is meaningful for the users)	Value for the market (why the design is better compared to the existing solutions)	Value for the stakeholders (why the design is attractive to the stakeholders)
Fun to play	Combining functions social play + fun + physically active	Educational value for schools
	Added value to play: developing motor skills	Guidelines on games to the teachers
Challenging	Trigger physical activity	Flexible play solution to the schools: possible to use indoors and outdoors
Suitable for both genders	Supporting the development of both motor and social skills of two different age groups through playful interaction	
Different levels of interaction		
Freedom of creating own game according to interests		
Building on game learned during gym class		
Develop motor skills		

age groups through playful interaction, which distinguishes the product from the existing gym equipment and education equipment.

Synthesizing

The synthesizing stage is the core of the method. At this stage, the design concept is detailed by considering the aspects in the design space, collaboration space and business space, in four steps.

Step 1: Define the (Dynamic) Use Scenario

As an initial step, a use scenario is developed by combining the value for the user and the value for the market, as specified in the identifying values stage. This initial use scenario can describe the main usage steps (scenario frames) with user actions in a typical use situation. It can be developed in a more extensive way by using available scenario development methods and techniques. If there are previously developed personas and use scenarios, they can be adopted in the process. The use scenario functions as the backbone of the discussion, to examine the consequences of pairwise comparisons between the three design perspectives. The scenario evolves as a part of the process, by adding or subtracting steps in the scenario gradually. It is possible that some steps are replaced by others, or new ones are added, as the use scenario gets more detailed through the process.

The Value Design Canvas provided in the design case included a basic use scenario (Fig. 4, area 4). The adjacent sections of the layout were provided to get the scenario developed through the paired comparisons of the two different age groups. First they evaluated which usage steps would constitute a typical play setting scenario and concluded that the scenarios for the two age groups and supervised versus unsupervised settings differ. Then they identified the product's interactive qualities by considering the differences between different types of play. Following this initial discussion, a general use scenario was made by identifying the user (children), the goal (physical play), the use context (gym class, outdoor/indoor), and the beginning (coming to the class) and ending moments (finalizing the play with the design) of the scenario. Then, a use scenario of 5–6 steps was constructed. In the evaluation stage, this scenario was detailed into two separate scenarios representing two conditions to examine the differences.

Step 2: Paired Consideration: Design and Collaboration Space

In the second step, the collaboration space is defined by identifying the possible stakeholders which can participate in the proposal. Based on the Value for the Stakeholders content in the Identifying Values stage, an initial set of stakeholders is defined. Following this step, based on the use context described in the use scenario, the stakeholders that are related with the use context are added to this initial set. This is done by identifying whether there are stakeholders that can influence the context of use or whether they can be affected by the solution.

In the design case, for instance, the role of the teachers regarding product use in the school setting was clarified. The experts pointed out that there could be two types of teachers depending on the school type: with and without a physical education background. They can both influence another stakeholder, the school director, on his/her decision on investing in the solution. Therefore the group identified that it is necessary to create a guide to support the teachers on the possible ways of using the product and benefits for the children. Some of the other identified stakeholders were the council that decides what can be installed/used in the schoolyard, and the educational equipment company who can be a knowledge partner in developing games.

Step 3: Paired Consideration: Collaboration and Business Space

After identifying the stakeholders, the motivations of these stakeholders to join in the proposal are evaluated by identifying relations. The stakeholder's involvement as a solution partner is elaborated based on defining the give and take relationships, in other words what a stakeholder can provide to the solution and what a stakeholder can obtain from the solution, respectively. These give and take relationships can be material (e.g. investment) or immaterial (e.g. exposure). If a stakeholder's give and take relations are not balanced, in other words the benefits do not meet their contribution, the commitment from that stakeholder can be considered weak or unrealistic. Based on the roles of the stakeholders in the proposal, some of the stakeholders can be considered to be left out of the proposal based on the imbalance of give-take relationships. By identifying the give and take relationships on a concrete level, new design challenges are discovered. The design concept is enriched based on the design opportunities by considering the stakeholders and what they can bring to the proposal. Attempting to balance the give and take relationships may also trigger hidden conflicting views to surface or may result in innovative ideas to emerge.

In the design case, the possible stakeholders that were defined in the earlier stages were placed on the related part of the canvas. For each stakeholder, give and take relations were identified. For instance, the teachers' motivation for providing expertise and feedback on the play types were evaluated. The experts gave insights based on their past experience in involving teachers in the design process, clarifying that the teachers are motivated to give input if there is a social benefit, if their input is acknowledged publicly in the solution.

Step 4: Paired Consideration: Business and Design Space

As the third step, the use scenario is detailed by considering possible roles of the stakeholders in realizing the proposal, by linking the give and take relations to the business model components, namely activities to realize the design and resources needed (Hakansson 1992; Osterwalder and Pigneur 2010). This step allows the participants to communicate what actions need to be taken, what resources are needed and which partners can contribute in the related steps. It also allows them to evaluate whether or not certain design features are necessary. The scenario parts are then detailed or changed based on the limitations and possibilities identified.

In the design case, the participants highlighted the fact that (1) creating new game opportunities can enrich the play experience of the children; (2) pedagogical advice is needed to develop the open-ended play concepts further.

The guidelines and play scenarios could possibly be created in collaboration with physical education teachers and the design company. So developing play rules was identified as one of the key activities to realize the design proposal. Teachers could be involved in this activity by providing physical activity knowledge as a key resource. In this activity, a game database would be developed, and the teachers could also give feedback about which games are more preferred by children.

Discussing the actions required to realize the concept also helped the group to spot the missing actions and challenging steps. For instance, to place the concept in the market, a certification needs to be obtained for school environments. This was an issue that the designer did not consider previously. In addition, the group also discussed the realization aspects based on the specific features of the product, such as the financial investment that would be needed at the initial stage and on which terms the company would invest in the idea, as well as the university's role as a shared idea owner in the case if the designer's idea would be commercialized.

Consolidating and Evaluating

At this last stage, the considerations discussed in the synthesing stage are brought together and evaluated. The finalized use scenario is combined with the discussions on how to realize the idea. The use scenario, the design challenges and important points of consideration for the next steps are documented on the layout and the concept is evaluated by considering the links between the decisions at different stages, through joint reflection.

The Consolidation and Evaluation layout (see Fig. 7) identifies the scenario with parallel layers that link the product specifications, use scenario, the business model concept and the roles of the stakeholders. Bringing the different aspects of the discussion together enables the participants to have the overview of the design decisions.

In the design case, the participants jointly defined the use scenarios in their final form. Due to the two types of users and differences in the supervised and unsupervised use context, the use scenario was detailed to form two scenarios, and

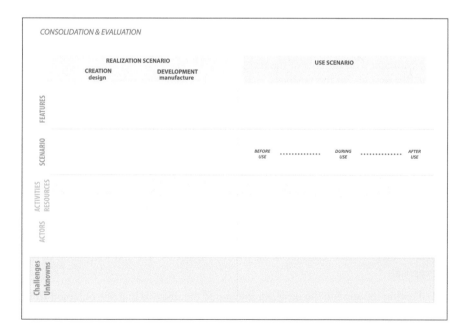

Fig. 7 A0 size consolidation and evaluation template of the value design canvas. The use scenario and the realization actions are brought together for final overview

the participants discussed the differences between the two cases to see whether the design decisions could meet the requirements of the two cases. In the final step, the use scenarios were brought together with the realization scenario, in which the roles of the stakeholders to realize the design proposal were also evaluated. Identifying the links between the design and realization decisions also clarified how the teachers would be involved in giving feedback on the design. Some unknown issues and challenges were also documented for the next stages of the design process.

Evaluation of the Value Design Method with the Designer

Following the design case we evaluated how the designer benefitted from applying the Value Design Method in her design process, with a semi-structured interview. She reported that the method helped the designer in aligning her understanding with the experts, in detailing some aspects of the design and in defining the company's role for introducing her design in the market. More specifically she reported that:

- The discussion process made her understanding of the design context more concrete with expert feedback; such as by differentiating the characteristics between two age groups and between the two use situations in supervised, indoors gym-class settings and in non-supervised, outdoors schoolyard settings.
- Use scenarios helped her to link this knowledge with what happens in the actual use situation, so that they gave the designer handles to evaluate whether some interaction rules would work or not. For instance, not only with a focus on how the children play with the product, but also by considering the classroom environment with the role of the teachers.
- The discussions on the business insights and stakeholder roles made her understand what the company and other stakeholders' expectations would be, if they take a role. For instance, the session made it clear for her that the company doesn't have an interest in developing the electronic components, but they have an access to market, so they could be interested if the design fits their portfolio. Therefore a third party for developing electronic components would be required. Also, the designer had an understanding over how far she should define the design specifications, so that the other stakeholders could participate in developing the concept.
- The structured process helped them to discuss on the subjects that were relevant, which they would have skipped without using the method, for example additional stakeholders were needed since they would affect the use and realisation of the proposal.

Following the session the designer integrated the insights that she obtained from the session into her final design iteration. She improved her design by adjusting the interaction rules based on the two scenarios developed duringthe session and used

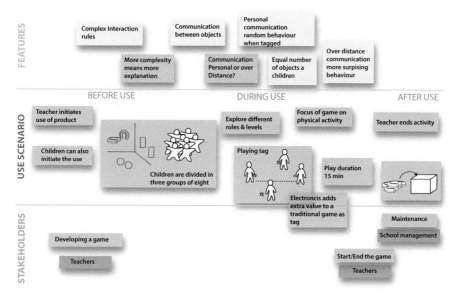

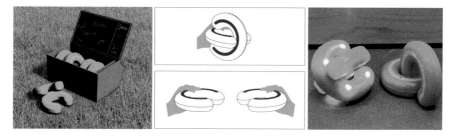

Fig. 8 The scenario template that the designer used in her final design report. The contents of the scenario were obtained during the session. The designer used the layout to digitize the post-its, in a scenario flow format with parallel rows of product-system features, use scenario and the stakeholders in relation to use instances

Fig. 9 Final concept of Lusio, with a storage box including Lusio set and game instructions on cards that the gym teacher can use during the classes. Two objects that can be separated or engaged together for different complexity levels in interaction

the scenarios to communicate about her design (see Fig. 8). She also chose to design a package in which all pieces of a play set were present and visible to the teacher. The package also contained practical guidelines to the teacher to use (see Fig. 9).

In the following part, we will evaluate the benefits and limitations of the method referring to the case presented above.

Discussion

The Value Design Method presented in this chapter is a method for designers to examine their design decisions with stakeholders, both to enable knowledge gathering and to integrate insights from several perspectives of the design concept.

The case presented in this chapter demonstrates how the method is applied. It represents a situation in which a designer (final year graduate student) evaluated her design with expert feedback and improved the design solution by considering the business and stakeholder perspectives. We discuss the benefits and limitations of the method and clarify the guidelines for method application in the following section.

Benefits and Limitations

The designer found the method useful as an instrument that supports her knowledge gathering action. In the case presented here, the concept did not diverge to a great extent from the beginning of the process, but rather got detailed with new interaction rules and expanded with service integration. This is partly due to the design concept being concrete with a concept prototype that was presented at the beginning of the workshop and how much this concept already fulfilled the user needs. It can be therefore expected that the resulted design concept can diverge from the original proposal to a larger extent if it is at a less mature state at the beginning of the session, or with less concrete user insights.

The method is also suitable to be applied to facilitate the involvement of stakeholders with different backgrounds and interests in the ideation. In the case presented here, the session was organized with experts from a company in the related market. These experts had a relevant understanding of the users and the market, therefore helped the designer to gather deeper knowledge regarding the business and market aspects, as requested by the designer herself. The session could also be organized by inviting school representatives or possible stakeholders who would be interested in solutions for social design challenges, such as motivating children to be physically active. Then the session would be useful for clarifying the requirements for involving schools as solution partners, or for evaluating the design solution based on the social impact.

As discussed in the previous paragraph, this case involved only one stakeholder in the earlier stage of the design (the company), although it was learned that there are additional stakeholders of importance. If the Value Design Method is applied with the direct involvement of all the identified stakeholders, the process can result in more conflicting views surfacing due to different perspectives on the solution. In those cases, the conflicting views can provide valuable insights, if handled in a constructive way. Therefore, while applying the method with direct involvement of the stakeholders, enough time should be provided for discussing the stakeholder roles. Alternative solutions can be developed with a deeper knowledge exchange, by elaborating on the balance between give and take relations or by considering other design solutions by with alternative actors, activities and resources. This also aligns stakeholders' understanding over each other's position.

Like many other participatory methods, Value Design Method has an open and participatory approach in which ideas are proposed and discussed openly and developed through joint reflection. It is important for the participants to share this mindset as well, and be ready for constructive discussions around challenging design issues.

The method process does not result in a fully verified business model, however it does support establishing the links between the design decisions with the business decisions and stakeholder roles, which will inspire the business model generation step for proposals with many actors. The Value Design Workshop can be applied as part of a series of workshops where the design decisions are developed into a more concrete business model.

We have already applied the earlier versions of the method in design and innovation workshops in which professional designers, experts and stakeholders came together for a social design challenge, for instance, motivating children to be active or designing Livinglab concepts to engage citizens in a more active lifestyle (Gultekin-Atasoy et al. 2013, 2014). These sessions were in the context of projects which aimed to bring universities, public sector and companies together to develop solutions with a commercial value. The participants reported that the method provided a useful process for discovering new design opportunities with the contribution of different stakeholders, and understanding each other's perspective towards the solution.

Integration with Existing Methods

The method utilizes use scenarios as a backbone of the discussion. Although it proposes basic steps of establishing and adjusting a use scenario, the process does not fully define how to develop a detailed use scenario. Detailed user personas and use scenarios are accepted as valuable tools for developing deep insights in the design process. We recommend the readers to refer to existing methods for scenario development (Carroll 1995; van der Bijl-Brouwer and van der Voort 2013) and story thinking, such as Co-constructing Stories (Ozcelik-Buskermolen and Terken Ch. 8) or Storyply (Atasoy and Martens Ch. 9) for developing scenarios and stories with a more focused approach. Value Design Method works in a compatible way with such scenario and story-based approaches, and will be a useful complementary method for further evaluation of the design concept.

The case presented in this chapter utilized the expert knowledge to clarify the business aspects and stakeholder expectations and possible roles, to support a compact session. More complex design problems may require focused studies to gather insights on stakeholder involvement and business aspects. For instance, stakeholder analysis methods (Bryson 2004) would be useful for the projects with many stakeholders, such as in the healthcare sector. Den Ouden's (2012) Value Flow Model provides a useful approach for identifying stakeholder relations in an innovation ecosystem. It can be utilized as a useful visualization tool defining stakeholder relations on the network level. The Business Model Generation approach (Osterwalder and Pigneur 2010) is a useful model in defining business aspects for a proposal, and can be useful when applied after the Value Design Method for further detailing the business model of the proposal.

Finally, readers are invited to focus on the required facilitation skills. The Value Design Workshop format lowers the facilitation requirements; however basic facilitation skills to organize the workshop and to manage the group process are required for the session to run smoothly. The effect of time pressure on ideation should be considered in the method application. More invested time will result in a more detailed concept and discussions; on the other hand more invested time is an obstacle for the participants to attend. In the previous applications of the method, it was observed that a minimum of 3 h is necessary for a compact session, while an extended session can easily take a full day. The time requirements of meeting with the other participants, warming-up, applying the method and evaluation, including breaks should all be considered in the session planning. The longest session segment is advised to be 45–50 min long to keep the participants motivated to contribute. Time pressure can be applied in the divergence stages. However, decision-making stages require sufficient time for discussion and reflection on the generated ideas. In the cases where challenging issues or conflicts surface, giving enough time for discussions may be necessary to explore alternative solutions or to resolve the issues.

Future Development of the Method

The layout prompts introduced in this chapter support the method application in a flexible way, by the use of post-its on the assigned parts of the layout. The layout informs the participants on the expected process and limits the number of comments with the space provided for each stage. Therefore it also communicates to what extent a topic is discussed. Although this approach has some advantages in easing up the facilitation, it also has some limitations. Firstly, the entries on the sticky notes can vary in depth, eventually giving less limitation and guidance. Secondly, the opportunities are only explored to a certain extent. The current version of the method is planned to be developed into a card-based toolkit with an integrated layout to support a flexible evaluation of existing opportunities, while motivating the participants to give more direct input with prompts on the cards. This is expected to support the creative discussion process to a higher extent. Another valuable direction for further development is to apply the method principles on a digital platform. This would allow taking more dimensions into consideration on the use scenario, and a more dynamic way of adding/subtracting scenario frames based on these considerations. It is also valuable for documentation, during and after the session, which allows sharing output of the session with other participants and for using the outcome in multiple sessions.

Conclusion

In complex design projects there are many interrelated components that should be taken into account during concept development. Therefore designers need to gather domain and expert knowledge at the early stages of the design process. It may be difficult for designers to envision all of the information that may be relevant. The challenges that are faced by designers in the design context at present can be answered by developing contemporary design methods that help evaluating design considerations in relation to the stakeholder expectations and how they can contribute to the proposal, by taking into account the tangible and intangible value exchange through the proposed concept. The design solution, business model and the roles of the stakeholders are advised to be designed together, through an explorative process of negotiation and participation.

The Value Design Method presented in this chapter helps designers to gather such information from experts early in the process, and hence to obtain an overview of the design problem at hand. This is likely to increase the designer's awareness when approaching a complex design problem. The method is observed to be especially useful for designers to consider design issues that may otherwise be missed in the ideation, such as business dimensions or stakeholder roles. It enables the designer to consider design concept in a broader context, beyond the typical focus on the user-product interaction.

Acknowledgments Value Design Method is developed within the PhD study of Pelin Gultekin in the context of the EU Interreg NWE ProFit Project. We would like to thank Hanneke Hooft van Huysduynen for providing the material that was used in the case study and for presenting it in this chapter. We would like to thank also to the session participants for joining in the study.

References

Atasoy B, Martens JB (2016) STORYPLY: designing for user experiences using storycraft. In: Markopoulos P, Martens JB, Mallins J, Coninx K, Liapis A (eds) Collaboration in creative design: methods and tools. Springer, New York
Basole RC, Rouse WB (2008) Complexity of service value networks: conceptualization and empirical investigation. IBM Syst J 47(1):53–70
Binder T, Brandt E, Gregory J (2008) Design participation(-s) – a creative commons for ongoing change. CoDesign 4(2):79–83
Brand R, Rocchi S (2011) Rethinking value in a changing landscape: a model for strategic reflection and business transformation. Philips Design, Eindhoven. Available: http://www.design.philips.com/philips/shared/assets/design_assets/pdf/nvbD/april2011/paradigms.pdf. Retrieved April 4, 2013

Brown T (2009) Change by design. Harper Collins Publishers, New York

Bryson JM (2004) What to do when stakeholders matter. Public Manag Rev 6(1):21–53

Buur J, Matthews B (2008) Participatory innovation. Int J Innov Manag 12(03):255–273

Carroll JM (1995) Scenario-based design: envisioning work and technology in system development. Wiley, New York

Den Ouden E (2012) Innovation design: creating value for people, organizations and society. Springer, London

Den Ouden E, Brankaert R (2013) Designing new ecosystems: the value flow model. In: de Bont, den Ouden, Schifferstein, Smulders, van der Voort (eds) Advanced design methods for successful innovation. Design United, Den Haag

Den Ouden E, Valkenburg R (2011) Balancing value in networked social innovation. Paper presented at the proceedings of the participatory innovation conference, Sønderborg, Denmark, pp 303–309

Drucker PF (1981) Toward the next economics and other essays. Harper & Row, New York

Gardien P, Djajadiningrat T, Hummels C, Brombacher A (2014) Changing your hammer: the implications of paradigmatic innovation for design practice. Int J Des 8(2):119–139

Gaver B, Dunne T, Pacenti E (1999) Design: cultural probes. Interactions 6(1):21–29

Gultekin-Atasoy P, Bekker T, Lu Y, Brombacher A, Eggen B (2013) Facilitating design and innovation workshops using the value design Canvas. In: The proceedings of participatory innovation conference, Helsinki, Finland

Gultekin-Atasoy P, Lu Y, Bekker T, Eggen B, Brombacher A (2014) Evaluating value design workshop in collaborative design sessions. In: The proceedings of the NordDesign 2014 conference, Helsinki, Finland

Gultekin-Atasoy P, Lu Y, Bekker MM, Brombacher AC, Berry JH (2015) Exploring the complex: method development by research through design. In: The proceedings of the 11th European academy of design conference: value of design research. Paris Descartes University, Paris, 22–24 April 2015

Hakansson H (1992) A model of industrial networks. In: Industrial networks. A new view of reality. Routledge, London, pp 28–34

Hooft van Huysduynen H (2014) Do you want to play? Final master project report, Eindhoven University of Technology, Department of Industrial Design. Eindhoven, The Netherlands

Ireland C (2003) Qualitative methods: from boring to brilliant. In: Laurel B (ed) Design research: methods and perspectives. MIT Press, Cambridge, pp 22–29

Mason K, Spring M (2011) The sites and practices of business models. Ind Mark Manag 40(6):1032–1041

Mattelmäki T, Visser FS (2011) Lost in Co-X: interpretations of co-design and co-creation. In: Diversity and unity, Proceedings of IASDR2011, the 4th world conference on design research (Vol 31), Delft, The Netherlands

Osterwalder A, Pigneur Y (2010) Business model generation. University of Warwick, Coventry

Osterwalder A, Pigneur Y (2013) Designing business models and similar strategic objects: the contribution of IS. J Assoc Inf Syst 14(5):237–244

Ozcelik-Buskermolen D, Terken J (2016) Co-constructing new concept stories with users. In: Markopoulos P, Martens JB, Mallins J, Coninx K, Liapis A (eds) Collaboration in creative design: methods and tools. Springer, New York

Pine BJ, Gilmore JH (1999) The experience economy: work is theatre & every business a stage. Harvard Business Press, Boston

Prahalad CK, Ramaswamy V (2004) Co-creation experiences: the next practice in value creation. J Interact Mark 18(3):5–14

Pruitt J, Adlin T (2010) The persona lifecycle: keeping people in mind throughout product design. Morgan Kaufmann, London

Roloff J (2008) Learning from multi-stakeholder networks: issue-focused stakeholder management. J Bus Ethics 82(1):233–250

Sanders EN (2000) Generative tools for co-designing. In: Collaborative design. Springer, London, pp 3–12

Sanders EB-N, Stappers PJ (2008) Co-creation and the new landscapes of design. CoDesign 4(1):5–18

Sanders L, Stappers PJ (2012) Convivial design toolbox: generative research for the front end of design. BIS, Amsterdam

Sanders EBN, Stappers PJ (2014) Probes, toolkits and prototypes: three approaches to making in codesigning. CoDesign 10(1):5–14

Tomico O, Lu Y, Baha E, Lehto P, Hirvikoski T (2010) Designers initiating open innovation with multi-stakeholder through co-reflection sessions. In: Proceedings of IASDR 2011. Delft, The Netherlands

Ulrich K, Eppinger SD (2004) Product design and development. McGraw-Hill, New York

van der Bijl-Brouwer M, van der Voort M (2013) Exploring future use: scenario based design. In: de Bont, den Ouden, Schifferstein, Smulders, van der Voort (eds) Advanced design methods for successful innovation. Design United, Den Haag

Visser FS, Stappers PJ, Van der Lugt R, Sanders EB (2005) Contextmapping: experiences from practice. CoDesign 1(2):119–149

Crowdsourcing User and Design Research

Vassilis Javed Khan, Gurjot Dhillon, Maarten Piso, and Kimberly Schelle

Abstract Crowdsourcing can be defined as a task, which is usually performed by an employee, that is given out as an open call to a crowd of users to be completed. Although crowdsourcing has been growing in recent years, its application to design research and education has only scratched the surface of its potential. In this chapter we first introduce the different types of crowdsourcing. Then, following the typical design cycle we present examples from literature and cases from an educational setting of how crowdsourcing can support designers. Based on these examples we provide a list of tips for utilizing crowdsourcing for design and user research activities.

Introduction

In recent years, crowdsourcing has exploded. Although most would be familiar with crowdfunding stories in the public media, crowdsourcing is, quietly but surely, increasingly growing. At the time of writing this chapter, Amazon's Mechanical Turk (AMT) has more than 300,000 tasks (known as HITs: Human Intelligence Tasks) available to be completed by workers. Elance combined with oDesk claim to have more than eight million workforce from more than 180 countries which have done $750 million worth of work done. It is that explosion of services which also affects user and design researchers and challenges in many ways to redefine what is the role of a design or user researcher amidst all of the currently available services. That need to explore how researchers and designers have made use of and can make use of such services is the driving motivation of this chapter.

One of the most cited articles in academic literature dealing with the topic of crowdsourcing is *"The dawn of the e-lance economy"* (elance as opposed to freelance) a paper by two academics from MIT's Sloan School of Management

V.J. Khan (✉) • G. Dhillon • M. Piso • K. Schelle
Industrial Design Department, Eindhoven University of Technology, P.O. Box 513,
5600 MB Eindhoven, The Netherlands
e-mail: v.j.khan@tue.nl

© Springer International Publishing Switzerland 2016
P. Markopoulos et al. (eds.), *Collaboration in Creative Design*,
DOI 10.1007/978-3-319-29155-0_7

in 1999 (Malone and Laubacher 1999). Although in that article the actual term crowdsourcing is not mentioned, the authors actually pretty much describe the core idea behind crowdsourcing. They use the term "temporary company" in which individuals from all corners of the globe, connecting through the Internet can form temporary organizations that work on a specific project and when that project is over they separate to form other organizations. The first time the actual term "crowdsourcing" appears is at Jeff Howe 2006 Wired article titled: "The rise of crowdsourcing" (Howe 2006). Howe juxtaposed the new term with the well-known IT-industry practice of "outsourcing". Unlike outsourcing, which requires sending out work to reduce costs, usually to a specific company from a developing country, with crowdsourcing one can solicit contributions from a large group of people all connected through the Internet rather than from a traditionally organized company. More specifically, in both cases work is commissioned outside of an organization, but outsourcing is directed at an employment relationship, often with fixed professional organizations or individuals, whereas crowdsourcing is directed at a public that is unknown to the organization and who are not necessarily professionals (Li 2011).

Others view crowdsourcing as something broader than the mere delegation of tasks. They view crowdsourcing as "*a strategic model to attract an interested, motivated crowd of individuals capable of providing solutions superior in quality and quantity to those that even traditional forms of business can*" (Brabham 2008). The model would be one that can serve both problem-solving and production by utilizing global distribution. Others argue that even though the term wasn't used before, crowdsourcing has a much longer history. Grier (2011) for example, relates modern crowdsourcing to mid-eighteenth century organized groups who completed computational processes together.

In its more recent, information-age history, it is telling that iStockphoto, one of the very first examples of a crowdsourcing service, belonged in the creative industries. Through iStockphoto, one can license high quality stock-photos submitted by a plethora of users with a fraction of the cost. iStockphoto is not an isolated case. Another early example in the beginning of the century is Threadless. Threadless created a website in which users uploaded designs for T-shirts. Then the users themselves would vote to decide which was the most popular design and the company would only then produce them and the users of the website would then become its customers. Thus, in this case the decision making process was essentially crowdsourced. Similar examples were followed by companies to created advertisements and decide which ones to air. Specific examples span from Converse, a sneakers company which welcomed homemade commercials from its customers at ConverseGallery.com, to potato chip giant Doritos who created a user-generated advertising contest with the winning ad airing during the Super Bowl, one of the most viewed televised events in the US (Brabham 2008).

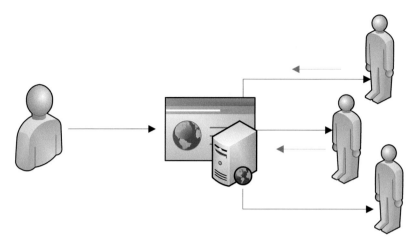

Fig. 1 The typical workflow in a crowdsourcing system includes a requester placing a task. That task is published to all workers. Note that all workers will not necessarily respond to the task

Typical Workflow

In the typical workflow of a crowdsourcing platform (Fig. 1), there are two groups of users: requesters and workers. The requester places a task and associates it with a reward, typically a monetary one, skills needed and a deadline. This task is published to all workers in the system. In some cases the requester can filter and select workers with certain characteristics, e.g. a certain skillset such as experience in graphic design. Typically a subset of workers will then compete to provide the best solution to the task. Workers can ask for clarifications to the initial task description. The requester then chooses the best one and can usually comment and ask the worker to make adjustments to the submitted solution. Once the solution is accepted the worker is rewarded.

The potential number and type of benefits for design have the prospective to alter the practice of design. The benefits are on multiple dimensions. Apparent dimensions include the type of task, the reward and the platform itself. There are already several examples that demonstrate that for typical design research tasks crowdsourcing can be: fast, affordable and engage a large, stable and diverse pool of participants. Those aforementioned advantages simplify several core principles of good design, such as: understanding user's needs, creating alternatives and iterating. Moreover, since design comprises of different and diverse activities, we strongly believe that crowdsourcing can help bridge knowledge and skill gaps for designers. Conducting a research study to better understanding users' needs is radically different when comparing it to sketching a paper prototype. Fundamental questions on user research deal with reliability and validity of findings whereas in the case of sketching paper prototypes the fundamental issue is to explore design alternatives. Nowadays, designers might have to perform several diverse tasks, that they do not necessarily master. Crowdsourcing can support designers in certain tasks as well as help them gather data to confirm or disconfirm certain decisions having as ultimate goal to reach better designs. With this chapter we wish to contribute in raising awareness about the services that already being used for design purposes, show what such services can yield for design and finally provide hints and tips to designers interested in using crowdsourcing. Before we review the chapter's sections and since there are many relevant systems we will present three classifications of crowdsourcing systems.

Types of Crowdsourcing

Industry-Driven Classification

A practical, industry-driven dissection of crowdsourcing can be found in the website Crowdsourcing.org. Crowdsourcing.org has divided the field of crowdsourcing into six major categories. These categories give an overview of companies and services related to crowdsourcing (Blattberg 2011). The six categories are (1) *crowdfunding*, (2) *crowd creativity*, (3) *tools*, (4) *distributed knowledge*, (5) *cloud labor* and (6) *open innovation*. The *crowdfunding* category lists services that help users gain online financial contributions from the crowd to fund non-profit and for profit initiatives and enterprises. The *crowd creativity* category lists services that make use of creative talent pools to design and develop original art, media or content. The *tools* category lists applications and software tools that support collaboration and sharing among distributed groups of people. The *distributed knowledge* category lists services directed at gathering knowledge or information from a distributed group of contributors. The *cloud labor* category lists services that provide online labor pools that fulfill both simple and complex tasks on-demand. Finally, the *open innovation* category lists sources that aid in the in the generation, development and implementation of ideas. Although this categorization is to a certain extent subjective, it does reflect an industry-driven perspective of existing services.

Organizational-Driven Classification

When crowdsourcing is looked at the relation of workers in connection to the requester-organization there are the following types of crowdsourcing: internal and external. In their taxonomy of crowdsourcing Corney et al. (2009) separate services and crowdsourcing tasks based on the nature of the crowd that completes them. They distinguish 'any individual', 'most people' and 'expert' tasks. Internal crowdsourcing is a specific instance of asking experts to work at the task to be crowdsourced, namely only employees of the company that aim for the conceptual design itself. The authors further divide external crowdsourcing in two categories: (1) crowdsourcing via companies' virtual communication environments and (2)

Fig. 2 Types of crowdsourcing systems based on a system-driven approach (Geiger et al. 2012)

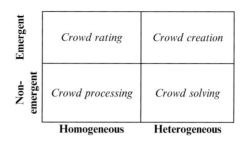

	Homogeneous	Heterogeneous
Emergent	*Crowd rating*	*Crowd creation*
Non-emergent	*Crowd processing*	*Crowd solving*

crowdsourcing via specialist service vendors. In the first category, companies create themselves a virtual collaboration platform through which end users are invited to create an account and share their ideas. Such an example is of Dell, which will be reviewed in the "Ideation" section. In the second category, an intermediary becomes part of the crowdsourcing activity as a party in between the client (company or individual requester) that sets the task and the crowd that is then asked to complete it. Some examples of this category will be presented in the following sections.

System-Driven Classification

When crowdsourcing is looked with a socio-technical systems perspective, a different classification is derived. Geiger et al. (2012), after analysing 50 systems derived a two-dimensional classification for crowdsourcing systems (Fig. 2). The first dimension is *"whether a system seeks homogeneous or heterogeneous contributions from the crowd"* and the second is *"whether it seeks an emergent or a non-emergent value from these contributions"*. These dimensions produce four types of systems: Crowd processing, Crowd rating, Crowd solving & Crowd creation systems.

In a nutshell, the main idea behind *crowd processing systems* is to divide large jobs into equal chunks of work, known as microtasks and then combine the individual contributions for a final result. A famous example of such a system is Recaptcha (von Ahn et al. 2008). The main idea behind *crowd rating systems* is that individual contributions represent votes on a given topic and only the aggregation of a sufficient number of these votes allows the deduction of a collective response. A famous example of such a system is IMDB or eBay's reputation system. The main idea behind *crowd solving systems* is that values of individual contribution's is determined with respect to evaluation criteria, either objective, well-defined criteria or subjective criteria. A famous example of a system with objective criteria is the Netflix Prize (Geiger et al. 2012) and with subjective criteria 99designs. The main idea behind *crowd creation systems* is that they are based on large, diverse crowds that aggregate a variety of contributions into a comprehensive artefact. A famous example of such a system is YouTube but also all kinds of other user-generated content systems.

Review of Sections

The organization of this chapter follows the design cycle. We decided to divide the chapter into four sections, that map to a design cycle: Establishing requirements, Ideating, Prototyping and Evaluating. We base this categorization of the design cycle partly from an interaction design perspective (Rogers et al. 2011) and partly from an innovation management perspective (Dow et al. 2013).

Since there is a plethora of terms used for the two main stakeholders of crowdsourcing platforms, throughout the paper we will use the terms worker to refer to the one that performs the task and requester to refer to the one that plans and sets the task that has to be completed.

Moreover, throughout the chapter, we will draw examples from the outcomes developed at a post-graduate, 2-week crowdsourcing course offered at the User-System Interaction (USI) program of Eindhoven University of Technology. The main learning objective for the course was how to make use of "the crowd" when trying to design an innovative solution. Five groups of three trainees with diverse academic backgrounds sought to leverage crowdsourcing platforms to help them in the process of coming up with a concept and prototyping features of it. Their design decisions were based on findings from several supporting crowdsourcing platforms.

Utilized Crowdsourcing Platforms of Case Studies

For better comprehension of the examples we will draw we first shortly describe the crowdsourcing platforms that were used. We would like to underline that the purpose of this is not to provide an exhaustive list of all available services. Instead, it makes an indicative selection of services that would be helpful for user and design researchers.

Mindswarms

Mindswarms (mindswarms.com) is an appealing crowdsourcing service because it opens a door into the user's world in the form of a video. Mindswarms allows researchers to gather short video feedback from consumers around the world using webcams and mobile devices.

This core property of the service can be very beneficial for questions which require the user to act in or talk about their natural environment. Using the service requires setting up an account for both requesters and workers and has its own worker community. The process starts with the creation of maximum five screening questions (200 characters each) in the form of four closed and one open question. Based on the results, one or several workers can be selected to record short videos in response to concepts or questions devised by the requester. The workers receive a payment of $10 for recording one video or $50 for recording multiple videos in the same study. Workers can be flagged and asked to redo a video if it is not up to the required standards, or a new worker can be selected. Response times are dependent on these factors.

Microworkers

Microworkers (microworkers.com) allows the collection of data through numerous channels like YouTube, Twitter, and Google. An account is required to use Microworkers for both requesters and workers and it has its own community of workers. Target users can be selected based on numerous criteria in vast regions of the world. Microworkers operates by giving workers instructions to use other services to perform a task, and provide the proof back to the requester. The service asks 7.5 % of money spent on each campaign plus 75 cents for campaign approval. Paying the workers is different for each task and varies between $0.10 and $2.50.

Workers receive a rating afterwards and are only paid if the collected answer is satisfactory to the requester's standards. Results though cannot be easily visualized in Microworkers itself because a lot of different services are often used outside the service and then linked back to it.

Crowdflower

Crowdflower (crowdflower.com) allows researchers to create surveys and distribute them among its workers. It requires an account to use the service as a requester or worker. Surveys are open to pre-selected groups based on a number of limited demographics and even in terms of rated Crowdflower contributions of workers. Crowdflower also has an extensive choice of countries from which workers can respond. Extra options are available only for premium users which vary from $100 to $12.000 a month. The free account allows the requester to input verbal data, images and external links. It includes rating, open questions, multiple choice, checkboxes and dropdown questions in several settings like survey, business data or even tweet analysis. Workers are paid for each survey they complete. Response times are often directly correlated with the amount of payment that is offered which varies from minutes to hours to even days. Workers, such as spammers, can also be flagged if the data is not satisfactory. The gathered data can be visualized during and after the campaign on the website itself.

DesignCrowd

DesignCrowd (designcrowd.com) is an open marketplace with close to half a million graphic designers across the world. Requesters can post their design jobs for the

DesignCrowd community and workers can create the designs from a wide variety of design categories. A posted project can get unlimited entries from the subscribed graphic designers. All entries receive a participation payment, and the winner and second place receive a higher amount. The winner and the second place entries are chosen by the requester. The requester can predetermine the amount of money for the whole project. The project can be posted as refundable or committed payment. As refundable projects, the requester can decide not to pay (service charges for DesignCrowd are still charged), if they do not like any design entry. As committed projects, the design team guarantees a payment for their posted project. DesignCrowd requires an account to post a project for the design team and to participate and submit design entries from the crowd. The chosen worker can be requested to make changes to their original design.

After having shortly presented the services that will be referred in the following sections, in the next section we address how crowdsourcing can help designers establish requirements.

Establishing Requirements Using Crowdsourcing

Rogers et al. (2011) describe the process of interaction design into four categories for which the first one is "*Identifying needs and establishing requirements*". An emphasis on this category is important because research suggests that incorrect use of requirements can lead to project failures in software development. The lack of user involvement is one of the top reasons for this failure (e.g. Taylor 2000; Viskovic et al. 2008; Standish Group 1999). Thus, having direct contact with users to better understand their context and needs is a salient part of the design cycle.

Although initially crowdsourcing was associated with work related tasks, the availability of users, in the form of "workers", allows them to complete a variety of data-collection tasks which serve this phase of the design cycle. Crowdsourcing allows requesters to reach a large and diverse group of people within a short time period, find out what their needs and preferences are and then translate these into requirements. Furthermore, crowdsourcing can help in involving workers in various stages of the design process and therefore help in setting and adjusting requirements.

In fact there are services that specialize for such purpose. Examples of such services include the execution of surveys, observations from posted videos and computer logging. In this section we review relevant examples in literature in which studies utilized crowdsourcing for eliciting user requirements. Since eliciting requirements is in essence applied user research we also cover studies that had a broader scope than that of a specific system. Finally, we draw an example from our case study.

Examples in Literature

Crowdsourcing is starting to be recognized for its increase in quality, comprehensiveness and economic feasibility of requirements elicitation (Hosseini et al. 2013). Despite crowdsourcing's potential for gathering requirements, few attempts have been made to explicitly use crowdsourcing in this way (e.g. Adepetu et al. 2012; Lim et al. 2010a; Glinz and Wieringa 2007). CrowdREquire (Adepetu et al. 2012) is such an endeavor in which a conceptualized crowdsourcing platform is explicitly formulated for requirements gathering. In this conceptualized service, the crowd contributes to tasks by coming up with explicit requirements as solutions themselves. This allows requirements to be formulated from different knowledge, perspectives and experiences which otherwise would not be easily available. As the authors state, the goal of this service is to give timely and complete response to clients who submit such a task definition.

Another interesting approach to requirements engineering is through noting the importance of stakeholders in a project. Stakeholders are an important source of requirements (Glinz and Wieringa 2007) and an incomplete stakeholders list will therefore lead to incomplete requirements. This view led to the development of the StakeNet tool (Lim et al. 2010a). In this tool, experts identify stakeholders by asking for recommendations to other stakeholders. This information is then used to build a social network. Network measures are then used to prioritize stakeholders. Using StakeNet in large software projects showed that it could accurately identify and prioritize stakeholders according to their importance to the project. Because experts are needed to ask for recommendations from stakeholders, it might become costly in large projects. To overcome this, the authors developed a tool called StakeSource to support the process (Lim et al. 2010b). StakeSource works in four ways: Identifying stakeholders, prioritizing stakeholders, identifying potential problems and displaying the stakeholder information, tasks that would otherwise be up to the experts in StakeNet. The tool proposes a way of using crowdsourcing, where stakeholders are essentially the crowd, to ensure an accurate list of stakeholders is being generated and prioritized. The crowd suggests new stakeholders and when new ones are recommended, StakeSource sends an email to those requesting recommendations about other stakeholders. This reduces the influence from experts and lowers the likelihood of overlooking stakeholders. This example elicits an interesting use of classifying stakeholders through crowdsourcing the stakeholders themselves. Knowing a complete list of stakeholder's results in more complete requirements and StakeNet is therefore useful in gathering requirements.

If one would broaden the scope of eliciting requirements to user research, crowdsourcing seems to be able to handle experimental type of studies as well. If done correctly, there seems to be consensus between controlled laboratory experiments and crowdsourcing experiments for a variety of tasks (e.g. Kittur et al. 2008; Heer and Bostock 2010). For example, Heer and Bostock (2010) have used crowdsourcing for graphical perception tasks. Results from perceptual tasks like participants making choices about the position, length, angle of shapes or

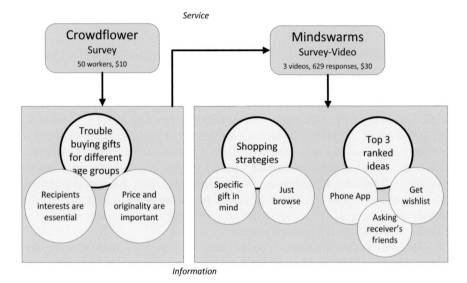

Fig. 3 The process followed in this case study about eliciting requirements for a gift-recommendation app. A survey in Crowdflower showed: (1) People have trouble buying gifts for people in different age groups than their own. (2) Price and originality are deemed important when buying a gift and the recipients' interests are essential. Based on this information, video responses were gathered through Mindswarms results revealed important shopping strategies. Some people buy something with a specific gift in mind, while others just go out and browse

graphs, concurs with controlled laboratory experimental results. This also holds for manipulations to find out acceptable contrast luminance settings. However, operating system and monitor details influence results and should be recorded when conducting such an experiment. Requirements can therefore also be set by crowdsourcing methods with more psychological factors in mind like perception.

Case Study

To specifically illustrate how crowdsourcing can be used to elicit user requirements we will draw an example from our case study (Fig. 3). The focus of our example is about gifts. Buying gifts is a task that many people have trouble with. Two crowdsourcing services were used to help the design team understand what aspects of giving gifts are perceived difficult.

The first one was Crowdflower. Fifty workers were selected and were asked to fill out a questionnaire. More specifically they were asked to report which things they found the hardest when buying gifts for other people. Results showed that workers had most trouble buying gifts for people in different age groups than their own and that a key aspect of gift selection is the recipient's personal interests. In addition, price and originality were deemed very important. Because of the worker population and survey options, Crowdflower provided a quick aid to the design team in gathering an initial set of user requirements.

The second one was MindSwarms. Based on the previous results, this service was used to elicit more focused responses. Out of 629 screening responses, three video responses were chosen. The number of chosen responses was primarily limited to the team's budget. Analysis of the video responses revealed behaviors and strategies such as: some people visit a shop with a specific gift in mind, and buy if it is there. If it is not there, they will return home think of a new gift idea. Others think and browse at the same time and visit several shops until they decide upon a gift. As part of this study, workers also rated some proposed ideas for improving the gift giving process. The top three ranked ideas were: (1) a smartphone application with local gift suggestions, (2) asking friends and family about the preferences of the receiver of the gift and (3) getting a wish list from recipients themselves. A subsequent brainstorm session with this information by the researchers led to a more specific direction. This resulted in specific requirements concerning item categories. For example it was decided that the system should contain sections with: books, flowers, the ability for ranking by price/popularity, demographic and social information about the gift recipient and information about where to buy the gifts.

Summary

This section attempted to clarify that crowdsourcing can be used to elicit user requirements. Although seemingly a natural fit, few attempts have been made to explicitly use crowdsourcing in this way. We presented examples of services developed specifically for the purpose of eliciting requirements as well as creating a network of stakeholders with the same purpose. Finally, a case study of the use of crowdsourcing for eliciting requirements was described in which two iterations using two crowdsourcing services were sufficient for extracting valuable information for a proposed system on gift recommendations. Once a design team has elicited requirements ideating about possible solutions usually follows.

Ideating Using Crowdsourcing

Dow et al. (2013) describe the second phase of the design cycle as 'ideating', where ideas are gathered and brainstormed based on the established requirements and needs. We present in this section how online crowds can support the typical brainstorming/ideating phase in design by increasing the number of ideas that come to the table. The structure of this section follows the organizational classification of crowdsourcing, which was mentioned in a previous section. As in the previous section we first review relevant examples in literature and then we draw an example from our case study.

Examples in Literature

Several companies have implemented the use of crowdsourcing during this phase of product and service design, such as Nokia, Starbucks, Dell, IBM and Lego Group. In their white paper Aitamurto et al. (2013) in cooperation with Nokia's idea crowdsourcing team distinguish several ways of using crowdsourcing during the 'ideating' phase. Firstly, it is possible to separate crowdsourcing with external collaborators from internal crowdsourcing, which describes initiatives for ideation among the employees of the company.

To the aforementioned distinction by Aitamurto et al. (2013) we will add idea contests and microtask services for idea generation. With this division this section describes in what way crowdsourcing can be of value during the ideating phase of the design cycle by giving examples of companies or relevant services for internal crowdsourcing, external crowdsourcing by the (company) clients themselves and external crowdsourcing via external services. The cases are selected to provide an insight in the use of crowdsourcing by large companies as well as to give a reflection from relevant academic literature. Almost all methods described by company examples can also be used on a smaller scale by anyone who wants to use crowdsourcing within the ideating phase of design.

Internal Crowdsourcing

An example of internal crowdsourcing for conceptual design can be found in IBM. Since 2001 IBM organizes 'jams', described by Bjelland and Wood (2008) as a group of interlinked bulletin boards and related Web pages on IBM's intranet, where tens of thousands of employees get involved over 3 days to generate new ideas and for example clarify IBM's values. In 2006 IBM opened up to employees' family members, business partners, clients and university researchers and organized an online "Innovation Jam" with over 150,000 participants. The innovation jam consisted of several steps, namely preparation, idea generation (Jam phase one), reviewing ideas of Jam 1, idea refinement (Jam phase two), reviewing ideas of Jam 2 and finally the proposal of new business units. This two-phase method is confirmed as a valuable method for gathering ideas by Yu and Nickerson (2013). They compared the results of a sequential combination system, in which participants had to combine good ideas from a first phase of idea generation during a second phase, to a Greenfield system in which ideas are generated by a crowd that is unaware of previously crowd-generated ideas. Participants in the experiment were asked to design an alarm clock that was easy and safe to use and inexpensive to manufacture. After applying either the sequential combination system or the Greenfield system the ideas generated by the crowd were evaluated with a measure of creativity developed by Ronald Finke (1990), based on the components of originality and practicality. This resulted in finding that the ideas made by participants following the sequential combination system were more creative.

External Crowdsourcing by Clients Themselves

An example of a company that creates new conceptual designs by crowdsourcing idea generation to individuals outside the company is Dell (Aitamurto et al. (2013); Bayus 2013). Dell has created a permanent virtual collaboration platform on which end users can create an account and share their ideas. Other users can comment on the topics raised and vote for their favourite ideas. In the first 4 years that Dell IdeaStorm was launched already 15,400 ideas were posted to the platform. However, Aitamurto et al. (2013) describe how the absorption of an idea by the company is dependent on the knowledge gap between the user and the company and the ability of the user community to reduce the ambiguity of the idea. Furthermore, they describe how the platform has, to a large end, become a place for regular customer feedback, instead of big business or innovation ideas. A more in depth study of the Dell IdeaStorm community demonstrated that ideators who did suggest an idea that was implemented by the company were not very likely to repeat their success. Instead they started proposing ideas similar to the ideas that were already implemented (Bayus 2013). Other companies that have embraced this approach are Starbucks' MyStarbucksIdea; Nokia's Ideasproject; and Lego Group's Lugnet.

In addition to permanent platforms that offer the opportunity for repeated idea generation several companies propose one time or yearly design contests. An example of a yearly competition among students is the Electrolux Design Lab, a global competition to create innovative ideas for future households. An example of a onetime challenge, is a pilot project of The Federal Transit Administration in the US focused on bus stop design in Salt Lake City, Utah (Aitamurto et al. (2013)). Over the course of the pilot 47 designs were submitted. From this pilot it became clear that amateur-participants did not see value in participating once the level of submissions reached a professional level. This is thus something to keep in mind when starting a contest for conceptual designs for a heterogeneous participant group.

External Crowdsourcing Via Other Services

Sometimes an intermediary becomes part of the crowdsourcing activity as a party in between the requester and the crowd asked to complete the task. Two types of intermediaries are distinguished, namely services that are focused specifically on gathering innovative ideas and services that enable clients to offer more broad types of microtasks.

Services that specifically focus on gathering ideas from the crowd are for example OpenIdeo, InnoCentive and NineSigma. They present themselves as open innovation services, where a community of solvers is created around a number of problems that clients of the service face. Whereas OpenIdeo creates a community where anybody can join, InnoCentive provides next to open challenges also the opportunity to create a closed environment for a challenge (for example only for employees) On their website InnoCentive describes how they had more than 1650 external challenges and thousands of employee-facing challenges, with awards ranging from $5000 to more than one million USD. The use of an open innovation service to find solvers for a problem can, to some extent, solve problems that companies face when putting challenges to the online crowd (Simula and Ahola 2014). For example, by not revealing the name of the initiator, sensitive information about future plans of a company can be kept secure. Furthermore, with their experience in posting challenges online, the services can assist companies to acquire the most relevant solutions as quickly as possible. This can for example relate to the amount of information given, the type of task or level of difficulty asked and the communication style to reach the right workers. See for more information the tips section at the end of this chapter.

A more accessible service is Tricider, an online brainstorming and decision making tool. It provides the option to list questions and invite people to answer them. Afterwards everybody can vote for the submitted ideas. The service is more accessible because no registration is needed, neither for proposing problems nor for proposing potential solutions. However, the community that can be reached without putting effort in spreading the word is small and projects need to be paid to receive input from more than twenty idea generators.

Services that enable clients to offer a wider type of tasks, among which idea generation can be one, are for example Crowdflower, Microworkers and Amazon Mechanical Turk (AMT). The microtask services provide anybody the option to set a task, e.g. to complete a survey, categorize data, and moderate or create content for a certain amount of money per completed task by a maximum amount of workers. After that, the workers registered on the service can choose whether or not to take on the task. Tasks can be completed on different websites by sending proof to the service, or by completing them on the virtual environment of the service itself. Some services, such as Microworkers, provide the option to reject unsatisfying work, until the necessary amount of workers with satisfying work is collected. An option for idea generation is creating a short survey in which the problem can be described as extensively as necessary and ideas can be gathered with one open question or multiple questions that give a more narrow description of the problem.

AMT was used by Dow et al. (2013) in the ideating phase of a student course on using crowdsourcing for design. In addition to a brainstorm within the student team, each team asked workers to generate five ideas each, leading to an average of about 140 ideas per group. Students were encouraged in their own brainstorm by the large number of ideas that could be gathered by the crowd, although they were concerned with the lack of quality and overlap between ideas of the crowd.

Case Study

We will continue with describing a project, from our case study, to demonstrate how young designers have used crowdsourcing for the ideating phase of their projects. The goal of this specific project was specified after crowdsourcing for user needs

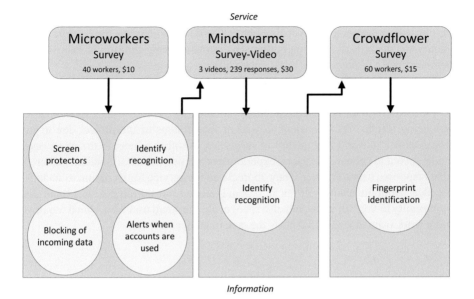

Fig. 4 The services used and information gained

around security and privacy and resulted in the aim to improve the feeling of security during the process of making online payments with a smartphone (Fig. 4).

Two services were used to gather ideas, based on the aforementioned description of crowdsourcing via external services. First, forty workers on Microworkers were recruited to come up with two solutions for problems that they could experience around online privacy. The task was introduced with little information to keep it as simple as possible:

> *"This question is about online privacy. We would like to ask you to think about solutions to create more online privacy while making payments (e.g. for shopping) via your smartphone. We are looking for ideas that make users more aware of and improve the feeling of online privacy. There are no wrong answers, don't let technical limitations constraint your ideas, be as creative as you want."*

Generated ideas ranged from popups with information about the security of one's Internet connection, to fingerprint and voice recognition for confirming payments. Requesters found the option to reject work that does not satisfy the task a big advantage of using Microworkers, because it was clear that not all entries were of similar quality and the amount of effort taken to respond to the task varied. Furthermore, they considered it a cheap way to crowdsource a survey to an international crowd, as only $0.56 was paid per entry of two ideas. While generating ideas via Microworkers the team also generated ideas by themselves, by an open brainstorm session on the topic. Finally, when comparing all ideas, substantial overlap was found between workers in the crowd, and between the ideas of the crowd and the team. However, both sources also provided unique ideas. Putting this in the perspective of Brabham's (2008) description on the superiority in quality and quantity of solutions gathered by crowdsourcing, the current case mostly demonstrated a higher quantity, but not quality.

The second service that was used was Tricider. Although the same task description as in Microworkers was used and a reward of €5 for the best idea on Tricider was set, no useful ideas were gathered via Tricider. Two entries were gathered, with one being a joke. Tricider was not used in future steps of the project due to its low response rate and low quality of responses. Another project in the same case study, set out to enhance the users' experience of visiting a football stadium gathered nine responses using Tricider. They appreciated the service for its option to include the ideas from the team itself (as an internal online brainstorming session) and the opportunity to vote for ideas, but also recognized the small user base which requires more work from the team that posts the task to gather workers.

Summary

This section attempted to clarify that crowdsourcing can be used in different ways to increase the number of ideas in a design cycle. We discussed examples of internal crowdsourcing, in which a company targets their own employees, external crowdsourcing by clients themselves, where companies or other organizations create virtual collaboration platforms or design contests, and external crowdsourcing, where intermediary services assist in communicating a task to a crowd. Finally, a case study of the use of crowdsourcing for ideation was described to give an example of a task, suitable services and output that can be expected. After having completed the ideation phase usually design teams will design their ideas.

Designing Alternative Ideas Using Crowdsourcing

Prototypes are concrete and tangible artifacts that act as a communication bridge for mutual understanding between the designers, developers, clients, and the end users. Prototypes act as a tool to aid the creative design process, wherein the designers can create and share ideas with other members of a design team. These can then be used by the designers to get feedback from the end users about the products being developed. For different Agile User Centered Design (UCD) methodologies with short and iterative design cycles, designers can feed the developers with complexities or details of a product easier with a prototype than a requirements text.

Depending on the required fidelity, prototypes can also be crowdsourced. Nevertheless, to the best of our knowledge there are no specific examples reported in the academic literature on utilizing crowdsourcing services for prototyping purposes

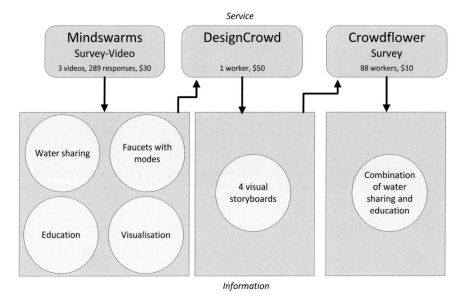

Fig. 5 Before using DesignCrowd for getting storyboards requirements were elicited through a survey and video responses on MindSwarms and the received storyboards were then evaluated with workers in Crowdflower

per se. Therefore, in this section we will report our case study experiences in crowdsourcing storyboards that could be considered as a form of a low-fi prototype.

Case Study

The goal of this specific project was to provide solutions for encouraging water conservation and reducing water wastage in the general population (Fig. 5).

Initially, workers (N = 88) on MindSwarms were asked how they use water in their homes. Three videos were selected to get in-depth information on water usage. Analysis of this data led to four main ideas: Water sharing, Faucet with various modes, Education and a water consumption visualization.

Subsequently, workers on DesignCrowd were asked to design a storyboard for each of these ideas. The final storyboard was designed by a skilled and favorably-priced (€50) designer from Indonesia. The requesters could see the graphic designer's portfolio and based their selection on this. The requesters directly contacted the graphic designer via email about the details of the project and to set deadlines. The graphic designer confirmed that she could complete the job within the deadline. A first draft of storyboards was delivered within 24 h of accepting the design job. The design team reviewed and provided some feedback for the draft storyboards. In another 24 h, the design team received the final version of their

You live together with your family in a suburban area.	In your community there is a platform for saving water. Here you can see your neighbors' water consumption.	You join the platform. Now you can see how much water your neighbors spend, and they can see yours as well.	This triggers your family to save water, because you want to be better than your neighbors.

Fig. 6 Storyboard Concept – Neighbor's usage of water affects and encourages other users' water usage in the neighborhood, figure courtesy of trainees: Zhiyuan Zheng, Areti Paziourou, Maria Gustafsson

My faucet have various modes, meaning that I can control the flow of the water depending on what I want to wash.	For the cutlery I'm choosing the "min" mode since I don't need much water to wash them.	For the regular plates I am choosing the "medium" mode.	'For the pots I need a lot of water, so I'm turning the mode into "max".

Fig. 7 Storyboard Concept – Faucet with pre-defined different discrete water flow control levels, figure courtesy of trainees: Zhiyuan Zheng, Areti Paziourou, Maria Gustafsson

prototypes. These four design concepts involving different ways of encouraging people to conserve and save water. Two of them can be seen in Figs. 6 and 7.

These prototyped storyboards were again posted on a crowdsourcing websites to get feedback on the most appealing concept for conserving water. The storyboards were presented to workers (N = 289) on the Crowdflower service which showed that a combination of water sharing and education would be preferable. Crowdsourcing evaluation will be further discussed in the following section.

The use of crowdsourcing services results delivered input to the design process starting from idea generation to actual concept selection. By crowdsourcing the generated ideas, researchers were ensured quality visualizations (outside their own skills) to be used as input for the actual concept selection.

There are many other crowdsourcing platforms like DesignCrowd that offer similar design options and designer skills such as: 99designs (99designs.com) Crowd-SPRING (crowdspring.com), Choosa (choosa.net), Hatchwise (hatchwise.com), Logomyway (logomyway.com), Eyeka (eyeka.com), just to list a few. Design teams should consider the time they have available, the quality of work required, and the minimum costs involved while selecting a crowdsourcing service for a prototyping assignment.

Summary

This section presented a case study in which crowdsourcing was used to support a design team in getting alternative design solutions from crowdsourcing services. Few storyboards were received from a crowdsourcing service in a matter of 2 days. After having design alternatives usually design teams will evaluate those.

Evaluation Using Crowdsourcing

While on the one hand evaluation depends on specific research objectives, there are certain questions that every designer needs to ask when being in such a phase. What to evaluate is the first one. That can vary from low-fi prototypes to a complete system. The previous section covered how crowdsourcing can actually help designers crowdsource such artifacts.

When to evaluate is the second question. Broadly speaking there are two types when it comes to this question: formative and summative evaluation (Lewis 2014). Formative evaluation would occur during design to check whether the current state of the product meets users' needs. Whereas summative evaluation would occur to assess the success of a finished product. Crowdsourcing platforms can certainly have a role to the timing issue of evaluation.

Examples in Literature

When it comes to summative evaluation, Liu et al. (2012) evaluated the usability of a graduate school website with the traditional lab usability test and with a crowdsourcing service. AMT in combination with Crowdflower was used as the crowdsourcing platform for recruiting participants. For the traditional (non-crowdsourcing) test they recruited five of the school's students, who were regular users of the website. Since those five students were not registered in the particular crowdsourcing service, there was a different group of participants. Moreover, the requested tasks were slightly different for the two groups. The tasks for the crowdsourcing group were designed to represent prospective students that had not visited the website. For the crowdsourced test first a pilot with 11 participants was conducted and then the actual test with 44 other participants took place. In the case of the traditional test, the think aloud protocol was enforced. The differences with the crowdsourcing test is that in the latter participants were compensated, their actions on the website were not recorded and there were of course no researchers present to observe the users' behavior.

Regarding the consequences for performing usability testing with the help of crowdsourcing platforms, since researchers are remotely located, instructions and tasks must be specific and unambiguous (Liu et al. 2012). Further, the survey to capture the user's views must take into account that users might cheat. Therefore it must be designed to make cheating difficult to occur. On that respect, Liu et al. (2012) did the following actions: (1) they informed participants that the study is for academic purposes (2) instead of multiple-choice questions they used blank-filling questions that required active searching for information on the website, the downside of that being increase of task completion time (3) participants giving substantial feedback were given a bonus while those giving random answers were not compensated.

Although it is difficult to generalize from Liu et al. (2012), since tasks and participants were different for the two conditions, some useful insights can be stated. To begin with, the time spent to collect the data was significantly less with the crowdsourcing platform. The results of 44 participants in the crowdsourcing case took less than 1 h while with the traditional approach 30 min per participant were required. Although there were differences in the actual usability problems that the tests uncovered, nevertheless a significant overlap, specifically of the major ones, was found. The issue of clarifying instructions was not possible in the crowdsourcing case. Moreover, asking for additional input from participants in case something was unclear was more difficult for the crowdsourced case.

In another example, Dow et al. (2013) were the first to publish their findings of utilizing crowdsourcing in the classroom. Among other crowdsourcing activities, students also gathered feedback for their prototypes from crowds. For example, one group of CMU students gathered feedback using MindSwarms for six storyboards that proposed design solutions, each yielding 28 video clips. Based on that feedback students presented their findings on how people reacted to those storyboards. In another part of the course students were asked to created working prototypes on the web, install Google analytics and perform A-B testing by recruiting at least thirty AMT workers.

Apart from the aforementioned platforms and when thinking of summative evaluation, one of the first aspects to evaluate is the usability of the finished product. Usability testing beyond the lab is well-known as remote usability testing (Albert et al. 2009). Nowadays, there is an explosion of companies offering such services. Just to list some: Loop11 (Loop11.com), UserZoom (UserZoom.co.uk), UserTesting (UserTesting.com), Youeye (youeye.com), Usabilla (Usabilla.com), Conceptfeedback (conceptfeedback.com), Crowdsourcedtesting (crowdsourcedtesting.com), Trymyui (Trymyui.com).

Going beyond the fact that some crowdsourcing platforms try to imitate the process of traditional usability testing there is an emergence of new, innovative services that take advantage of their digital nature. For example Usabilityhub.com has three possible tests that differ significantly from the traditional think-aloud protocol. The first one is titled: "five second test". In that, workers see the design to be tested for just five seconds. Then workers are asked a series of questions, such as "What product do you think this company sells?", or "What was the company name?". This test particularly aims to test first impressions and how easy is a certain design to understand. The second one is titled: "click test". In that test a design is shown and the system records where workers click on it. The aim of this is to test navigation placement and prominence as well as how clear calls-to-action are in the design. The third one is titled: "nav flow test". In that test navigation of a design is tested. The requester uploads a series of page designs, and specifies where workers have to click to proceed. The success and failure rate of workers is then recorded at each step. The aim of this test is to test multi-step navigation flows and where users will leave a certain screen.

Beyond the different types of tasks Usabilityhub.com has a virtual currency model. A user of the platform can complete tasks and for each task gets credits. Those credits can then be used in the platform itself to request for tasks. One can also purchase credits. This is a novel and interesting alternative to the traditional compensation model of crowdsourcing systems. In most systems the compensation is financial. In this platform the compensation is essentially data from design tests.

Case Study

For illustrating how crowdsourcing platforms can support designers in the case of formative evaluation we will refer to two projects from our case study.

For the first project the objective was to enhance the football stadium experience. Input from a questionnaire administered through Qualtrics and an idea generation round on Tricider led to the selection of three ideas: Replays on phone, Vote for the player of the match, and showing replays on an extended screen. One hundred workers of the Crowdflower service were engaged to investigate whether people would agree with the interpretation of the previous results. Responses showed that almost half of the workers would download the app, few would not, and the remaining workers were in doubt. Most functionality in the app scored four out of five on a Likert scale. Moreover, MindSwarms was used (200 screening workers, six video responses chosen) to gain insights into what people think about the app and to get real-time video response. The conclusion was a positive response confirming that this concept is on the right track in developing an application to enhance the experience of visiting a football match.

For the second project, the storyboards to encourage water conservation that were mentioned in the previous section, were submitted back into the crowd, for evaluation purposes, using MindSwarms (86 screening responses, four video responses) and Crowdflower (50 workers). The evaluation objective was on the one hand to get more descriptive feedback on the storyboards and on the other hand to reach a larger group of workers. Workers were asked to select the most appealing storyboard and come up with suggestions. These results showed that two of the four storyboards were liked the most by workers and two combinations of storyboards were proposed. The final concept was a faucet that a user can adjust so that water can be regulated according to their current water usage needs (seen in Fig. 7).

Summary

This section presented examples reported in literature and a case study in which crowdsourcing was used to support a design team in evaluating design solutions. There is a growing number of services that goes beyond remote usability testing which include short design tests as well as video responses from remote workers.

Tips for Utilizing Crowdsourcing Platforms for User and Design Research

In this section we present a list of issues that designers will probably face when trying to crowdsource phases of the design cycle. Furthermore, we elaborate on tips to tackle those issues.

For Checking the Quality of Responses, Have Verifiable Questions

Getting quality responses and generally controlling the quality of the crowdsourced tasks is a recurring issue. One solution to this salient issue is to have verifiable questions in addition to the tasks of interest. However, Kittur et al. (2008) mention that the quality of results seem to be dependent on carefully setting up the task presented to workers. By making participants rate, on a Likert scale, several Wikipedia articles on factors such as whether the article was well written the comparison of these ratings with expert ratings showed that there is very weak evidence for the two being consistent. However, in a second experiment, Kittur et al. (2008) explicitly added verifiable questions and found that the consistency between worker and expert reviews was much higher and that there were less meaningless responses. They conclude that it is important to: (1) have explicitly verifiable questions, (2) that tasks are designed so that they take least effort possible and (3) that tasks have multiple ways to detect suspect responses. Like Kittur et al. (2008), Heer and Bostock (2010) emphasize the importance of using qualification tasks and verifiable questions in the process to increase the likelihood of quality responses. An example of verifiable a question could be a simple one like "How much is $5 + 5$?". If the answer is senseless it means the worker is not taking the task seriously and can be excluded from analysis.

Put Extra Effort in Describing the Requested Task

Poor quality of output can also be attributed to lack of understanding of the task's context. Geographic, time and cultural differences are bound to occur when using crowdsourcing. Therefore, one needs to be more sensitive when formulating task objectives or survey questions. One needs to also think very carefully of the context implied with the study and make sure to capture it if necessary. For example, when it comes to usability testing since workers are in their own environment, they might not completely focus on the task. The results, such as for example the time taken to navigate through a website might vary if a participant is simultaneously doing something else. In that case one would need to carefully instruct workers to make sure to have booked the time and place where distractions will be avoided. Thus, instructions and tasks in the crowdsourced usability testing must be described specifically and totally unambiguously (Liu et al. 2012).

Dealing with Money

Crowdsourcing platforms offer what outsourcing offered as a monetary advantage. Higher currency conversion rates might act as an advantage for some countries. In lesser money, design teams might be able to obtain similar quality work from a crowd member in a country with lower currency value.

Nevertheless, there is always the question of how much to actually pay for a task. Monetary compensation for participation plays a major role in the performance of the tasks by the crowds on different crowdsourcing platforms. Nelson and Stavrou (2011) noted that the performance of users on a usability evaluation task on AMT increased as the compensation was increased till a certain amount. The performance dropped again when the amount of the compensation was further increased. Furthermore, for user requirements elicitation, Mason and Suri (2012) showed that the amount of money paid for each task directly influences the time in which surveys are completed. Payments of $0.01 showed a drastic increase in response time compared to prices of $0.03 and $0.05 for each task.

Furthermore, to the issue of payment and output of work, it has been shown that increased financial incentives increase the quantity, but not the quality of work performed by participants (Mason and Watts 2009). The recommendation for optimal pricing would be to experiment with pricing and to lower the payment to workers down to the price where the speed of picking up the published tasks was still acceptable.

Finally, one needs to be aware that there are more reasons for workers to participate in a crowdsourcing platform than money. Reasons such as altruism, enjoyment, reputation and socialization have been shown to exist among workers (Quinn and Bederson 2011). Another reason is found in the case of the open-source web browser Firefox. Improving the performance of the product is a salient reason for people to actively contribute to the project (Ko and Chilana 2010).

Be Aware of Selection-Bias

Although there are increasingly more people connecting to the Internet and although it has been shown that specific user groups can be reached that might be difficult to reach otherwise, such as senior citizens (Tates et al. 2009), one needs to be aware of the selection bias that inevitably occurs in crowdsourcing platforms. For example, in a 2010 study Ross et al. (2010) showed that AMT had essentially two major populations: well-educated, moderate-income Americans, and young, well-educated but less wealthy workers from India.

Around the same time period Brabham (2008) stated that an online crowd does not guarantee diversity of opinion and might thus not be able to generate many high quality ideas. The typical web user, at least in 2008, was still likely to be white, middle- or upper-class, English speaking, higher educated and having a high-speed Internet connection. When it comes to ideating this creates the dangers of not finding useful ideas at all, but also of only creating solutions that fit white middleclass young individuals. In addition, Aitamurto et al. (2013) describe how the best idea generators may prefer closed networks instead of open online communities, because their ideas might then more likely be developed. It is therefore clear that selecting the right workers for an idea generation task can be challenging and take time and effort.

Furthermore, it is obvious that workers would need to have access to a computer connected to the Internet as well as basic skills of computer usage. Moreover, one also needs to be sensitive to the fact that a lot of workers participate in crowdsourcing platforms due to the need of extra income.

On the other hand, if the design team lacks some skills, then the crowd can make up for that skill thereby making usage of the crowds a practically viable option for teams that need a skill temporarily for one of the projects.

Great Tool for Ideation

Several benefits have been recognized by authors reflecting on the use of crowd-sourcing for ideation. The benefits can be easily described in the large amount of ideas that can be generated (Aitamurto et al. 2013; Bayus 2013; Brabham 2008; Dow et al. 2013), in the fact that it is cheaper when compared to internal R&D (Bayus 2013; Brabham 2008) and that results can be seen in a very short amount of time (Bayus 2013; Brabham 2008). In contrast to the cases presented from our case study and the course by Dow et al. (2013), Poetz and Schreier (2012) found that a crowd of users could provide ideas scoring higher on novelty and the attribution of customer benefit than ideas from professionals, indicating that sometimes ideas could also be more useful for a design team or company. However, ideas from professionals turned out to score slightly higher on feasibility.

Nevertheless the large number of ideas that can be yielded from crowds might have the downside that design teams sometimes have to face a tough selection process to come to the most relevant or useful ideas that can be developed and tested further. Screening all ideas may take high costs, and a lot of time and effort (Aitamurto et al. (2013)). A solution to this problem can be found in crowdsourcing itself! One can give back the task of ranking of the submitted ideas (applied e.g. by OpenIdeo).

Be Aware of Ethical and Intellectual Property (IP) Issues

Finally, companies and teams who aim to use crowdsourcing for idea generation should consider legal and ethical questions that rise during the development of a project. An important legal matter is the distribution of the intellectual property of the ideas (Aitamurto et al. 2013; Simula and Ahola 2014). In almost all cases when one reviews the terms and conditions of crowdsourcing services on the matter of IP, one will find that the complete IP rights lie with the requester. Nevertheless, the global worker base of such services raises both issue of international laws and practical issue with actually being able to check whether the requester's IP rights are not actually infringed. If a requester is in a European country and a worker in far-east Asia uses the design or parts of the concept it would take a lot of resources for the requester to find out about it and then to legal pursue the worker. A straightforward solution to such an issue would be to split the task at hand in smaller pieces and then request those smaller pieces to be completed by different workers. On the other hand, issues of quality might arise. Form the workers' perceptive, an ethical issue to keep in mind is whether workers get fair payment for their ideas, or crowdsourcing should be seen as part of a slave-like-economy (Brabham 2008; Felstiner 2011). Due to the novel nature of such services this is still an open question that needs to be addressed.

Conclusion

It has become clear that although relatively new, crowdsourcing is often used in all phases of the design cycle. Examples are widespread among profit and non-profit companies, government agencies and researchers. Crowdsourcing can be used in several distinct ways, varying in the nature of the crowd to whom a task is send, to the type of task and the nature of the reward.

For eliciting requirements crowdsourcing can provide access to a global user base. Different platforms provide a number of ways to query and get data from the users to better understand their needs and the context in which those needs arise.

When one wants to use crowdsourcing to gather ideas, it is important to keep in mind that a diverse crowd that fits the complexity of the topic is selected, that the selection of ideas is prepared and that legal and ethical issues surrounding crowdsourcing are properly addressed. Overall, crowdsourcing can be a useful addition to the phase of ideation in the design cycle, by cheaply and quickly adding a large number of ideas in the creative process.

If the design team lacks some skills or is under time pressure to create design alternatives or prototypes, then reaching out to crowds can be a pragmatic choice. Crowds can complement the strengths of the design team with strong visual design, programming, technical writing, and other skills. Crowds can provide multiple prototypical versions of a design idea for the design team to choose from in short time. Crowdsourced prototypes can vary from whole website design to just creating a color them for the website or the landing page of a website. Unlike testing with crowds, prototyping from crowds ensures a certain quality of the prototyped product according to price offered by the design team.

When one wants to use crowdsourcing to evaluate design artifacts, again a plethora of services exist. Traditional usability evaluation services are better known as remote usability services whereas re-appropriation of several existing services, such as asking users' thoughts after having interacted with a prototype on Mindswarms, can serve system evaluation purposes. New crowdsourcing services open up the design evaluation options by offering novel micro-evaluation tasks.

Acknowledgements We would like to extend our gratitude to all trainees of the User-system interaction (USI) program of Eindhoven University of Technology (generation 2013) for their insights into using crowdsourcing for design and user research. Moreover, we would like to specially thank the Mindswarms team for their continuous support in helping us to use their service for our educational objectives and the other aforementioned services.

Annotated Bibliography

Kittur A, Chi EH, Suh B (2008) Crowdsourcing user studies with Mechanical Turk. In: Proceedings of the SIGCHI conference on human factors in computing systems, April. ACM, New York, USA, pp 453–456
The authors describe an experiment in which they investigate Amazon's Mechanical Turk and how they can get quality responses from workers
Mason W, Suri S (2012) Conducting behavioral research on Amazon's Mechanical Turk. Behav Res Methods 44(1):1–23
The paper demonstrates how to use Amazon's Mechanical Turk website for conducting behavioral research and tries to lower the entry barrier for researchers who could benefit from this platform
Description of how Google Consumer Surveys is superior to current probability based Internet panels by using what is known as a "surveywall" to attract respondents
Aitamurto T, Leiponen A, Tee R (2013). The promise of idea crowdsourcing: benefits, contexts, limitations, September 25. Retrieved from http://www.academia.edu/963662/The_Promise_of_Idea_Crowdsourcing_Benefits_Contexts_Limitations
The authors review crowdsourcing for idea generation ('idea crowdsourcing') both from the perspective of academic literature and actual cases from businesses to understand how and when to use crowdsourcing and with which benefits and costs

References

4 Reasons you should consider crowdsourced design for your next big project (n.d.) Retrieved September 30, 2014, from http://michaelhyatt.com/crowd-sourced-design.html

Adepetu A, Ahmed KA, Al Abd, Y, Al Zaabi A, Svetinovic D (2012) CrowdREquire: a requirements engineering crowdsourcing platform. In: AAAI spring symposium: wisdom of the crowd, March

Albert W, Tullis T, Tedesco D (2009) Beyond the usability lab: conducting large-scale online user experience studies. Morgan Kaufmann

Bayus BL (2013) Crowdsourcing new product ideas over time: an analysis of the dell IdeaStorm community. Manag Sci 59(1):226–244, http://doi.org/10.1287/mnsc.1120.1599

Bjelland O, Wood R (2008) An inside view of IBM's 'innovation jam. MIT Sloan Manag Rev 50(1):32–40

Blattberg E (2011) Crowdsourcing industry landscape. Retrieved from http://www.crowdsourcing.org/editorial/november-2011-crowdsourcing-industry-landscape-infographic/7680

Brabham DC (2008) Crowdsourcing as a model for problem solving: an introduction and cases. Converg Int J Res New Media Technol 14(1):75–90, http://doi.org/10.1177/1354856507084420

Corney JR, Sanchez CT, Jagadeesan AP, Regli WC (2009) Outsourcing labour to the cloud. Int J Innov Sustain Dev 4(4):294, http://doi.org/10.1504/IJISD.2009.033083

Crowdsourcing your brand design: the math just doesn't work out (n.d.) Retrieved from http://forty.co/crowdsourcing-your-brand-design-the-math-just-doesnt-work-out

Dow S, Gerber E, Wong A (2013) A pilot study of using crowds in the classroom. ACM Press, New York, USA, p 227, http://doi.org/10.1145/2470654.2470686

Felstiner A (2011) Working the crowd: employment and labor law in the crowdsourcing industry (SSRN scholarly paper No. ID 1593853). Social Science Research Network, Rochester. Retrieved from http://papers.ssrn.com/abstract=1593853

Finke RA (1990) Creative imagery: discoveries and inventions in visualization. L. Erlbaum Associates, Hillsdale

Geiger D, Rosemann M, Fielt E, Schader M (2012) Crowdsourcing information systems – definition, typology, and design. In: ICIS 2012 proceedings. Retrieved from http://aisel.aisnet.org/icis2012/proceedings/ResearchInProgress/53

Glinz M, Wieringa RJ (2007) Guest editors' introduction: stakeholders in requirements engineering. Software IEEE 24(2):18–20

Grier DA (2011) Foundational issues in human computing and crowdsourcing. In: CHI 2011

Heer J, Bostock M (2010). Crowdsourcing graphical perception: using mechanical turk to assess visualization design. In: Proceedings of the SIGCHI conference on human factors in computing systems, March. ACM, pp 203–212

Hosseini M, Phalp K, Taylor J, Ali R (2013) Towards crowdsourcing for requirements engineering. In: Joint proceedings of REFSQ-2014 workshops, doctoral symposium, empirical track, and posters, co-located with the 20th international conference on requirements engineering: foundation for software quality (REFSQ 2014). Retrieved from http://ceur-ws.org/Vol-1138/et2.pdf

Howe J (2006) The rise of crowdsourcing. North 14:1–5, http://doi.org/10.1086/599595

Ko AJ, Chilana PK (2010) How power users help and hinder open bug reporting. ACM Press, New York, USA, p 1665. http://doi.org/10.1145/1753326.1753576

Lewis JR (2014) Usability: lessons learned ... yet to be learned. Int J Hum Comput Interact 30(9):663–684, http://doi.org/10.1080/10447318.2014.930311

Lim SL, Quercia D, Finkelstein A (2010a) StakeNet: using social networks to analyse the stakeholders of large-scale software projects. In: Proceedings of the 32Nd ACM/IEEE international conference on software engineering – Volume 1. ACM, New York, pp 295–304. http://doi.org/10.1145/1806799.1806844

Lim SL, Quercia D, Finkelstein A (2010b) StakeSource: harnessing the power of crowdsourcing and social networks in stakeholder analysis, vol 2. ACM Press, p 239. http://doi.org/10.1145/1810295.1810340

Liu D, Bias RG, Lease M, Kuipers R (2012) Crowdsourcing for usability testing. Proc Am Soc Inf Sci Technol 49(1):1–10, http://doi.org/10.1002/meet.14504901100

Li Z (2011) Research of crowdsourcing model based on case study. Manag Sci 1–5. http://doi.org/10.1109/ICSSSM.2011.5959456

Malone TW, Laubacher RJ (1999) The dawn of the E-Lance economy. In: Nüttgens M, Scheer A-W (eds) Electronic business engineering. Physica-Verlag HD, Heidelberg, pp 13–24. Retrieved from http://link.springer.com/10.1007/978-3-642-58663-7_2

Mason W, Watts DJ (2009) Financial incentives and the "performance of crowds". In: Bennett P, Chandrasekar R, Chickering M, Ipeirotis P, Law E, Mityagin A, Provost F, von Ahn L (eds) Proceedings of the ACM SIGKDD workshop on human computation (HCOMP '09). ACM, New York, pp 77–85, http://dx.doi.org/10.1145/1600150.1600175

Nelson ET, Stavrou A (2011) Advantages and disadvantages of remote asynchronous usability testing using Amazon Mechanical Turk. Proc Hum Fact Ergon Soc Annu Meet 55(1):1080–1084, http://doi.org/10.1177/1071181311551226

Poetz MK, Schreier M (2012) The value of crowdsourcing: can users really compete with professionals in generating new product ideas?: the value of crowdsourcing. J Prod Innov Manag 29(2):245–256, http://doi.org/10.1111/j.1540-5885.2011.00893.x

Quinn AJ, Bederson BB (2011) Human computation: a survey and taxonomy of a growing field. ACM Press, New York, USA, p 1403. http://doi.org/10.1145/1978942.1979148

Rogers Y, Sharp H, Preece J (2011) Interaction design: beyond human computer interaction, 3rd edn. Wiley

Ross J, Irani L, Silberman MS, Zaldivar A, Tomlinson B (2010) Who are the crowdworkers?: shifting demographics in mechanical turk. ACM Press, p 2863. http://doi.org/10.1145/1753846.1753873

Simula H, Ahola T (2014) A network perspective on idea and innovation crowdsourcing in industrial firms. Ind Mark Manag 43(3):400–408, http://doi.org/10.1016/j.indmarman.2013.12.008

Standish Group (1999) Chaos: a recipe for success. Standish Group International

Tates K, Zwaanswijk M, Otten R, van Dulmen S, Hoogerbrugge PM, Kamps WA, Bensing JM (2009) Online focus groups as a tool to collect data in hard-to-include populations: examples from paediatric oncology. BMC Med Res Methodol 9(1):15, http://doi.org/10.1186/1471-2288-9-15

Taylor A (2000) IT projects: sink or swim. Comput Bull January, 24–26

Viskovic D, Varga M, Curko K (2008) Bad practices in complex IT projects. In: ITI 2008 – 30th international conference on information technology interfaces, p. 301

von Ahn L, Maurer B, McMillen C, Abraham D, Blum M (2008) reCAPTCHA: human-based character recognition via web security measures. Science 321(5895):1465–1468, http://doi.org/10.1126/science.1160379

Why Designers Hate Crowdsourcing (n.d.) Retrieved September 30, 2014, from http://www.forbes.com/2010/07/09/99designs-spec-graphic-technology-future-design-crowdsourcing.html

Yu L, Nickerson JV (2013) An internet-scale idea generation system. ACM Trans Interact Intell Syst 3(1):1–24, http://doi.org/10.1145/2448116.2448118

Cardboard Modeling: Exploring, Experiencing and Communicating

Joep (J.W.) Frens

Abstract This chapter presents Cardboard modeling as a tool for design that allows for simultaneous exploration, experiencing, and communication of design proposals. It introduces basic techniques and exercises to build skill and speed in Cardboard modeling and then demonstrates how it can be used as a tool for exploration. It ends with presenting two Cardboard models that were made to give a sense of the fidelity level and type of design that is possible with the technique.

Introduction

This chapter presents Cardboard modeling as a tool for the design of interactive products. It aims to show how Cardboard modeling can be used to create lo-fi, early phase design explorations, and how these explorations lead to mid-fi experiential models and hi-fi presentation models. This chapter is meant to act as a first foray into Cardboard modeling. Before I dive into the techniques of Cardboard modeling I briefly contextualize the technique in design, design education and in design tools.

Design has changed during the last few decades and with it design education. We come from a practice where design was focused on the design for appearance but this has changed towards the design of interactive products and even interactive systems (Frens and Overbeeke 2009). Design education has followed and aimed at providing students with new skills and new tools. The computer has made inroads into design, first as a literal replacement of the tools of the trade, technical drawings moved from paper to the computer, but later also as a new tool for the exploration of, for instance, interaction (e.g., Obrenovic and Martens 2011). Generally speaking this has been a blessing for design and for design education; the introduction of the computer gave us new inroads into the solution domain of design challenges and has been instrumental in spawning a range of new maker-tools like laser-cutters, 3D printers and CNC milling machines. Yet, the computer also comes with drawbacks and I belief that those are neither recognized nor adequately acted upon. Computer tools

Joep (J.W.) Frens (✉)
Department of Industrial Design, Designing Quality in Interaction Group,
Eindhoven University of Technology, Eindhoven, The Netherlands
e-mail: j.w.frens@tue.nl

© Springer International Publishing Switzerland 2016
P. Markopoulos et al. (eds.), *Collaboration in Creative Design*,
DOI 10.1007/978-3-319-29155-0_8

149

are operating in an abstract world and they require strictly defined input before they can operate (Frens and Hengeveld 2013). They, more often than not, leave little room for ambiguity and do not allow for a visceral experience of the designed artifact. In design sketching, where ambiguity is a strong resource (Fish and Scrivener 1990) and experiential evaluation of your actions is the starting point for grasping the complexity of design challenges this is an important drawback, particularly in the early phase of the design process. While computers seem to offer more efficient ways of working, it is important to realize that the material you design with co-shapes the solution domain. When entering the design space from the abstract, this is carried through into the solutions that are created. This is as much an argument for having a broad repertoire of skills, as it is an argument for ideating with the hands. When engaging in making and doing you start 'thinking with your hands', responding to what happens in the material that you are working with. Your hands catalyze your thoughts (Frens and Hengeveld 2013). You will be surprised by the turn your exploration takes and it is this surprise, this serendipitous experience that I feel is one of the most powerful aspects of making. And it is this experience that Cardboard modeling tries to capitalize on.

The use of foam-core and cardboard as modeling material is not new. In architecture but also in design there is a long tradition of making maquettes and models using those materials for reasons of exploring aspects of form or presenting ideas and concepts. The techniques that are presented in this chapter build on this practice but bring it to the domain of designing for (physical) interaction, as such Cardboard modeling stands in a rich tradition of tools for design (Buxton 2007). The techniques are developed to integrate into the design process from the early phase onwards and range from early lo-fi explorative models to hi-fi experiential presentation models.

A final remark before moving to the core of this chapter concerns terminology. I refer to the techniques as 'Cardboard modeling' techniques but the material that is used is pre-dominantly foam-core; cardboard is only used as a secondary material. The very mundane reason for this odd choice of terms is that in my mind '*cardboard modeling*' paints a more recognizable picture of the type of models that are created through these techniques than '*foam-core modeling*' would. The term 'Cardboard modeling' stuck and is now associated with many of my activities, I have no desire to change the term.

Cardboard Modeling

I have been teaching courses on Cardboard modeling for over 10 years and have developed a website on the topic to support my teaching. This chapter adapts and refers to the original material from the website, particularly in this first section. Where appropriate I refer the reader to the website which covers many of the topics discussed in this section in the language of video. The URL of the Cardboard modeling website is http://www.cardboardmodeling.com

Fig. 1 Cardboard modeling toolset

Cardboard Modeling Tools

To start with Cardboard modeling a toolset is necessary, see Fig. 1. This toolset can be as extensive as you want and it is subject to personal preference. Below I outline a basic toolset that I have found pleasant to work with, I explain my reasoning and I make suggestions for how you could extend your toolset. [website: basics/tools]

Basic Toolset

- Refillable 0.3 mm HB pencil: It allows for drawing defined lines that are visible but neither too contrasty nor smudgy. Alternatives are ordinary wooden HB pencils provided that they are kept sharp or thin fine-liners provided that the ink dries quickly.
- Scalpel (no. 3) with a straight, pointy blade (no. 11): It is sharp, maneuverable and sits well in the hand as it has a flat handle. Hobby-knives with pointy blades (e.g., x-acto handle with blade no. 11) are an alternative but are generally less sharp and in my experience tend to bend. This makes it more difficult to cut straight down also because the handle is round. Break knives and Stanley knives are great for making straight cuts, but do not work well when cutting (small) circles.
- Flat 30 cm stainless steel ruler with millimeter markings: It leaves no gap between it and the material and resists cutting into. Aluminum rulers are an alternative but avoid plastic.

- Flat transparent plastic set triangle with an integrated protractor: It is well suited for constructing perpendicular lines and angles.
- A4 (or letter) size 'self-healing' cutting mat: It protects the work surface and because it does not (easily) scar it does not misdirect the blade.
- Simple bow compass.
- Glue: I use a solvent-based glue that does not dissolve the polystyrene middle layer of the foam-core material. This glue is not readily available in every part of the world (e.g., it is difficult to source in the USA). A quick-setting tacky glue is a good alternative.
- Paper masking tape (18–25 mm): The cheapest variant of paper masking tape (18–25 mm width) will do.
- Double sided tape (15–18 mm): It is important to source a (roll of) tape with a non-stick backing as it allows you to cut it to size on the cutting mat without it sticking to it.
- Round toothpicks: Round, not square.
- Steel pins with a small steel head.

Possible Extensions to Your Toolset

An extra scalpel (no. 4) with a round blade (no. 24) will allow you to delaminate and scrape clean foam-core more easily than with a pointed blade. It could be a worthwhile extension if you do much delaminating (this is done to create curved surfaces). A circle cutter is another extension that could be considered. Good circle cutters are expensive and require some skill to cut through foam-core specifically, but if you want to cut many circles it might be worth the investment. Finally, a useful extension is a pair of flush-cutting pliers to cut toothpicks to length. This is an investment that makes sense when you are diving into the world of mechanisms.

Material

Here I describe the materials that are used in the Cardboard modeling techniques and the considerations for using other materials [website: basics/materials].

Foam-Core Foam-core is the main material that is used in the techniques described in this chapter. Foam-core is a paper-foam-paper sandwich that comes in many sorts: they vary in color, glossiness, and thickness. The foam layer is mostly polystyrene and the paper layers come in different weights. Cardboard modeling focuses on models on a product scale and for that a 3 mm thickness is ideal, larger models might require a thicker foam-core. Another consideration concerns the glossiness of the paper layers: if the paper is too glossy and smooth it becomes difficult or even impossible to draw legible layouts on the material, making it impossible to make precise cuts. Often a mid glossy foam-core yields good results but it is to the preference of the individual to make final decisions on the glossiness of material and the same goes for color. A final consideration concerns the weight of the paper layers, they range from very flimsy to rather heavy. The paper layer needs to be flexible enough to curve (when making a cylinder out of delaminated foam-core) and it needs to be heavy enough so that it does not easily break when making joints. Moreover, the heavier the paper, the more difficult it is to cut the material and the faster the scalpel blades dull.

Cardboard Stock Cardboard stock is the secondary material that is used in the techniques describes in this chapter, mainly to create mechanisms or curved surfaces that need to have some structural strength. The challenge is to find a cardboard stock that is easy to cut and that resembles the foam-core material in terms of color and glossiness on both sides. As the nomenclature of cardboard stock is different in every country it is difficult to pinpoint a specific type, nevertheless, Bristol board seems to be quite universally available and depending on the application 350–500 g works well.

Other Materials While most of the models that I show in this chapter or on my website are made of plain white foam-core and cardboard, colored paper can add accents to your models. It is often easier (and less risky) to create these out of colored paper. Another material that is often used is transparent plastic sheets for example to simulate screens on models of interactive products.

Further Considerations on Material Use A prime consideration for materials use is how they fit to each other: always consider the color and the shininess and try to match those or contrast those if it suits your purpose better. The other consideration is how it handles. If it is impossible to draw on a material it is difficult to construct a cutting layout on it. Finally, if it is hard to cut a material it will be very difficult to use it for your models. Try out materials before buying them in bulk.

Building Skill in Static Models

A first step in using Cardboard modeling as a tool for design is to build skill. Crucial to building skill is to master a series of techniques but also to be able to plan models and to have speed in building models. Cardboard modeling is about striking a balance between modeling speed and aesthetics. In what follows I present first some basic skills and outline a series of exercises to help build these skills further.

Basic Techniques

Many foam-core constructions are based on square beam structures for rigidity and strength. A second recurring construction element concerns single curved shapes.

Fig. 2 A 40 × 40 × 40 mm cube

These basic constructions are exemplified in two shapes: a cube and a cylinder. In making a cube one encounters the need to construct and draw layouts, to cut along a ruler and to make simple but strong joints. In making the cylinder one encounters the need to cut freehand, to delaminate material and to practice creating made to measure pieces. If these basic techniques are mastered, demonstrated by the building of a good cube and cylinder, the world of Cardboard modeling opens and follow-up steps can be made.

Creating a Cube Cut of Foam-Core

In this sub-section I show how to make a 40 × 40 × 40 mm cube, see Fig. 2. Before doing that I go through the basic techniques that are needed to make a cube.

Constructing and Drawing Layouts The basis of each layout is a primary and a secondary construction line drawn in pencil, perpendicular to each other, see Fig. 3. These reference lines are used as the basis for each further measurement and construction line (if possible). It is generally a bad idea to use a protractor to construct perpendicular lines onto perpendicular lines as the small measurement errors tend to multiply into large errors. Try to use light but legible lines and never erase your construction lines as the cardboard and foam-core material will acquire a nasty shine and the line will still be visible by the dimple that is left by the pressure of the pencil [website: basics/basics].

Cutting Along a Ruler One of the more difficult things to do right when cutting foam-core is cutting straight into the material, it is, however, necessary as it is the basis of truly square models. The reason for this difficulty is that when cutting along

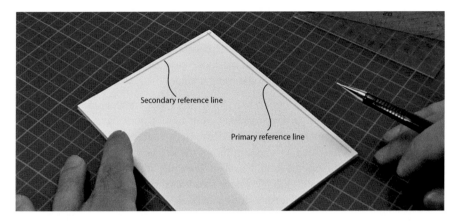

Fig. 3 Constructing a primary and secondary reference line

Fig. 4 The blade bending against the ruler

a steel ruler the blade of the scalpel bends sideways as a force is exerted on the blade to keep it aligned with the ruler: the handle seems to be pointing perpendicular to the material but the blade is actually not, see Fig. 4. To remedy this a cut is made in stages: the first cut only goes through the top layer of the foam-core sandwich to make a guidance cut. This cut is then used as a guide for the next cuts, because of this guidance less sideways pressure is necessary and it is easier to go straight down. It should be done in two more strokes, going all the way through the material without exerting excessive force on the scalpel. As a rule of thumb smaller details are cut before the bigger details. And cuts are made from the edge of the material towards the center of the material. These two strategies make sure that there is support for the ruler when cutting and the material keeps its integrity while cutting [website: basics/basics].

A word of caution: scalpels are sharp tools designed to cut human flesh. Even when a scalpel is no longer sharp enough to cut cardboard or foam-core it is still sharp enough to cut your flesh. Be extremely careful with a scalpel. Be careful when cutting: make sure your fingers are out of its path. Be careful when placing the scalpel on your work surface: it should not be able to fall and when moving your hands and arms it should not be possible to hit the blade. Be careful when walking with a scalpel: remove or cover the blade. Be careful when taking off the blade: use a pair of pliers if the blade does not come loose easy and dispose of a blade in a responsible manner [website: basics/changing a blade].

The Side with the Pencil Marks Is the Good Side The more complicated your models will be, the more construction lines your layouts will have (e.g., the axes of a cylinder). These construction lines will remain visible on the material after cutting and often this is seen as a reason to put the drawn on side on the inside of your models. While this seems to make sense, in reality it is a remarkably bad idea that often ruins the looks of a model. There are at least three good reasons to leave the drawn lines visible on the outside of your models [website: basics/basics]:

- The construction lines are information that is read by a person handling your models to understand its construction. Even when a circle is not completely tightly cut the drawn axes show that it was meant to be a perfect circle, it is even perceived to be more circular. It is therefore a good idea to not only leave the construction lines visible but even to show them in completeness: the center-point of a circle can be marked by a small cross or even a dot but by marking the complete axes it is perceived as more circular and more finished.
- When cutting the lines the blade sometimes goes through the material at an angle. This results in slightly slanted edges of the material and each edge has a slightly different angle. This makes the front of the material much more defined than the back. The front, with the pencil lines, has the shape that you wanted and the back is often slightly off. Another argument for the front of the material to be on the outside of your model.
- The side that you were cutting from, the front, is slightly rounded inwards making a pleasant edge to the touch. The back is similarly rounded but outwards, it has a slight burr, unpleasant to the touch. Putting the burr on the inside of the model makes a nicer model.

Creating Simple But Strong Joints There are several ways to create square joints in foam-core: for example miter joints, butt joints or rabbet joints. Miter joints are hard to do as it requires cutting the foam-core at an exact angle and butt joints leave the foam layer exposed. I find, however, that rabbet joints provide a good balance between speed and aesthetics. They are constructed by making a cutout in one of the pieces of a width of one material thickness and gluing the other piece into the created rabbet perpendicularly making sure that the perpendicular piece sits flush with the edge of the rabbet, see Fig. 5. The cutout goes all the way through the first and second layer of the material but leaves the third layer intact, care has to be taken to not touch the final layer with the sharp blade as this weakens the material and often leads to the paper layer to tear or delaminate. While the glue sets, the joint needs to be supported by means of small strips of paper masking tape [website: techniques/creating a joint].

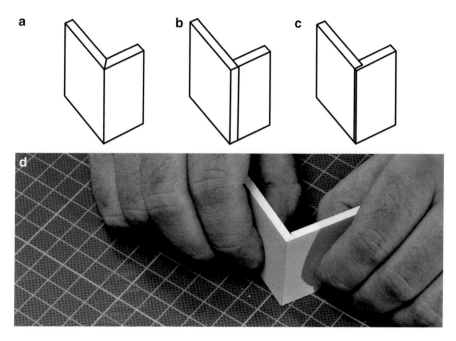

Fig. 5 (**a**) Miter joint – (**b**) Butt joint – (**c**) Rabbet Joint – (**d**) Rabbet joint in foam core

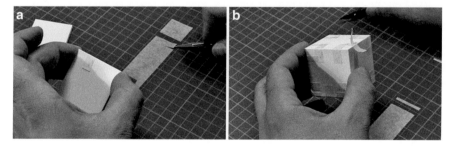

Fig. 6 (**a**) Bracing a glued joint with strips of tape – (**b**) Removing strips of tape

On Using Paper Masking Tape When using paper masking tape to brace joints or glued connections it is important to prepare the tape before starting to apply glue. The paper masking tape is designed to be removable but when it is used on paper surfaces it oftentimes is so sticky that it is hard to remove without tearing the paper. This is problematic as it ruins your model. It is therefore important to make the tape less sticky before applying it to your models. Sticking the tape to your clothes a few times so that it loses most of its stickiness can do this. After this is done the tape is stuck to the cutting mat and cut into little strips, see Fig. 6a. This makes it easy to apply to your models.

When you are finished gluing and the glue has set it is time to remove the strips of tape again. Never leave the tape on your models for too long (remove within a day or so) as the glue in the tape will at some point react with the surface of your models, making it very hard to remove. When removing the tape, care has to be taken that the model is not harmed. Particularly important is to not simply rip off the strips of tape in one go. The tape reaches over the edges of material, where it is most prone to tear. The way to go is to carefully peel both sides of the tape towards the joint, and then removing it parallel to the edge, see Fig. 6b. Thus the risk of tearing the surface of the model is minimized.

Creating the Cube Finally it is time to create a $40 \times 40 \times 40$ mm cube. There are four stages to this: (1) creating your layout, (2) cutting your material, (3) preparing the cut squares and (4) assembling your cube [website: simple models/creating a cube].

In the first stage six squares are constructed. First the two perpendicular reference lines are drawn, a parallel line is constructed 40 mm out from the first reference line and those two parallel lines are both marked in 40 mm intervals, creating six squares after the marks are connected, see Fig. 7.

During the second stage all lines are cut using the method that was discussed above.

The third stage concerns preparation for creating the joints. Three pairs of two squares are created, one pair is left intact and is put aside, the other two pairs are marked on the back for creating the cutouts. One pair needs cutouts on two of its sides, opposing each other. The final pair needs cutouts on all of its sides, see Fig. 8a.

During the fourth and last stage the six squares are assembled and glued to form a cube, see Fig. 8b. There is no particular order to do this; you can start with any square and work from there. After each piece is glued to the assembly small strips of paper masking tape are put on to hold the joint together. After each square is glued into the cube and the glue has set, the pieces of paper masking tape can be carefully removed.

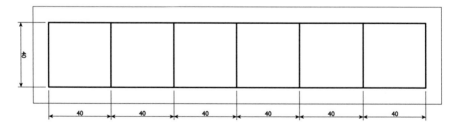

Fig. 7 Layout of the cube on the material

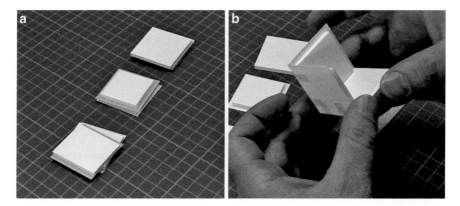

Fig. 8 (**a**) Pairs of squares: 2, 4 and no cutouts – (**b**) Assembling the cube

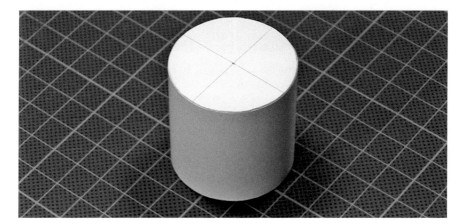

Fig. 9 A $40 \times \varnothing 40$ mm cylinder

Creating a Cylinder Out of Foam-Core

In this subsection I show how to make a $40 \times \varnothing 40$ mm cylinder, see Fig. 9. First I present the basic techniques that are necessary to accomplish this.

Cutting Freehand To create circles or other curves in foam-core one needs to be able to cut freehand. Circles, of course, can also be cut using a circle cutter. That requires skill in itself and a circle cutter is not ideal when only partial circles are required. So even when you think that you'll just be cutting circles it is good to have free-hand skills next to skills with the circle-cutter.

Cutting a circle starts with drawing a circle on foam-core using a simple bow compass (make sure not to forget to draw the axes). Then a scalpel (preferably with a fresh blade) is used to make a first guidance cut through the surface following

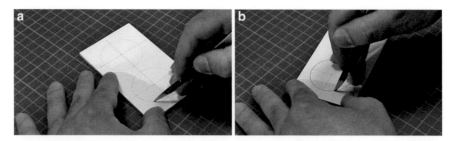

Fig. 10 (**a**) Shallow cut through the surface – (**b**) Through cut, bending the blade into the curve

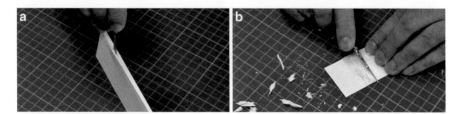

Fig. 11 (**a**) Cutting into the side of the foam-core piece – (**b**) Delaminating the foam-core piece

the circular pencil line, see Fig. 10a. After this, a second cut is made that goes all the way through the material. This is a difficult cut to make as the blade is straight and a circular cut needs to be made. When the scalpel is normally inserted into the material the tip of the blade is trailing outside of the circular path of the cutting edge that is visible at the pencil line, creating a cone shape rather than a clean circle with perpendicular edges. To remedy this the blade needs to be given a curve by putting sideways tension on it. Letting the tip cut into the cutting mat and then bend it sideways does this. The cut is made while bending the blade, see Fig. 10b. To make the complete cut the scalpel needs to remain inserted into the material and the material, together with the scalpel needs to be lifted and rotated to have optimal control over the cutting. This is complex and difficult and needs (a lot of) practice [website: techniques/cutting a circle].

Delaminating Foam-Core If one wants to create curved surfaces in foam-core models while using a material of the same color it is necessary to delaminate the foam-core and use only the (bendy) paper surface. To do this a piece of foam-core is cut to size and then cut sideways into the foam layer going as deep as possible without hurting the paper layer at the front, see Fig. 11a. After this the material is put on the cutting mat with the front facing down and the paper layer at the back is peeled away using the leeway that the sideways cut-ins give. This needs to be done carefully as the paper layer at the front needs to remain intact. After removal of the paper layer at the back the foam layer is peeled away as far as possible using your nails and scraping with the scalpel, see Fig. 11b (also the side of your ruler can be used as a scraper). Make sure that the scrapings do not find their way under

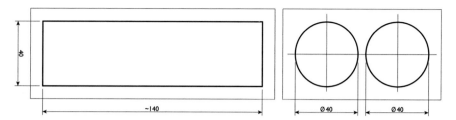

Fig. 12 Layout of the cylinder on the material

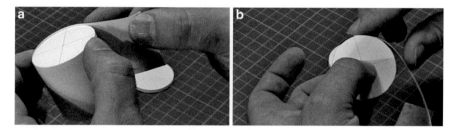

Fig. 13 (**a**) Measuring in the exact length of the mantle – (**b**) Starting at an axis endpoint

the material that you are delaminating, as it will mark the surface. Final cleaning up can be done by scraping with the scalpel. This will result in a flexible piece of paper of the same color as the rest of the model [website: techniques/delaminating foam-core].

Creating the Cylinder Finally it is time to make a 40 × ⌀40 mm cylinder. This is again done in four stages: (1) drawing the layout, (2) cutting, (3) preparing the material and (4) assembly [website: simple models/creating a cylinder].

The first stage concerns the drawing of two circles (together with the axes) and the outline of the cylinder mantle, see Fig. 12. The cylinder mantle needs to be oversized in length, as the true size needs to be made to measure. Math gives us the circumference of the circle ($2\pi r$) but foam-core modeling is not an exact science, particularly when cutting circles: take 20–40 mm extra length.

Secondly the circles are cut following the instructions outlined above. Also the mantle is cut out.

Next the mantle is delaminated and measured in. In order to make the mantle to size first the cut circles are marked number 1 and 2 on the back. The same is done on the back of the delaminated piece on the long edges. This is to ensure that the right length corresponds to the right circle. By wrapping the marked edge around the similarly marked circle the exact size is found, it is marked on the surface of the mantle, see Fig. 13a. This is repeated with the second edge and circle after which the mantle is cut to size.

The final stage concerns gluing the cylinder together. Glue is applied to the side of the first circle. In my eyes the cylinder is most beautiful when the endpoints of the mantle material are aligned with an endpoint of an axis, see Fig. 13b. When the piece is fully wrapped around the first circle a small strip of paper masking tape is applied to hold it in place while the glue sets. Next the other circle is glued. As the mantle of the cylinder is now glued around the first circle, positioning the second cylinder is more difficult. The circle needs to be carefully dropped in, making sure that the glue does not find its way onto the mantle. As this is quite tricky, aligning the axis endpoints to the point where the mantle endpoints now meet is easily forgotten. So plan your moves and be careful that the circle does not drop in too far. Another strip of tape is applied to hold the mantle in place while the glue sets. Finally the tape is removed after the glue has set.

Towards Mechanisms

Cardboard modeling is a tool for designing for (physical) interaction. To design for interaction is to explore rich physical action possibilities and this cannot be done without having skill in creating moving parts. Building on the previous skills I present ways of enriching models with action possibilities.

Mechanisms come in great variety and it is not my intention to present an overview of all possible mechanisms. Instead I wish to show some common ways of making parts of a model slide, or hinge or rotate. I use the properties of the materials and go from there, hopefully opening up a world of possibilities.

Creating a Slider Out of Cardboard Stock

To create a discrete and contained sliding motion is to create layers of material that limit degrees of freedom and create defined stops for the motion to start and end. Adding layers adds a lot of volume to a model and this is why I often use cardboard stock for such mechanisms rather than foam-core. Here I show a four layer 60×40 mm cardboard slider with an action of 15 mm to exemplify the possibilities of cardboard stock when creating mechanisms, see Fig. 14.

Friction Is the Enemy When creating mechanisms in foam-core and cardboard engineering challenges are enlarged as you are working with a fairly weak material that is not very forgiving in terms of overstraining it. In my experience most of this boils down to gaining an understanding of how friction works in terms of material characteristics, forces and geometry and acting accordingly when designing a mechanism. Below I outline some considerations.

When approaching friction from the perspective of material characteristics consider that hard and smooth surfaces create less friction than soft and pliable materials: two layers of cardboard have less friction when slid over each other than two edges of foam-core (with the soft and pliable foam layer exposed) have. Avoid

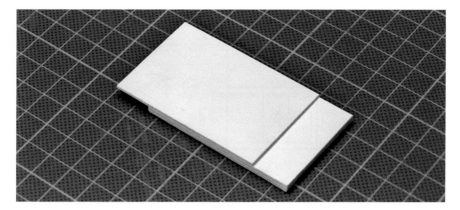

Fig. 14 A 40 × 60 mm slider mechanism with an action of 15 mm

constructions where materials 'eat' into each other. Surface characteristics can be tweaked by adding tape layers to make a surface smoother or less prone to deform.

When approaching friction from the perspective of forces it is important to avoid concentrations of forces where the material cannot deal with it and to reinforce at places where there are concentrations of force. To make mechanisms work is to design how they are engaged. As a rule of thumb you try to minimize torque by carefully considering the position of the engagement point.

When approaching friction from the perspective of geometry, consider the amount of play that a mechanism has. The intuitive assumption that more play in a mechanism creates less friction is faulty: a drawer with a lot of play in its sliding mechanism gets stuck rather than that it works smoothly (the infamous 'sticky drawer' effect). The 'art' of making a mechanism work is in finding the balance between play and surface friction. That is, the amount of play in a mechanism needs to be minimized to the point that it still runs smoothly in terms of surface friction between touch points of material.

This all means that you have to work cleanly and precisely and need to understand (by trial and error or by calculation if you prefer) the friction in your mechanisms and the forces that work on the mechanisms that you are creating.

Creating the Slider This slider mechanism measures 40 × 60 mm and consists of four layers of cardboard stock. To prevent the mechanism from bending a heavier gauge 500 g cardboard stock is used. The process of building again has four stages. [website: mechanisms/creating a slider]

The first stage is creating the layout of the slider on the cardboard stock, drawing the four layers, see Fig. 15a. The layers have different functions in the mechanism. Counting from the top down, the first layer is the sliding part, the second layer controls the sliding by means of slots, the third layer contains the connection of the slider to the base and the fourth layer closes of the mechanism for aesthetical purposes, see Fig. 15b.

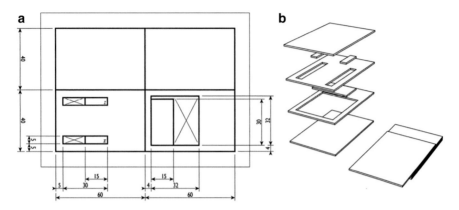

Fig. 15 (a) Layout of the mechanism on the material – (b) The four layers of the mechanism

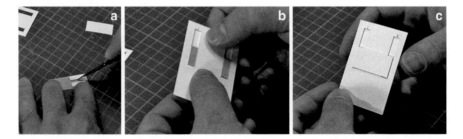

Fig. 16 (a) Cutting double-sided tape to size – (b) Attaching runners – (c) Attaching bar

The second stage is cutting the material. Here it is important to remember to cut the details before the larger lines and to keep track of which lines to cut and which to leave alone, be aware of your construction lines that should not be cut. Also mark the runners so that you know from which slot they originate.

The third stage is to prepare the material for assembly. In the case of this layered mechanism we use double-sided tape as glue can seep into the mechanism easily. The tape is applied to both sides of the runners and third layer. After application the tape is carefully trimmed to the size of the parts, using their contours as guide, see Fig. 16a.

Finally the mechanism is assembled. This is not difficult but it needs to be right in one go as the tape immediately and irreversibly bonds to the cardboard. First the runners are attached to the first layer using the second layer as a template for positioning. Make sure to match the runners to the slots, also make sure to position the runners at the right place, doing this wrong will result in exposing the slots when the mechanism is opened and that is unwanted, see Fig. 16b. If the runners are fixed the bar is attached over the runners locking the second layer to the mechanism, leaving it the freedom to slide but not disconnect, see Fig. 16c. Next, the third layer is attached, and finally the fourth layer is mounted.

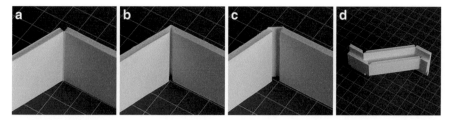

Fig. 17 (**a**) Simple cut-in hinge – (**b**) Shaped hinge – (**c**) Double hinge – (**d**) Four bar linkage

Hinges and Pivots

Next to sliding parts are often made to hinge or rotate when adding action-possibilities to a model. Foam-core is very suited to construct hinges that bring defined motion and by adding pins or toothpicks parts can easily be made to rotate. In what follows I explain basic techniques.

Hinges When foam-core is partially cut-in one of the paper layers can be used as a film hinge. By shaping the cut-in the hinge can be given nicely defined endpoints, see Fig. 17.

The simplest hinge is made by first scoring the surface of the foam-core at the place where the hinge is planned. The hinge line is copied to the back and the material is turned over. Next the foam-core is cut-in from the back, cutting through the back paper layer and through the foam layer but without touching the front paper layer. The material is folded and the result is a neatly defined hinge that is one-way rigid and folds over 180 °, see Fig. 17a [website: mechanisms/simple hinge].

A more complex hinge is made by shaping the cut-in in a 45 ° angle. Hinges of this type are particularly useful in creating a four bar linkage out of foam-core, see Figs. 17b, d [website: techniques/creating an angled cut].

The creation of a double hinge makes it possible to have the material fold over itself while hiding the foam layer. Making a slot with score lines at either end creates a hinge that folds over itself, see Fig. 17c [website: mechanisms/double hinge].

Pivots To make parts of a model rotate, a steel pin or a wooden toothpick can be used as pivot, see Fig. 18. Steel pins are easy as they are simply stuck into the material. The down side is that they get loose quickly, easily rip the material and cannot be cut to size as they are exceptionally hard. Although sometimes using pins makes sense, most of the problems that are summed up above are mitigated by the use of wooden toothpicks. Toothpicks can be glued in place, can be fitted with stops made from paper masking tape and can be easily cut to size. Moreover, because they are relatively thick (ca. 2 mm diameter) they do not tear the material easily. Yet, their thickness also makes them tricky to stick through your material. There are a few rules of thumb to take into account when using toothpicks. (1) Create the holes for a toothpick before the parts of a model are cut from the drawn layout. You'll exert some force on the parts and thin parts are likely to deform or break under this

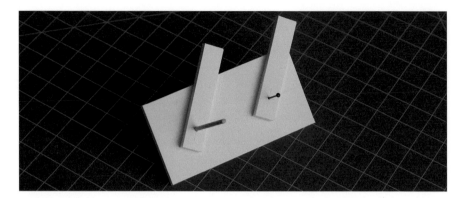

Fig. 18 Using a round toothpick (*left*) or a steel pin (*right*) as pivot

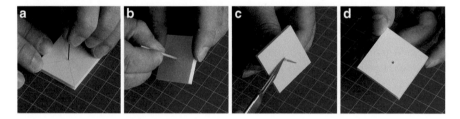

Fig. 19 (**a**) Position the hole – (**b**) Widening the hole – (**c**) Removing the burr – (**d**) Clean hole

force. (2) Use a pin to punch a hole (Fig. 19a) and follow it with a toothpick. Insert the toothpick while rotating it, as if you were drilling a hole, see Fig. 19b. (3) Make sure to remove the large burr that forms at the back of the material when a toothpick is pushed through, leaving it in place means that you cannot mount material flush to the surface otherwise, see Fig. 19c, d.

Other Techniques: The Use of Graphics

As a final 'basic technique' I want to highlight the use of graphics on cardboard and foam-core models. Adding drawn graphics on your models is a quick way to give a lot of expressivity to your models, see Fig. 20. If your model has a lot of small details it often makes sense to draw rather than to model those. A grippy surface on a camera lens, a button layout or an icon on a button is often more convincingly drawn than modeled. Yet, it is also easy to ruin a model when it fails. There are a few rules of thumb that I want to present here to guide the use of graphics. (1) Graphics need to be crisp and clean and (in my opinion) should not have too harsh of a contrast with the material. Draw your graphics with as much care as possible and use a ruler, guide or set triangle to do so. Construct your graphics rather than

Fig. 20 Adding graphics

draw them freehand: crisp graphics will elevate your models to a higher level of fidelity, even when there are a few flaws in the construction of your models. Sloppy graphics will pull down the quality and there is no skill in modeling that can pull that back up. As it is easier and less time consuming to add graphics than it is to model it seems to be a shame to ruin the modeling effort with a sloppy graphics job. Be particularly careful when adding lettering to you models, letters need to be constructed and drawn, they should not look 'scribbled' on. (2) Graphics are added when the model is laid out on the foam-core or cardboard material, when it is still flat and uncut. This way you can use rulers and guides to accurately construct your graphics. Moreover, it is difficult, if not impossible, to draw on a three-dimensional model with any precision, as you'll find that you cannot freely position a ruler or guide.

Another way to add graphics is to add (color) printed graphics. I have had mixed success with that. Often it seems to be difficult to have the model and the graphics form a unity. This is because there is a mismatch in the 'form language' between the model and the printed graphics. Care has to be taken to add some detailing on the model that makes the printed elements blend in. Also, printed elements are best attached after modeling, particularly on curved surfaces.

A last consideration goes to painting your models. I myself am not in favor of this practice as I feel that painting a model oftentimes highlights the little flaws in for example a glued seam rather than hiding them. Yet, when you do want to paint your model take care of what kind of paint you use, for example a water-based paint might damage the surface of the material.

Concluding

In this section I have discussed a variety of basic techniques and presented three models aimed at building skill both in static models and in models with moving parts. Of course one cannot expect to be skilled in Cardboard modeling after building only the three models that were presented here, but they provide a good starting point. To build further skill it is important to get away from the step-by-step guides that are presented here and on the Cardboard modeling website; it is important to learn how to plan your own models and mechanisms. A good way to practice this is to choose an existing product and to copy it. Start by picking an object that has clearly recognizable basic forms without too much double-curved surfaces and try to make a copy that simplifies the form where necessary but that captures the essence of the product that you are trying to copy. If you feel adventurous you can include mechanisms to practice those as well. After having done a fair amount of models on your own it is time to take the next step and start using Cardboard modeling as a tool to design with.

Cardboard Modeling as a Tool to Design with

After a basic level of skill and speed in Cardboard modeling is established it can be used as a tool for exploration rather than as a tool for the creation of models alone. This section is a more abstract take on what it means to explore a design challenge after which I present two exercises that can be used to acquire skill in using Cardboard modeling as a tool to design with.

A Framework for Exploration

Design exploration is often concentrated in the early phases of the design process. In the context of this chapter I'll define it as a synthesizing activity with the aim to: (1) generate and refine ideas and concepts, (2) experience and validate ideas and concepts and (3) demonstrate and present ideas and concepts. Cardboard modeling is a technique that scales in fidelity throughout all of these purposes of design exploration. Cardboard modeling, like other design techniques, leverages on ambiguity. The ambiguity of a sketch opens new paths for exploration (Fish and Scrivener 1990) and the same goes for the lo-fi models that are created when using Cardboard modeling as a tool for exploration. The models are made with a certain intention but as they are manipulated to evaluate the value of the idea a designer can come to the conclusion that a different orientation of the model, or an additional piece of material, can be meaningful as well. This mechanism is very similar to the 'interaction relabeling method' (Djajadiningrat et al. 2000) and it opens new paths for exploration.

Lo-Fi/Hi-Fi

A criticism that is often brought forward when discussing Cardboard modeling as a tool for designerly exploration is that it is too time consuming. This is based on the assumption that all Cardboard models need to be of the best quality possible Yet, in a design process a model needs to be good enough to answer the questions that you are asking in a given exploration (Frens and Hengeveld 2013); any more effort in polishing a model is wasting time.

Exploration generally starts with lo-fi models going through mid-fi models as insights grow and better answers are needed and ends up in hi-fi models that can be used for more than exploration. To make matters complicated the concept of lo-fi/hi-fi is somewhat fractal-like, characterizing the evolution of prototyping from early to final stages of the design process, but for explorations within a specific phase it is true as well. To give an example: a hi-fi cardboard model, that is the endpoint of a certain exploration is, in the grander scheme of things, a mid-fi prototype at most as it leaves unanswered important questions such as which material to use, or how to exactly fit components,

The bottom line is that exploratory models need not be of higher fidelity than the questions that they answer. When Cardboard modeling is used as a tool to design with, a conscious effort needs to be made to match fidelity levels and make effective use of the techniques.

Depth and Broadness

Design explorations can be characterized by the concepts of *depth* and *broadness*. These characteristics can be used as guiding principles of exploration, see Fig. 21.

Depth An exploration can be focused on a single issue and explore that in *depth*. For example, the focus can be on creating a beam shape that can be held at the long sides of the beam shape between thumb and one finger. The first concept (left) shows strips cut out of the sides, in the second concept (middle) these straight cutaways changed into slanted cutaways and in the third concept (right) there are large cutaways. This is an exploration in *depth* where the exploration is conceptually similar but changes in implementation, see Fig. 22.

Fig. 21 Model for exploration: deep exploration versus broad exploration

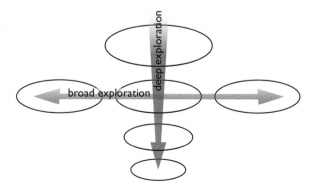

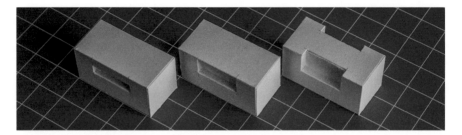

Fig. 22 A deep exploration: three variations of a side grip

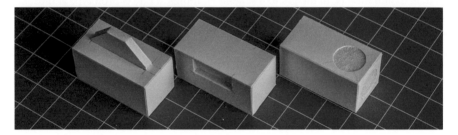

Fig. 23 A broad exploration: three variations of a grip

Broadness Other explorations have less focus; the concepts are very diverse. For example, the first concept (left) adds a handle to the top surface of the beam, a second concept (middle) has two slanted cutaways in its long sides and a third concept (right) tries out how a beam can be picked up at one of its corners by adding two circular cutouts. This is a *broad* exploration where the extent of the solution domain is mapped through variation in concept, see Fig. 23.

While the approach to design exploration is ultimately a matter of the individual designer, their preference and the design challenge at hand, it is essential to try out and mix both approaches towards exploration: a *broad* exploration is used to map the solution domain to a given design challenge, a *deep* exploration is a means to optimize specific implementations for a conceptual direction. Both are necessary to gain confidence in the 'rightness' of the solution that one works towards in a design process. In the reality of a design exploration both approaches mix and oftentimes are repeated.

Three More Things to Take into Account

Our Body Does Not Scale When making design explorations one can work on a different scale than the actual size of the design. This can be the result of the context of application: product designers often make sketches and models that resemble the actual size of the product but architects are more prone to make models on a smaller scale. The choice for working on a different scale than the final concept can have pragmatic reasons or it can be because the scale of the exploration does not influence the quality of the answer: to explore a mechanical principle the exploration can be smaller or bigger than actual size without harming the insights that are derived from the exploration. Yet, when we talk about *design for interaction* then it is important to understand that scaling the size of a design exploration is oftentimes undesirable, as it tends to influence the experiential properties of the model and thus the quality of the insights. This is because our hands and our bodies do not scale. Models that are considerably smaller or bigger than the actual size that is aimed for defeat the purpose of making the explorations physical. Exploring in the physical allows for experiential exploration of the concepts at hand. When they are scaled up or down this road is closed.

To Be Stuck A common complaint in using Cardboard modeling as a tool for exploration is that it is hard to start and easy to get stuck after a few models made. This is particularly true for people with little experience in the technique. Oftentimes people know better which conceptual direction they wish to avoid than that they know where they want to take their exploration. A good strategy to kick-start the process of exploration is to actually make a model in the conceptual direction that is not desired. The purpose of this strategy is to start the process of *making* because it opens the roads towards ambiguity thus unfolding pathways for new explorations.

Limitations of the Material Foam-core and cardboard have material limitations that constrain the forms and the mechanisms that can be modeled. For example, it is virtually impossible to create double curved surfaces and there are limits to miniaturizing models and mechanisms. When used as a tool for exploration, Cardboard modeling gives access to a very specific form language that follows its material limitations. If it is necessary or desired to go beyond the possibilities of foam-core and cardboard more complex forms can be suggested (e.g., by means of facets, contour-lines or graphics) or modeled using materials like (shaped) Styrofoam to enable more free form sculpting.

Two Exercises

To practice the use of Cardboard modeling as a tool for design I have developed two exercises. I outline them below.

Exercise 1: Exploring Expressive Horm – How to Pick Up a Beam
Explore how to pick up a beam of $20 \times 20 \times 40$ mm comfortably with your fingers and modify the shape such that it optimizes comfort. In this exercise at least three explorations (more is better) need to be made in a limited amount of time (45 min). This exploration should follow a cyclical approach of creating, evaluating and again

creating and so on. In other words, the concepts should not be drawn out on the foam-core material in one go (which would be the most efficient way to construct a series of models) but in succession with evaluation steps in between.

Exercise 2: Exploring Form and Interaction – How to Answer a Phone
Explore how to answer a phone call, or how to pick up the phone. The intention here is not to create full-scale models of telephones but to explore principles of interaction in combining form and interaction. Again the challenge is to create at least three, but preferably more, concepts in a very limited amount of time (45 min) also the use of physical actions is encouraged over the use of screen interaction.

Concluding

In this section Cardboard modeling is presented as a tool for exploration. It is clear that using Cardboard modeling as such requires fluency in the techniques, only when the technique is conquered does it become an instrument. The exercises in the 'Cardboard modeling' section are designed to accomplish this fluency. This section discusses 'exploration' and presents a framework for exploration and two patterns that are commonly observed. While all this provides a starting point for using Cardboard modeling as a tool for exploration, it still needs to be practiced in a design process in order to be internalized as a valuable instrument.

Beyond Cardboard and Foam-Core

When you are satisfied with your skill in Cardboard modeling and you can use the technique to explore you can go beyond the models with mechanical action possibilities. This can be done either by the use of video scenarios or by adding sensors and actuators to the models. Below I briefly discuss both.

Video Scenarios Cardboard models are a good base material for the production of video scenarios. Using postproduction software (e.g., Adobe After Effects or Apple Motion), your models can be brought to life, either by demonstrating the action-reaction loops or by showing it in context. If your models contain physical action possibilities that have physical reactions when utilized sometimes demonstrating this needs some planning. Try to think of how to temporarily actuate pieces of your model by adding soft springs or soft foam material or by actuating it by means of fish wire for example. Also the smart use of cuts and camera angles can be used to suggest much more interactivity than the model is actually capable of. To bring screen content to your models or to bring your models into different contexts you can make use of chroma keying. Through attaching a blue or green colored piece of cardboard to your model at the place where the screen is and replacing the blue or

Fig. 24 A model equipped with two servo motors

green screen with your screen content in postproduction. If you film your footage in front of a blue or green screen it is easy to free the video content from the background and place it in a different background. Always try to think ahead of what it is that you wish to show. Sometimes you need a few tries on the cutting table to really know how to create the right shots. Whenever you are shooting video make sure that enough attention is given to lighting your scene and model. In my experience that makes the postproduction faster and better.

Adding Sensors and Actuators A different route that can be followed to make your Cardboard models even more life-like is by adding actual interactivity to them, by adding sensors and actuators, see Fig. 24. This is an activity that has the same range of fidelity as the Cardboard modeling technique itself has. It adds new routes for exploration, experiencing and demonstration that have to do with exploring and fine-tuning product behavior. While it goes too far to exhaustively discuss everything necessary to do this I wanted to at least point to the possibilities this offers.

Example Models and Designs

This section presents two camera designs that I designed in foam-core and cardboard using the Cardboard modeling techniques. Please note that the models that are shown here are hi-fi end results of a design process. Each model builds on a box-full of early sketch models. I try to showcase both the models and their interaction styles but I refrain from giving the complete rationale of the designs. Should you be interested in reading more on the cameras and the design rationale I refer the reader to my PhD thesis: Designing for Rich Interaction: Integrating Form, Interaction, and Function (Frens 2006).

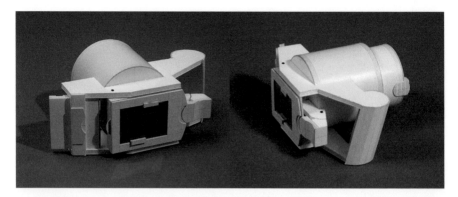

Fig. 25 Two views of the Rich Actions camera

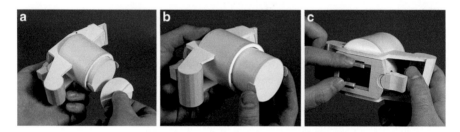

Fig. 26 (**a**) Switching the camera on – (**b**) Zooming in and out – (**c**) Setting the resolution

Rich Actions Camera

The 'Rich Actions' camera has controls that show what they are for and how you can act on them, see Fig. 25. The camera is switched on by taking off the lens cap, see Fig. 26a. To zoom the two handles at the front of the lens are grabbed to move the lens, see Fig. 26b. The resolution of the images is set by making the screen 'smaller' or 'bigger' through controlling two small sliders, see Fig. 26c. To take pictures the 'trigger' at the side of the screen is pushed, see Fig. 27a. If the trigger is pushed all the way in, the screen releases from the lens and shows the image that was taken, see Fig. 27b. This image can be saved by pushing the screen towards the memory card at the side of the camera, see Fig. 27c.

Control per Function Camera

The 'One Control per Function' camera was designed to have a dedicated control for each of the functions of the camera, see Fig. 28. It employs three layers of controls

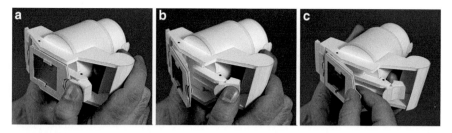

Fig. 27 (**a**) Ready to take a picture – (**b**) Picture is taken, screen separates – (**c**) Save image

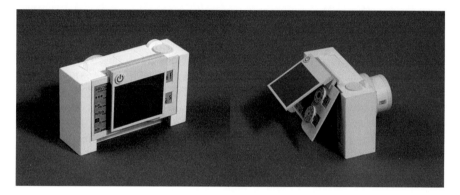

Fig. 28 Two views of the One Control per Function camera

Fig. 29 (**a**) Changing settings – (**b**) Make available a control– (**c**) Manually controlling aperture

that each contains controls that are associated with distinct modes of use. The top layer and outer shell of the camera contains the controls that are needed to access the primary functionality of the camera (i.e., framing and taking photos). Deeper into the camera controls are placed that have to do with changing the settings of the camera, see Fig. 29a, b. In certain cases changing settings in the inner layer of the camera makes available new controls in the outer layer of the camera as is shown in Fig. 29c.

Concluding

In this final section of this chapter I want to come back to the title of this chapter: Cardboard modeling: exploring, experiencing and communicating.

Exploring In section "Cardboard modeling as a tool to design with" I thoroughly explain how Cardboard modeling can be used as a tool for exploration. When used for exploration it is important to tune the fidelity level of your modeling to the phase of the design process for reasons of speed and scope. Trust your hands to catalyze your thinking.

Experiencing One of the reasons to bring your creative process into the physical world is that you can try out your ideas, you can evaluate the size and proportion, you can get a feel for matters of use and best of all, you can give your model to another person to gauge their feelings about your design. A physical model that was created by hand already gives insights while it is being made. The designer has immediate access to the properties of the design while making it. I want to stress that this type of access is not afforded by current computer software. Next to the experiential qualities during the process of making, of course it also offers opportunities for experiencing after it is finished, for the designer, but importantly for others as well. Allowing for the validation of the design exploration on the fly as well as after it is finished, by the designer and by others. The latter point brings me to the final part of the title: communicating.

Communicating Hi-fi models afford to be experienced, but there is more, they communicate the values of a design and the intentions of the designer. These hi-fi models represent key moments in a design process where assumptions, decisions, insights and design intent come together and can be evaluated. What is interesting to note is that Cardboard modeling scales through all these levels of fidelity and can be used to explore, experience and communicate. What is more, you can do it from the comfort of your desk, no machinery is needed.

References

Buxton B (2007) Sketching user experiences: getting the design right and the right design. Morgan Kaufmann, Boston
Djajadiningrat JP, Gaver WW, Frens JW (2000) Interaction relabelling and extreme characters: methods for exploring aesthetic interactions, In: Proceedings of the DIS'00, New York City, New York, pp 66–71, August 17–19
Fish J, Scrivener S (1990) Amplifying the mind's eye: sketching and visual cognition. Leonardo 23:117–126
Frens JW (2006) Designing for rich interaction: integrating form, interaction, and function. Unpublished doctoral dissertation. Eindhoven University of Technology, Eindhoven, The Netherlands. Retrievable from http://www.richinteraction.nl

Frens JW, Overbeeke CJ (2009) Setting the stage for the design of highly interactive systems. In: Proceedings of the IASDR'09, Seoul, South-Korea, pp 1–10, October 2009
Frens JW, Hengeveld BJ (2013) To make is to grasp. In: Proceedings of the IASDR'13, Tokyo, Japan, pp 26–30, August 2013
Obrenovic Z, Martens JBOS (2011) Sketching interactive systems with Sketchify. In ACM transactions on computer-human interaction, TOCHI 18, 1, Article 4, May 2011

Part III
Designing with Stories

STORYPLY: Designing for User Experiences Using Storycraft

Berke Atasoy and Jean-Bernard Martens

Abstract The role of design shifts from designing objects towards designing for experiences. The design profession has to follow this trend but the current skill-set of designers focuses mainly on objects; their form, function, manufacturing and interaction. However, contemporary methods and tools that support the designers' creative efforts provide little help in addressing the subjective, context-dependent and temporal nature of experiences. Designers hence need to learn by trial and error how to place experiences at the center of their creative intentions. We are convinced that there is room for new tools and methods that can assist them in this process. In this chapter, we argue that storycraft can offer part of the guidance that designers require to put experiences before products right from the very start of the design process. First, we establish the background behind the shift from products to experiences and explain the challenges it poses for the designer's creative process. Next we explore the contemporary conceptual design process to understand its shortcomings, point out the opportunity that storycraft offers and propose our approach to take on this challenge. Last but not least, we propose a specific method called Storyply that we have designed and developed iteratively by testing it in conceptual design workshops with students and professionals.

From Products to Experiences

Design is fundamentally an exploratory process. It starts with an idea. Inspirations motivate designers to approach a problem and/or opportunity by generating, developing and testing ideas (Brown 2009). This is a process in which creative minds extend the boundaries of what is possible (Osterwalder 2010). An average training in design teaches you what you need to know about how to systematically try a design challenge, and provides you with the skill-set to explore your options. This skilled practice of conceptualizing and appropriating innovative ideas into people's lives on a daily basis is in the designer's generic job description.

B. Atasoy • J.-B. Martens (✉)
Department of Industrial Design, Eindhoven University of Technology, Eindhoven, Noord-Brabant, The Netherlands
e-mail: j.b.o.s.martens@tue.nl

© Springer International Publishing Switzerland 2016
P. Markopoulos et al. (eds.), *Collaboration in Creative Design*,
DOI 10.1007/978-3-319-29155-0_9

181

However, the designer's specific roles in industry along with their impact on society have evolved tremendously in the last couple of decades. If you put the popular designer titles in a chronological order of appearance you would get something like: (1) Industrial Designers, (2) Product Designers, (3) Interaction Designers, (4) User Experience Designers and (5) Service Designers. This is due to the shift of focus from 'manufacturing-centered form-giving' towards 'human centered creation of experiences and services.'

To explain this evolution, Brown points at the shift of economic activity from industrial manufacturing to knowledge creation and service delivery in the developing world (Brown 2009). Irwin, suggest that this is the century for design since it is the only discipline that is equipped to address today's problems, envision a new future and shift the conventions around how people think, behave and understand the world (Baskinger 2012). Sibbet sees this situation as a consequence of the emerging economies interest in creativity and innovation (Sibbet 2011).

Surely technological developments play a big role here since design is more relevant through technology than ever before. The extensive integration of technology into people's lives translates into the design domain as a new generation of requirements. Useful, comfortable and beautiful is not enough anymore. Design efforts need to explore solutions beyond useful, usable, efficient and effective towards universal (inclusive), sustainable, socially responsible, emotionally desirable and meaningful. Consequently, the technical and behavioral knowledge that is expected from a designer has significantly increased. The variety of the people that need to collaborate has multiplied. Designers are less independent and the design profession is more inter-dependent than ever before.

In order to conceptualize and appropriate innovative ideas into our lives, designers first need to work through those ideas. An ideation process is just that. Ideation starts with diverging by generating and elaborating a number of options (ideas) (Buxton 2007; Osterwalder 2010), followed by converging and synthesizing those options into decisions. The outcome is concepts to follow through. To live this process properly designers need to be able to appropriately discuss ideas. To discuss, they need a common ground to communicate and collaborate with each other. Even a one-man-design job requires a self-reflective process where a similar discussion occurs in the form of a self-conversation by externalizing thoughts through sketching (Buxton 2007; Cross 2011; Lawson 2005).

Discussing experiences is quite different from discussing products and interactions. User experiences are subjective, context-dependent and dynamic over time (Moggridge 2006) and the usual operating principles that work for products do not allow to bring the emotional, contextual and temporal aspects of an experience into discussion (Buxton 2007). When envisioning a tangible product, you can always make a two dimensional (2D) sketch to see the reflection of the idea you had in your mind on paper. It helps you to understand a little more about what you imagined. You can play with that idea just using pen and paper. Even then you can only envision a certain level of complexity and if you are convinced of the potentials of your idea then you need to advance to a tangible model that enables you to envision by exploring your options in the third dimension (3D). This routine becomes even more complicated when the challenge shifts from product to experience.

Allow us to elaborate with the story about The Mechanic and the Heart Surgeon. A renowned heart surgeon brings his car to the garage. To pass time he starts watching the mechanic while he removes a cylinder head from the engine. The mechanic notices him and teases the doctor saying: "Well Doc, check this out. I just opened up the heart of your car, found the problem, fixed it, closed it up and it works like new. So, tell me, how come you earn so much more than me while we are basically doing the same work?" The cardiologist glanced at the mechanic, leaned over smiling and said: "Try doing the same work with the engine running".

In a designerly fashion, what is expected of an experience designer is similar to what is expected of the heart surgeon in the story. He needs to design for an ongoing sequence of events and be aware of the implications that constant change brings. Designing an experience means generating, developing and testing ideas on the fourth dimension (4D); time. The challenge exceeds the physical product and he needs to consider the time factor, as this is an essential aspect of experiences. How can he discuss and weigh his options about an event that has not occurred yet? How does he assess an emotional experience that belongs to a future context long before it becomes an established component of his design?

Designing-for-experiences means designing for the state of changing. Moggridge, described this situation as designing verbs instead of nouns (Brown 2009). Irwin seconds this thought by pointing at the shift of articulation from designing as an entity to designing as a behavior (Baskinger 2012). We understand that designing verbs or behaviors requires different principles, knowledge, skill-set and tools than designing nouns or entities.

In the next section, we first provide an overview of contemporary conceptual design methods and tools that are currently in use with designers. Subsequently, we introduce storycraft as a means to incorporate experience into the designer's creative thinking, and propose a method and a framework to achieve this goal.

Contemporary Conceptual Design Process

As we mentioned in the first section, design work starts by exploring new ideas. Implementation of those ideas (design actions taking place, supported by design skills) is guided by decisions made in the conceptual design process. Therefore, the best opportunity to establish an experiential influence over the whole project lies in the conceptual design stage. However, as design professionals find out repeatedly, this is easier said than done. To explain the complications that an experience focus brings to the envisioning process, we first need to introduce the contemporary state of the art.

Conceptual Design

Conceptual design is the act of sketching, outlining, and drafting key characteristics of a product, interaction, service or experience early on in the design process, with the goal of initiating creative reflection and planning subsequent phases (Atasoy and Martens 2011). In order to test, evaluate and refine ideas, designers externalize and represent ideas into tangibles (Cross 2011). It can be a sketch on a piece of paper or a bundle of sticky notes. It is an essential step in any creative process where the mind(s) needs to keep track of the thought process and reflect on the options under consideration. Sketching also functions as an external memory for the designer and supports the creative process by reducing the mental effort on the working memory (Bilda and Gero 2007).

Often the conceptual design process starts with a brief, which is an initial summary of project objectives. The team is likely to adapt the brief as the project develops. It is hence an open-ended document that helps the team to creatively explore without losing perspective.

The first step is collecting information where the team explores the project background with the objective of revealing opportunities. The aim is to cultivate insight from the project domain, relevant user and context information. A quick approach is to consult secondary information sources (websites, wikis, books, etc.) but the team may also decide to collect primary user information through interviews, direct observation or self-documentation (such as diaries, probes, inquiries, etc.). Interpreting the collected information into a shared understanding can help to envision present and future user behaviors. Subsequently, the 'concepting' starts with the team collaboratively processing content and creating alternative approaches that lead into propositions for new concepts. The outcome is shared with outside stakeholders by presenting illustrations of the new concepts such as product sketches, system flow charts, storyboards of user scenarios, etc. Next, the selected concepts are compared mutually and with competing systems in the market by reviewing their potential with respect to the project objectives. Depending on the scale of the project, qualitative research methods might be used, either without user involvement, such as team evaluation, expert evaluation, heuristic evaluation, or with user involvement, such as focus groups, contextual inquiry, cultural probes and low-fidelity prototype testing. Finally a decision is made and the chosen concept starts its long journey towards the market (Keinonen and Takala 2006).

Contemporary Conceptual Design Tools

The conceptual design process has incorporated diverse tools to support the complete range of design activities. Their common function is to structure conversations and to assist design teams to understand and empathize with people in their pursuit of envisioning meaningful products (Marin and Hanington 2012) and experiences. For the purpose of this chapter, we focus on a selected group of tools that are widely used in concept generation to generate and illustrate concepts. The common purpose of these tools is to enable designers to explore and discuss the variety of propositions that they generate. These tools complement each other and their goal is to promote effective communication between team members by providing an opportunity for explicit reflection.

Mind Mapping, Affinity Diagramming and User Journey Mapping are prominent examples of tools that support the collaborative thinking process by visually mapping out the discussion topics. Mapping out the thinking process encourages designers to collaboratively explore and recognize new patterns by revisiting and reorganizing fragments of information and elaborating new ideas.

Mind Mapping is used to visually organize data by representing relations between dynamic elements of a topic or a problem (Marin and Hanington 2012). It helps to make connections for providing a general sense of a whole (Brown 2009). Affinity Diagramming is used to find common aspects between ideas and to meaningfully cluster them according to identified themes (Bonacorsi 2008; IDEO 2003). It helps the team to constantly ground their decisions on the collected data (Marin and Hanington 2012). User (or Customer) Journey Mapping aims to visually represent the steps taken by the user in order to envision key events and interactions during the process (Hagen and Gilmore 2009). It helps the team to exclusively focus on and evaluate key moments of an experience (Marin and Hanington 2012) (See Fig. 1).

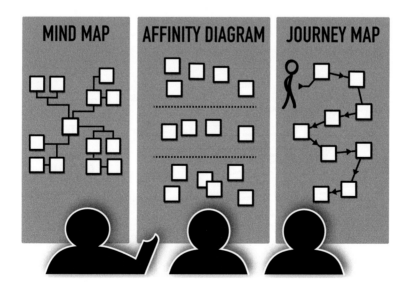

Fig. 1 Mind mapping, affinity diagramming and user journey mapping

Personas condense actual user information into the general characteristics of a user group and represent it in the form of the profile of an (fictional) individual. Personas help the design team to align their design intentions with the people for whom they are designing and to ground their decisions on actual user data (Long 2009).

Scenarios are believable narratives about imaginary courses of actions while people engage with a product or service. They are usually written for a future setting from a Persona's perspective to describe the goals, behaviors and experiences of users in a specific context to test the design assumptions (Cooper 2004; Marin and Hanington 2012).

A Poster (Gray et al. 2010) is a fictional advertisement of a product and/or service while it is still being designed. It is a visual representation aimed at elaborating the design vision and "understanding the link between the service idea and the existing reality" (Norman 2009). Next Year's Headlines a.k.a Cover Story (Gray et al. 2010) have a similar goal as the Poster but focus more on the possible impact of a design idea on society in an imagined future (IDEO 2003). Designers use these tools to project themselves into the future and to ideate around future scenarios in an imaginary but believable context (IDEO 2003) (See Fig. 2).

Mood boards (a.k.a. Image boards or Collages) are visual compositions made out of a collection of pictures and materials to communicate impressions in order to share values between collaborators that are difficult to express in words (Gray et al. 2010; Lucero 2009; Tassi 2010) (See Fig. 3). While a picture of a calm lake in the middle of a pine forest in the fall could infer 'serenity', a ripe green apple with a drop of morning rime could represent 'freshness' and 'health'. Mood boards involve imagery with a focus on the impressions rather than interpretations. They

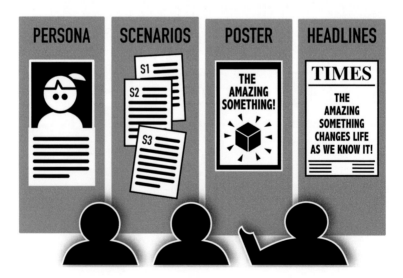

Fig. 2 Persona, scenarios, poster and next year's headlines

Fig. 3 Mood board, Storyboard

aim to create a unified perception of inspirational materials in order to guide the team towards a shared direction. Discussing such abstract impressions in the early stages of the design process helps the team to build a shared understanding about values on which decisions are made.

Storytelling, Storyboard, and A day in the Life are used to ground the conceptualization process by attaching design ideas to their experientor in a real context. These tools help designers to establish an empathy with users on a level where designers can participate in the feelings and ideas of their target user group (Fritsch et al. 2009). Storytelling is an articulation tool that connects ideas within a flexible context and stimulates discussion towards a unified understanding on a level where all stakeholders can contribute regardless of their background (Chastain 2009; Rees 2010). 'Storytelling' is a notoriously generic term and used in various disciplines (including design) for various purposes. This causes a cacophony around the term and we will try to address this issue in the next section to avoid any misunderstandings.

Storyboarding (a.k.a Continuity Board) is a visual story-planning tool that was originally used in the pre-production of movies to plan and communicate scenes between members of a film crew (director, cinematographer, cameraman, sound designer, etc.). It is telling a story through a series of pictures (Glebas 2008). In the case of a design process, storyboards help design teams to capture key events of an experience by illustrating and viewing them in a sequential order (See Fig. 3).

A day in the Life is a tool to imagine the potential experience of a user in the course of a day. Focusing on routine actions during the day helps disclosing unnoticed details that might reveal essential issues (IDEO 2003) (See Fig. 4).

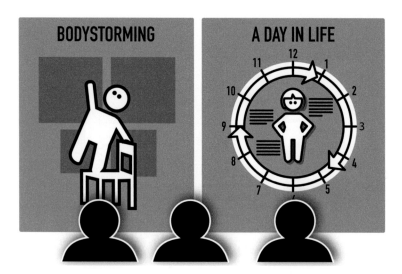

Fig. 4 Body storming, a day in life

Role Playing and Body Storming assist in igniting fresh inspiration through improvisation and physical involvement. Role Playing is used as a tool in which collaborators act out potential users within a real or imagined context. Assigning roles to the team members emphasizes empathy with different users and strengthens the perception of context, hence providing insight into the impact of circumstances on the user experience (IDEO 2010). Body Storming is a type of role-playing where the focus is on the physical interaction with the objects and the environment. Props are generally involved in the process as product placeholders and this gives the team the opportunity to observe potential responses and behaviors in an imaginary context (Tassi 2010). Improvisation tools provide valuable opportunities for designers to envision the embodied qualities of an experience (See Fig. 4).

The mentioned methods and tools all support designers in the conceptual design process to collaboratively envision hypothetical situations where imagined users act in an imagined context. The task of the design team is to pull new and yet believable concepts out of this process and describe them in a convincing way to influence the overall design process. All these tools represent a clear demand for organizing the creative thinking process in order to promote collaborative creativity through explicit communication. In the hands of experienced design teams, these tools can significantly improve the process of conceptualizing for tangible products and the interactions they provide.

Envisioning experiences in their temporality on the other hand requires the ability to communicate and reflect on the dynamic change of the emotional experience over time according to the design intentions and none of the above methods looks like it is adequately providing just that. We will elaborate further on this topic in the following section.

Storycraft

Although Storytelling is the popular term when there is a need to address story-related issues in the design domain, it does not fully support the meaning that we are trying to infer within our work. Storytelling takes many forms. The traditional storytelling forms are oral storycraft, written storycraft and in the last century; film storycraft (McClean 2007). Firstly, we are interested in film storycraft over its predecessors for several reasons, which we will explain in the next section. Secondly, our work is more about the act of "crafting" instead of "telling". Therefore, we are going to use the term storycraft when we need to emphasize our focus and intentions.

Why Storycraft?

The similarity between the key properties of a story and an experience is evident. They are both made up of people, places and objects and they both emerge from the interrelations of all three over time (Buxton 2007). This implies that both are subjective, context-dependent and dynamic in nature. They share a sequential structure with a beginning, middle and end that can be crafted and influenced through design. However, most importantly, both stories and experiences evoke and influence the emotions of their experientors.

Our interest in storycraft is motivated by its proven ability to create emotionally satisfying and meaningful experiences (Glebas 2008). Storycraft has a well-established tradition, and makes use of an appropriate set of tools to influence the experiences of its consumers (audiences). Certain branches, such as film storycraft, also have extensive expertise in managing the complex orchestration of multidisciplinary production teams towards a common purpose. We look at the communication and orchestrations required in the conceptual stage of designing for user experiences and investigate if it can be improved by adopting tools used in film storycraft. Glebas suggests that we organize our experiences in the form of a story and we use stories to structurally communicate our experiences to others (Glebas 2008).

Although the film and design domains produce different kinds of outputs, the early design process where they create their fictional/conceptual backgrounds share common tools such as Scenarios, Storyboards, Role Playing and Personas to envision future experiences. The reason for this similarity between disciplines comes from a common need for a quick, inexpensive and flexible way to explore and communicate ideas between collaborators.

Despite their shared benefits, the filming process seems to be more advanced in its explicit awareness and clearly defined strategies to engage the audience emotionally (Chastain 2009). Filmmakers tell stories in a professional way. Stories have the power of engaging people into an experience, and screenwriters (a.k.a. scriptwriters or scenarists) are experts at crafting story structures and plots based on story ideas. It is therefore relevant to try and understand the structural strategies behind this craft that deliberately aims at influencing the emotions of its audience.

Story Structures

A story structure is the structural framework that determines the order and fashion in which a story is presented. A plot is the storyteller's pick of events and their arrangement in time from numerous interrelated possibilities of how things could unfold (McKee 2010). They are used as a strategic guide to pull the audience emotionally and hold their attention (Inchauste 2010). Explicit knowledge and use of story structures is essential for a successful transition from script to scenes (visual planning or storyboarding).

There are prominent studies that suggest empirical evidence for a common structure that is at work in the type of stories that transcend through time, despite cultural and geographical differences: Propp's Morphology of the Folktale analyzes numerous Russian folk/fairy tales that indicate the same structure with 31 commonly occurring themes. He called them 'story functions' – classifiable actions that characters can take that occur in a consistent order (Hammond Sean 2011). Campbell studied myths, fables, folktales and stories across cultures and various periods in history. He pointed out that there is a common structure to the journey of the hero in all the great stories that have transcended throughout history (Duarte 2010) (See Fig. 5).

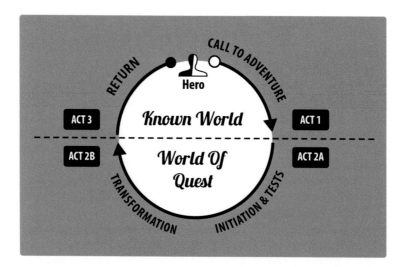

Fig. 5 A visual representation of the structure of a hero's journey

This basic formula that he called "Hero's Journey" suggests a process where the story can be summarized as follows; first the hero is introduced, then he is called to an adventure. Though the hero resists the call at first, he is encouraged to take the challenge by the mentor with whom he crosses a threshold from the ordinary world to the new world of the quest. Then the hero is tested and events take a climax where the hero is forced to take an ordeal. He collects the reward of the ordeal that also transforms him. Finally, he starts the journey back home, enriched by the experience (See Fig. 5). Gilgamesh, Beowulf, Odyssey, Hobbit, Star Wars, Matrix are few examples sharing the same basic formula (Inchauste 2010; Schlesinger 2010).

Evidently, the history of story structures started long before screenwriting. In Poetics, Aristotle observed and analyzed the connection between storytelling techniques and emotional experiences. His drama theory inquiries into the capacity of the storytelling technique to evoke certain emotional experiences in the audience (Hiltunen 2002). The Aristotelian system maps the structure of actions as a unified plot where beginning, middle and end describe the order of actions. Hiltunen interpreted Aristotle's approach as a strategy of understanding the mechanics of creating emotional experience through storytelling and suggests that it may lead to the ability of predicting probable success of stories beforehand (Hiltunen 2002). The Aristotelian plot has been expanded by Freytag into a five-act structure that mostly applies to ancient Greek and Shakespearean drama (Freytag 1900).

In Freytag's Curve, Exposition provides the context by introducing the hero and the villain along with the conflict at hand. Rising Action is the stage where the tension increases through complications and uncertainties towards an identified goal. Climax (or Crisis) is the point in time at which tension and uncertainty in the story peaks while pulling up the audience engagement to the maximum. Falling Action is the point where the conflict unravels in favor of the hero over the villain. Dénouement (or Resolution) is the conclusion where the suspense ends and the complications are resolved (Wheeler 2004) (See Fig. 6).

A more contemporary story structure is proposed in Field's Paradigm (Duarte 2010). It is more commonly known as the three-act structure and is a simplified and compressed version of Freytag's five-act structure.

According to Field, what moves a beginning to the middle and the middle to an end are called plot points which are definitive moments where an event happens that changes the direction of the story (Duarte 2010). Act One introduces the reader to the setting, the characters, and the conflict and establishes relationship between them along with the hero's unfulfilled desire (Duarte 2010; Quesenbery and Brooks 2010). Act Two develops through a series of complications where characters encounter obstacles that keep them from achieving their goal (Duarte 2010). Though each of these crises are temporarily resolved, the story leads inevitably to an ultimate crisis, which is the global climax (Quesenbery and Brooks 2010). Resolution ties together the loose ends of the story, offers a solution rather than an end and allows the reader to see the outcome of the main character's decision or action at the time of the climax (Quesenbery and Brooks 2010) (See Fig. 7).

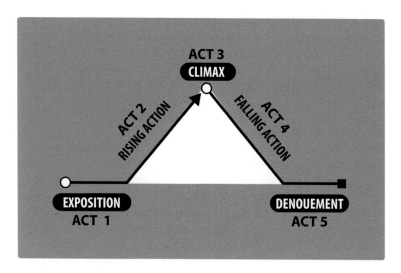

Fig. 6 Freytag's (five act structure) curve

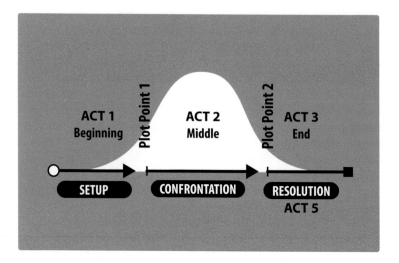

Fig. 7 Three act structure

Freeman, interpreted the three-act structure as an energy curve, called "Aristotle's Plot Curve" (Sparknotes 2011), that visually communicates the relationship between time (horizontally) and dramatic intensity (vertically) (See Fig. 8). Glebas took a step further and interpreted dramatic intensity as emotional involvement. In his interpretation, the vertical axis depicts how much the audience is involved or "lost" in the story (Glebas 2008).

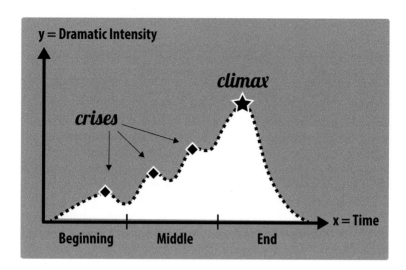

Fig. 8 Aristotle's plot curve

Quesenbery and Brooks suggests that a well-structured story with appropriate plot points can provide a storyline with qualities such as coverage, fit, coherence, plausibility, uniqueness and audience imagination (Quesenbery and Brooks 2010). Coverage means that the story addresses all the necessary facts. Coherence and plausibility assure that the story makes sense (i.e., doesn't create confusion). A story that fits well with the facts of the context feels in place. Uniqueness is required to intrigue the audience (i.e., avoids boredom). A good story always leaves room for the audience to fill in-between the lines, i.e., to imagine details (Quesenbery and Brooks 2010).

The efforts around analyzing stories that 'work' originate from the common need to distill operating principles for good storytelling. However, it is important to realize that the strategies above do not provide a formula or recipe that guarantees success but can only serve as guidelines to think in a structured way.

We provided an overview of the prominent crafting tools that support a story-teller's creative process. From a User Experience Design point of view the identified patterns of story structures may inspire designers in several ways: (1) they provide an explicit representation of the strategy behind creating emotionally satisfying and meaningful experiences, and can help to raise common awareness, (2) they reveal that storytelling is not entirely an intuitive technique; and that usage of pre-defined guides to support the process is common practice, (3) the change in pattern of the intended user response, when visually represented, can assist the storyteller in the orchestration and communication of the story.

Storyboarding in Film and Design

Another aspect of crafting stories is visual planning, or storyboarding, where storytellers interpret written scripts into a visual sequence of actions. Storyboarding has its origin in Disney's animation studios out of a need to track and organize the massive amount of drawings in a feature-length animation. It was then and still is a tool that enables collaborators to view the sequence of frames in order to reflect, discuss and detect problems and make changes quickly and easily (Glebas 2008). Storyboarding starts with a script that is "a verbal plan for a story" (Glebas 2008) or loose treatment that is a sketch of a script. From this, the visual artist builds the actual look of the world in the movie under the supervision of the art director and production designer (Glebas 2008).

In live action, Alfred Hitchcock was one of the first who implemented storyboarding in order to project and assure the seamless flow between shots (continuity) beforehand (Glebas 2008). Storyboards provide visual assistance to the director and feeds information to the planning phase with respect to cinematography, camera angles, choreography and the positioning of actors (blocking) and props. It is a cheap way of exploring and experimenting options before shooting live action (See Fig. 9) (Glebas 2008). It also assists the producer in searching for financers, casting and locations (Lelie 2005).

Storyboards became especially important with the development of special effects in live action movies where Computer Generated Images (CGI) need to be combined with real footage (Glebas 2008). Peter Jackson is one of the contemporary directors who utilize storyboarding extensively in his process, which he refers to as "version

Fig. 9 An example of how a storyboard for film looks like

Fig. 10 An example of how a storyboard for design looks like

zero of the movie" (Botes 2003). During the filming of The Lords of the Rings Trilogy, in addition to storyboarding firsthand he commissioned a professional storyboard artist to redraw his own sketches. He also used Story Reels, which are a version of storyboards with voice-over, sound and music (Botes 2003). Story Reels are not only useful to help detect and solve story problems but they also provide an emotional road map for the film (Glebas 2008).

Storyboarding in the design domain is used to help designers understand and communicate the interaction between a user and product/service in context and over time (See Fig. 10) (Lelie 2005). Designers are visual thinkers. Representing an idea in the form of a story is a challenging but lucrative task. It forces designers to consider the context of use in the temporality of action. The challenges that target users might face gain believable circumstances even though they are hypothetical. The reader of a storyboard can empathize with the design issue at hand and can reflect upon it regardless of his/her background or discipline (Lelie 2005).

Several of the visual planning and communication strategies used in the early moviemaking process are already being adopted partially into the design practice. We want to take this interest further and focus more explicitly on the cross-section of user experience design and film storycraft. We believe that understanding and assimilating the principles of storycraft will enable designers to "aim for the heart by working at a structural level" (Glebas 2008). In order to test this proposition, we have adopted a research through design approach in which we iteratively design and develop a method that incorporates many of the key storytelling techniques mentioned above, and subsequently apply this tool in the user experience design process in order to establish opportunities, problems, etc.

Storyply

Designing for experiences requires design teams to empathize with the people for whom they design on an emotional level. The challenge is to envision not only the 'dynamic qualities of experiences' but also 'the constantly changing emotional response to those changes'. This calls for an approach that allows exploring and discussing on the fourth dimension starting from the very beginning of a project.

We argue that this could be achieved by designers expanding their creative process and cultivating visual story-crafting skills. Especially film storycraft provides unequaled strategic guidance in addressing emotional change over time. We try to harvest this potential and integrate it in the designer's creative process using a conceptual design method called Storyply, accompanied by a toolkit and implemented through a workshop.

In this section we start by describing the research process we followed and then introduce the method, toolkit and workshop we designed, tested and developed through successive case studies. We conclude with a reflection on what was learned along the way.

Research Approach

Our research approach is to generate knowledge to enhance design practices by linking the theory to practice through investigating the processes and tools of thinking and making in design projects (Marin and Hanington 2012). This is frequently referred to as a Research Through Design Approach (Archer 1995; Marin and Hanington 2012). It involves a creative and critically reflective process in which literature survey and case studies are used to discover insights that are subsequently incorporated into the (next version of a) design of a conceptual design method.

We previously mentioned that storytelling takes many forms such as oral, written and film storycraft. We focused on film storycraft as our role model for the qualities it provides for visual communication and orchestration of a collaboratively creative process. We also pointed out the limitations of contemporary tools that support creative thinking when it comes to envisioning experiences. To overcome these limitations we seek guidance in storycraft as a means to assimilate experience into designer's creative thinking. Storycraft is the skilled practice of generating/building stories. Experience -driven design requires design teams to embrace storycrafting skills in order to ideate comfortably in the fourth dimension: time.

We built a method called Storyply (formerly known as Storify (Atasoy and Martens 2011)) that tries to merge the skilled practice of generating stories with the skilled practice of design thinking. Following a year of literature studies (backed by 15 years of professional design consultancy experience) we started designing and testing the framework through workshops that we conducted with BA, MSc

Fig. 11 Snapshots from several storyply workshops conducted with students, trainees and professionals

and PhD students, professional design teams, professional storytellers, R&D and Innovation teams at firms from various industries in Europe (See Fig. 11).

Throughout the process we were guided by the following research questions:

- Does incorporating storycraft within conceptual design provide an improvement on the process of designing for experiences?
- Does Storyply help designers to focus on and prioritize the experiential aspects of a design project?
- Does Storyply help designers to address the subjective, context-dependent and temporal nature of experiences?
- Does Storyply help designers to envision user experiences in a better (more profound) way?
- Does Storyply help to envision – better (more profound) user experiences?

Addressing these questions required a relevant and realistic project context were the value that is generated by the user experience focus can clearly manifest itself (in order to increase external validity). Moreover, to observe real design teams trying the method we needed a coherent framework and appropriate setting to apply, observe and document. For communication purposes we first built a dummy case project scenario called *Re-designing the Waking-up Experience* and presented everything designer (creative workshop) style; on a (white) board using a collage of drawings, sticky paper notes and printed materials (See Fig. 12). The dummy scenario raised an experience design challenge where we could employ the proposed method, try out and prepare for the real project challenges. This presentation style complemented our explanation and provided not only the experts to follow a visual track of how process flows but also allowed them to comment on our thinking and decision making process. We utilize the same "Re-designing Waking-up Experience" case also in this chapter to explain the method in the following section.

Fig. 12 Mock-up presenting the dummy case project scenario called *Re-designing the waking-up experience* in which a renowned alarm clock producer is rapidly losing its customers to personal mobile devices. The scenario presents this imaginary company as desperate and open to revolutionary ideas. They hire a team of designers (the workshop participants) to come up with an innovative idea/solution that will change the game in their favor. The creative brief interprets the goal as generating new ideas and concepts by focusing on the experience such as sleeping, waking up, the need for an alarm, etc. It encourages prioritizing experience over 'product'

Observing the success of communication in these sessions encouraged us to follow a similar format to test our initial design. We built a 'meaningful workspace' (Gray et al. 2010) out of pre-made templates that served as the canvas for the participating team members to post artifacts/idea containers (Osterwalder 2010) (print outs, sticky notes) and to sketch and write on. Our goal was to allow the design team to quickly and easily explore relationships between the artifacts they generated while keeping a visual track of their progress. Artifacts/idea containers acted as a carrier of meaning that makes the processed information tangible, explicit, portable and persistent (Gray et al. 2010; Osterwalder 2010). The combination of our framework within the specified workspace allowed a mixture of conceptual design tools to be utilized during the workshop such as Brainstorming, Scenarios, Persona generation, Post Up and Storyboarding to support the creative process. In fact, these early mock-ups laid the foundations of the Storyply Toolkit that we use currently for academic and professional purposes in practice. The details of the method, process and toolkit are explained in the following section.

Storyply Method, Toolkit and Workshop

The Storyply Method combines 'conceptual design' and 'story planning techniques' to help the design team visualize the solutions for themselves as well as for the potential audience of the project such as users, clients and other stakeholders. The method suggests two main layers: Backstage and Onstage. Each stage is guided

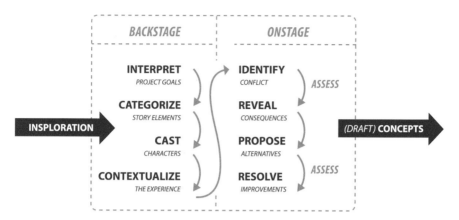

Fig. 13 Sub-activities in the Storyply method

Fig. 14 In order to make the insploration process more accessible for the whole team, participants are asked to prepare six images (two images of people, two images of places and two images of objects) of their own choice that belong to the project context according to their perspective before the session. In the beginning of each session, these images are added to a collection of images that is called "insplorers deck" that consists of 90 (5x5 cm) readymade generic images to support the creative process

by ready-made templates and instructions to work with the provided template. The templates divide the total effort into more manageable sub-activities, next to offering support for the sub-activities themselves (Fig. 13).

Insploration is a pre-workshop phase that suggests a preliminary effort where team members individually prepare for the conceptual design process by exploring the sources of inspirations that will fuel the creative process. Design is a creative endeavor and designers are explorers of inspiration: Ins-plorers. Insploration works as a conscious and systematic act of searching for and capturing stimulants that may inspire new ideas (Atasoy and Martens 2011). The project brief sets the focus of interest which influences the creative minds' selective perception to start highlighting relevant and potentially inspirational materials according to the subject matter. Designers are visual-thinkers so the majority of insplorational material is likely to consist of visual collections (Fig. 14) (Keller 2007).

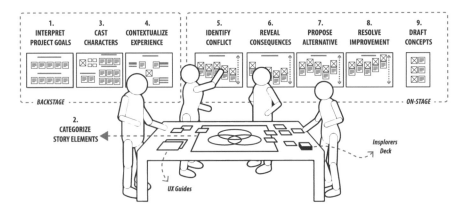

Fig. 15 Storyply workspace and toolkit. The toolkit consists of nine various A0-sized paper templates with instructions, insplorers deck: 90 pieces of 5x5 cm pictures (30 people, 30 places and 30 objects), UX guides: a compact collection of practical and relevant information from user experience research, sticky papers and markers for each member to write and draw. Moreover, the workshop experience requires a workspace with a wall to hang eight of the templates side by side and a desk to lay one template on and gather around while using other components such as insplorers deck, stick papers, sharpies, etc. throughout the session. The storyply toolkit will be available at www.Storyply.com soon

Storyply provides a toolkit with clear instructions to assist the creative process in which participants can apply the method by simply following the instructions on the foreground while they actually implement a combination of story-thinking and design-thinking principles in the background.

Backstage is the first layer in which the collaborative process starts at the workspace. The first template of this section is the *Interpret Project Goals* which guides the team to first self-reflect and then to share and discuss their understanding of the project brief at hand. The format and instructions are designed to encourage the participants to briefly look at the initial project brief from an experience-focused perspective (See Fig. 15).

Our observations showed that Template1 serves the following purposes; (1) The chance to write down a personal interpretation of the challenge at hand is especially valuable not-only for self-reflection but also to detect the diversity of perspectives and to start building a shared vision at the very beginning. (2) The wording of the instructions sets a tone that agrees with the experiential focus from the start. (3) The opportunity to individually title the project not only reinforces participant's ownership of the task but also sometimes helps revealing their priorities (Fig. 16).

The second template that guides the backstage is called *Categorize Story Elements*. Here the design team organizes and interprets the collected materials to assist them in the design process. The template is laid flat on a desk and each team member categorizes their inspirational collections individually under three themes; People, Places and Objects.

Describe The Project Goal As An Experience

*Individually think about what this experience is essentially about and summarise what
should be the main experiential goal according to you on one sticky paper & post them up side by side*

"STOP THE RAPID LOSS OF CUSTOMERS TO THE PERSONAL MOBILE DEVICES.	"...EVEN IF THEIR MOBILE DEVICES CAN MATCH EVERY SINGLE FUNCTIONAL- ITY THAT WE OFFER, PEOPLE STILL SHOULD GO FOR OUR PRODUCT!" - CEO	GENERATE NEW IDEAS & CONCEPTS THAT COULD LEAD TO INNOVATIVE SOLUTIONS	FRESH START WITH A FOCUS ON THE WAKING-UP & LAYING DOWN EXPERIENCE	CONNECT WITH THE USER ON AN EMOTIONAL LEVEL!

Fig. 16 A section of Template1: interpret project goals for re-designing the waking-up experience project

Fig. 17 A section of Template2: categorize story elements

The themes are represented by three intersecting Venn diagrams to be filled with images according to their relevance (See Fig. 17). Then, as a group, the elements placed on the canvas are sorted for potential building blocks of a context of experience. We call this template *Categorize Story Elements* since it provides a pool of potential elements for the participants to choose from in later stages. During our studies, we observed that Template2 serves the following purposes; (1) it helps individual designers to organize their visual collections and focus on the project space. (2) Sorting collections according to people, places and objects appropriates materials to be used as story elements. (3) Participants see each other's collections and get inspired by each other while they try to pick appropriate materials together, which also encourages discussion. (4) Filling the canvas serves as a good warming-up exercise for the team members to familiarize themselves with the process and each other.

The third template of this section is *Cast Characters* that allows the team to build characters that will be actors in the user experience that is aimed for. Similar to a Persona creation, the team is expected to envision believable archetypes to represent a target group of potential users. Step 1 is giving a face, a name, age, occupation and location to a main character then, giving a face and a name to the most prominent supporting characters who have a direct interaction with the main character and/ or the experience. One important difference of Casting from a Persona is that the main interest is the conflict that drives the character in a story, rather than their consuming behaviors and/or decision-making strategies. Hence, Step 3 and 4 encourage participants to think about the motivations and potential points of tension in a characters life, which helps to identify the conflict that could provide the team with an understanding of the character on an emotional level, which they can in turn build the experience upon.

We learned from our observations that Template3 serves the following purposes; (1) since the process resembles casting a main-character for a film story, it is familiar to every participant regardless of their background and this familiarity adds confidence to their actions. It is especially important at this early stage of the process to obtain contributions from all participants, even those who claim an allegedly non-creative background. (2) Casting provides a means of identification as in the film. Identification cultivates emotional attachment with the experience that is the fundamental goal of the process. (3) Conflict is a professional expression in storycrafting with rather unfamiliar vocational interpretations for a non-story crafter. The most meaningful instances often arise when one can imagine a conflict from personal experience. Some sort of guidance tends to be required at this stage, and is provided in the Storyply Method by two quick questions about Motivations (i.e., Drivers) and Points of Tension (i.e., vulnerabilities) (See Fig. 18).

The fourth template is called *Contextualize Experience*. It helps the team to build a visual representation of the space where the experience takes place. A set of guiding questions about the spatial properties of the experience supports their imagination about touch-points in context (See Fig. 19). The task is to imagine the space where the experience could take place as vividly as possible and then sketch

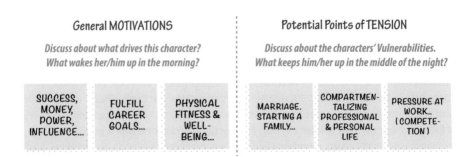

Fig. 18 A section of Template3: cast characters; Step 3 and step 4

Fig. 19 A compacted version of Template4: contextualize experience

a very simple visual representation of that scene as if you are a fly on the ceiling or a bird flying over the scene.

We learned from our observations that Template4 serves the following purposes: (1) Simple but concrete questions prompt participants' imagination to a richer setting than they would usually think without guidance. (2) The visual representation of the setting establishes a mutual agreement about the spatial qualities of the setting and grounds the discussion in a more believable context that leads to concrete suggestions.

Onstage is the second layer where the design team generates and assesses the content and creates alternative approaches that can lead into propositions for new concepts. The outcome is one or more concepts that can subsequently be assessed with users and other stakeholders. The process is very similar to storyboarding, knowing that the biggest challenge of storyboarding, even for the experienced storytellers, is where and how to start. Story-craft becomes an essential tool to assist them in this stage. The procedure guides and facilitates the team to build five key frames for each step that encompasses one event with a beginning, middle and an end.

The first step of the Onstage section is the fifth template of the overall framework and initiates the visual story-building process itself. It is called *Identify Conflict* and it is the most crucial of the following steps in which the participants are asked to imagine the story space (See Fig. 20). We learned from our observations that Template5 serves the following purposes: (1) Introduction of the quest for conflict feels utterly refreshing and mind opening even to the most seasoned designers since it clearly promotes a smooth transition from design thinking to story-thinking. (2) Each previous step proves to be indispensable at supporting this first step into the story space.

The sixth portion is *Reveal Consequences* and follows the same procedure as the previous stage but this time the framework guides the team to envision the impact of the conflict on the character and thus the consequences on the experience (See Fig. 21). We learned from our observations that Template6 serves the following purposes: (1) pushing participants into envisioning further into the consequences of their imagined experience, encouraging them to bring daring and rich insights from their own personal life into the story. (2) The severity of consequences fuels empathy with the main character and thus the reality and believability of the situation.

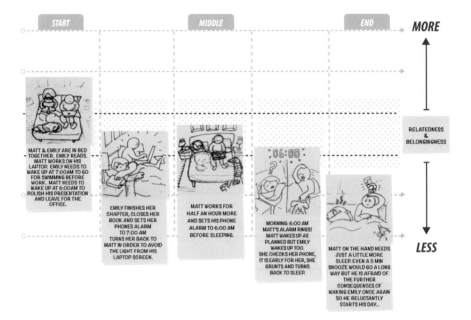

Fig. 20 A compacted version of Template5: identify conflict. The event is assessed frame by frame over time by moving sticky paper frames up and down with respect to the chosen value on the right. Relatedness/belongingness is picked out of UX guides, as the emotional need to appraise the experience in this instant. It is described by Hassenzahl et al. (2010) as "the feeling that you have regular intimate contact with people who care about you rather than feeling lonely and uncared for"

The seventh portion is called *Propose Alternative* and this time the framework guides the team to make their own proposal using the conflict and its consequences that they identified in previous stages (See Fig. 22). Once again, they are expected to use the same five key frame structure to build their proposal and evaluate the situation with the same values.

We learned from our observations that Template7 serves the following purposes: (1) feeling the pressure of the consequences from the previous stage adds an extra sense of responsibility towards the opportunity to propose the experience that they would rather like to have. (2) Participants start to utilize the dramatic qualities in their story more confidently than in previous steps.

Finally, it is time to see the impact of your contribution and *Resolve Improvement* is the template just for that. For the last time the team creates five frames that illustrate how the events could unfold with their intervention to the situation at hand. Additionally they still have the scale of emotional needs that they can reflect on and see how the envisioned experience could be better aligned with those values over time (See Fig. 23).

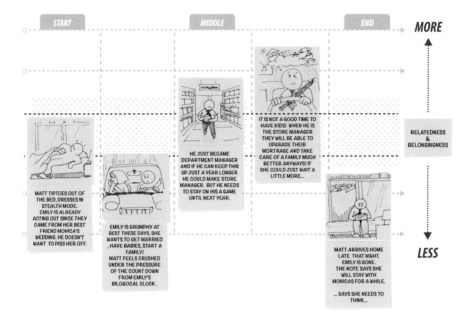

Fig. 21 A compacted version of Template6: reveal consequences

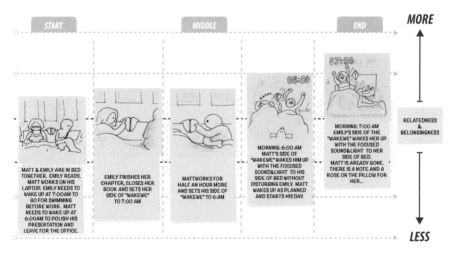

Fig. 22 A compacted version of Template7: propose alternative

We learned from our observations that Template8 serves the following purposes: (1) The resolution in the story provides a proper closure for the process (2) participants starts to see the further potentials of the method in the future.

Fig. 23 A compacted version of Template8: resolve improvement

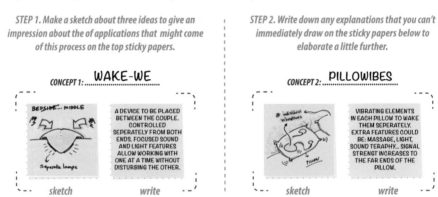

Fig. 24 A compacted version of Template9: draft concepts

The very last step is to capture initial ideas as a tangible outcome of the process. That is called *Draft Concepts* and its purpose is to scribble quick and dirty visual reminders of the initial ideas and directions that the project may take in the future (See Fig. 24).

We learned from our observations that Template9 serves the following purposes: (1) it provides the opportunity for the team to take note of the small ideas that they had during the process but did not have time to focus on in the heat of discussion. (2) The small concept sketches work as good reminders of the intensive process and earns a whole new level of respect for the small sketch that previously looked easy to achieve. Hence, the real outcome of the process is the whole template filled with content, which functions as a visual map of the envisioning process and the discussions initiated by the framework.

Reflections on Designing Storyply

We tested the Storyply Method in 15 workshops in Italy, Sweden, Turkey and the Netherlands with 150 participants (59 Professionals and 81Students/Trainees) from diverse backgrounds such as: designers (industrial, product, visual, interaction, service, strategy, software, hardware, UX.), researchers, engineers, managers, film makers, R&D specialists and CEO's. Throughout these studies we focused our attention on the aspects that coincided with our expectations, identifying the general expectations that didn't work out as expected and identifying unexpected obser-vations. Each session provided feedback for the next iteration and the framework iteratively matured while we collected, analyzed and synthesized findings into the overall design.

Some significant aspects that coincide with our overall expectations were: (1) the guiding principles of storycraft extended the boundaries of explorative sketching towards story thinking. The framework allowed the participants – with no specialized knowledge in storycraft – to conceptualize applying principles of storycraft. Breaking down the conceptual design process into clearly visible, meaningful partitions where they can freely navigate and track their progress helped the participants to organize and share their thoughts clearly and confidently. (2) The narrative competence (Pink 2005) liberated creativity from the bondage of "must-have". Story context provided a safe zone where the participants could take a vacation from the boundaries of technical rationality. The fewer participants worry about manufacturing and/or marketing, the more creative they get and started discussing experiences through stories rather than features and functions. (3) Non-designers and designers fancied and valued different qualities of the Storyply Method. Designers appreciated the framework where they can combine a competence they already have (conceptual design) with a competence they would like to have (storycraft) in a practical and visual style. Meanwhile, non-designers (engineers, managers, researchers, etc.) were more interested in the qualities that promote cross-disciplinary communication, which provide grounds to discuss values and helps the decision making process.

Some significant aspects where our general expectations didn't work out as expected: (4) the intensive one day workshop format had certain shortcomings. Ideally, the participants are supposed to have sufficient time to process and digest the user research data before starting the session. This requires a longer time frame than a one-day workshop where the participants learn and apply the method simulta-neously. On certain occasions, the workshop became an overwhelming experience and we observed that most of the participants performed less energetically in the afternoon sessions. In the light of this observation, we spread the workshop across two days when we can and observed significant improvement on participant's energy levels and ideation performance. Ideally, the framework is designed to accompany the conceptual design process for as long as the project goes. (5) We also witnessed that the participation of an experienced facilitator who understands story-thinking and design-thinking principles to the process made a much bigger difference than we initially expected.

Some significant unexpected observations were the following: (6) it was surprising to see that the skepticism on non-designers wore off faster than their designer teammates when it came to think with stories. (7) the Storyply Method was appreciated eagerly by participants who are more business-minded and/or who are in a decision maker role. They were overly enthusiastic about being able to express what they mean and discuss values with 'creatives' on equal grounds.

Conclusion

In this chapter, we laid out the groundwork for an approach where user experience design is guided by storycraft. We started by pointing at the change of focus from objects to experiences in the design domain, and explained the background and implications of this shift on how designers operate. We placed the conceptual design phase at the center of our attention and provided an overview of how designers currently cope with the conceptual design process. We introduced storycraft, as a means to incorporate experience into the designer's creative thinking and briefly introduced the practical outcome of our evidencing process as a framework to support our claims. The first goal of this chapter is to provide readers with the motivational background behind introducing storycraft into conceptual design and introduce relevant literature regarding our approach to the subject matter. The second goal is to back our claims with evidence gathered through extensive testing in the field with real users and projects.

Good design thrives where creativity can be channeled through skilled practice. Our studies show that the real-time visual mapping of the thinking process under the guidance of story crafting principles has several benefits. First, embedding narrative competence into visual thinking drives the discussion towards experiences. Consequently, the ideas that spin out of such discussions are more likely to serve the purpose of designing for user experiences. Second, the mystical creative process is opened up to the contribution of users and non-designer project stakeholders whilst the concepts are still under consideration. Third, the document, which gradually appears in front of the design team, provides a blueprint of the ideation process which can be iterated back and forth on different occasions with various participants who were not present at the time of the generation process. Above all, the design team can discuss and iterate new concepts in a platform that offers a structure which allows sketching experiences true to their temporal, emotive and contextual nature.

References

Archer B (1995) The nature of research. Co-Design J 2(11):6–13
Atasoy B, Martens J-B (2011) Crafting user experiences by incorporating dramaturgical techniques of storytelling. In: Proceedings of the second conference on creativity and innovation in design (DESIRE'11). ACM, New York, pp 91–102

Baskinger M (2012) From industrial design to user experience. Available via UX Magazine. http://
uxmag.com/articles/from-industrial-design-to-user-experience. Accessed 3 Dec 2012

Bilda Z, Gero J (2007) The impact of working memory limitations on the design process during
conceptualization. Des Stud J 28(4):343–367

Bonacorsi S (2008) What is... an affinity diagram? Available via Improvement and
Innovation. com. http://www.improvementandinnovation.com/features/articles/what-affinity-
diagram. Accessed 16 Mar 2010

Botes C (2003) Making of lord of the rings. WingNut Films, New Zealand

Brown T (2009) Change by design: how design thinking transforms organizations and inspires
innovation. HarperCollins e-books, New York

Buxton B (2007) Sketching user experiences: getting the design right and the right design,
Interactive technologies. Morgan Kaufmann, San Francisco

Chastain C (2009) Experience themes – boxes and arrows: the design behind the design http://
www.boxesandarrows.com/view/experience-themes. Accessed 6 Oct 2009

Cooper A (2004) The inmates are running the asylum: why high tech products drive us crazy and
how to restore the sanity, 2nd edn. Sams, Indianapolis

Cross N (2011) Design thinking: understanding how designers think and work. Berg, New York

Duarte N (2010) Resonate: present visual stories that transform audiences. Wiley, New Jersey

Freytag G (1900) Freytag's technique of the drama: an exposition of dramatic composition and art.
Scott Foresman, Chicago

Fritsch J et al (2009) Storytelling and repetitive narratives for design empathy: case Suomenlinna.
Nordes 2:1–6

Glebas F (2008) Directing the story: professional storytelling and storyboarding techniques for live
action and animation. Focal Press, Boston

Gray D, Brown S, Macanufo J (2010) Game storming: a playbook for innovators, rulebreakers,
and changemakers. O'Reilly Media, Inc, Sebastopol

Hagen P, Gilmore M (2009) Stories: a strategic design tool. Available via Johnny Holland. http://
johnnyholland.org/2009/08/13/user-stories-a-strategic-design-tool/. Accessed 13 Aug 2009

Hammond SP (2011) Children's story authoring with Propp's morphology. Dissertation, The
University of Edinburgh

Hassenzahl M, Diefenbach S, Göritz A (2010) Needs, affect, and interactive products – facets of
user experience. Interact Comput 22(5):353–362

Hiltunen A (2002) Aristotle in Hollywood: the anatomy of successful story telling. Intellect Books,
Bristol

IDEO (2003) IDEO method cards: 51 ways to inspire design. W. Stout Architectural Books, San
Francisco

IDEO (2010) Work-human centered design toolkit. Available via http://www.ideo.com/work/item/
human-centered-design-toolkit/. Accessed 16 Dec 2010

Inchauste F (2010) Better user experience with storytelling – Part one. Available
via http://www.smashingmagazine.com/2010/01/29/better-user-experience-using-storytelling-
part-one/. Accessed 30 Jan 2009

Keinonen T, Takala R (2006) Product concept design: a review of the conceptual design of products
in industry. Springer, New York

Keller I (2007) For inspiration only. Design Research Now, pp 119–132

Lawson B (2005) How designers think: the design process demystified. Architectural Press-
Elsevier, Burlington

Lelie C (2005) The value of storyboards in the product design process. Pers Ubiquit Comput
10(2–3):159–162

Long F (2009) Real or imaginary; the effectiveness of using personas in product design. In:
Proceedings of the Irish Ergonomics Society annual conference, pp 1–10

Lucero A (2009) Co-designing interactive spaces for and with designers: supporting mood-board
making. Dissertation, Eindhoven University of Technology

Marin B, Hanington B (2012) Universal methods of design: 100 ways to research complex problems, develop innovative ideas, and design effective solutions. Rockport Publishers, Beverly

McClean ST (2007) Digital storytelling: the narrative power of visual effects in film. The MIT Press, Cambridge

McKee R (2010) Story: substance, structure, style, and the principles of screenwriting. Harper-Collins, New York

Moggridge B (2006) Designing Interactions. MIT Press, Cambridge

Norman DA (2009) THE WAY I SEE IT signifiers, not affordances. Interactions 15(6):18–19

Osterwalder A (2010) Business model generation. Wiley, Hoboken

Pink D (2005) A whole new mind: why right-brainers will rule the world. Riverhead Books, New York

Quesenbery W, Brooks K (2010) Storytelling for user experience: crafting stories for better design. Rosenfeld Media, Brooklyn/New York

Rees D (2010) User journey mapping. Available via Articlesbase. http://www.articlesbase.com/international-business-articles/user-journey-mapping-521154.html. Accessed 16 Mar 2010

Schlesinger T (2010) Screenwriting seminars and script consultations with Tom Schlesinger. Available via http://www.writingfilms.com/continue.html. Accessed 29 Mar 2010

Sibbet D (2011) Visual teams: graphic tools for commitment, innovation, and high performance. Wiley, Hoboken

Sparknotes Editors (2011) Themes, motifs, and symbols. Available via http://www.sparknotes.com/film/starwars/themes.html. Accessed 10 Mar 2011

Tassi R (2010) Moodboard. Available via Service Design Tools. http://servicedesigntools.org/tools/17. Accessed 16 Mar 2010

Wheeler K (2004) Freytag ' s Pyramid adapted from Gustav Freytag ' s Technik des Dramas (1863) The structure of tragedy. Available via https://web.cn.edu/kwheeler/documents/Freytag.pdf. Accessed 20 Mar 2010

Storyboards as a Lingua Franca in Multidisciplinary Design Teams

Mieke Haesen, Davy Vanacken, Kris Luyten, and Karin Coninx

Abstract Design, and in particular user-centered design processes for interactive systems, typically involve multidisciplinary teams. The different and complementary perspectives of the team members enrich the design ideas and decisions, and the involvement of all team members is needed to achieve a user interface for a system that carefully considers all aspects, ranging from user needs to technical requirements. The difficulty is getting all team members involved in the early stages of design and communicating design ideas and decisions in a way that all team members can understand them and use them in an appropriate way in later stages of the process. This chapter describes the COMuICSer storyboarding technique, which presents the scenario of use of a future system in a way that is understandable for each team member, regardless of their background. Based on an observational study in which multidisciplinary teams collaboratively created storyboards during a co-located session, we present recommendations for the facilitation of co-located collaborative storyboarding sessions for multidisciplinary teams and digital tool support for this type of group work.

Introduction

Creativity and collaboration are inevitable in user-centered design (UCD), which considers end-user needs from the beginning of the design and development process of a user interface (UI) for an interactive system. The teams that are responsible for UCD of interactive systems ideally involve members with various backgrounds. As considered in the ISO standard for usability, a multidisciplinary UCD team should include team members with expertise in, among others, human-computer interaction (HCI)/human factors/ergonomics, user interface/visual/product design and systems engineering/software engineering/programming, as well as end-users/stakeholder groups and application domain specialists/subject matter specialists (International

M. Haesen (✉) • D. Vanacken • K. Luyten • K. Coninx
Hasselt University – tUL – iMinds, Expertise Centre for Digital Media, Wetenschapspark 2, 3590 Diepenbeek, Belgium
e-mail: mieke.haesen@uhasselt.be; davy.vanacken@uhasselt.be; kris.luyten@uhasselt.be; karin.coninx@uhasselt.be

© Springer International Publishing Switzerland 2016
P. Markopoulos et al. (eds.), *Collaboration in Creative Design*,
DOI 10.1007/978-3-319-29155-0_10

Standards Organization 2010). Nowadays, several creative and participatory design techniques support collaboration within such multidisciplinary teams for the design and development of interactive systems. However, since the team members often have different expectations about the representations and transformations of end-user needs and concepts for a future software system, it is a challenging task to ensure that they all reach a common understanding in the first stages of a UCD process.

An additional challenge when collaborating within a multidisciplinary UCD team, is communication within the team without information loss. One missing link in most user-centered processes is an approach and accompanying tool to progress from informal design artifacts (e.g. scenarios) toward more structured and formal design artifacts (e.g. task models, abstract user interface designs) without losing any information. Existing tools and techniques often require specific knowledge about specialized notations or models, and thus exclude team members not familiar with these notations or models. Furthermore, functional information may be missing in informal design artifacts, while structured design artifacts may not always contain all non-functional information. We put forward storyboards as a comprehensible notation to overcome these shortcomings.

In the next section, we present a storyboarding notation that specifically considers UCD practices in multidisciplinary teams. This notation mainly focuses on collaborative storyboarding to facilitate multidisciplinary teams during different stages of the UCD process. Collaborative storyboarding activities also involve the contributions of team members with different skills, perspectives and goals in the creation of storyboards. We present an observational study that presents insights into important aspects of collaborative co-located storyboarding in multidisciplinary teams, which lead to recommendations for facilitating this type of storyboarding sessions. In addition, these recommendations can inform the design of digital storyboarding tools that support this type of group work.

Storyboards for User-Centered Design

The early design stages of a UCD process include a user needs analysis and generally result in several artifacts that contain the user needs, such as usability requirements documents (Redmond-Pyle and Moore 1995), scenarios that represent how a future system is used in a certain context (Carroll 2000) and personas concerning hypothetical archetypes of key users of the future system (Pruitt and Adlin 2006). These artifacts are written in natural language, usually have a narrative style and are typically created by team members with expertise in UCD or HCI, but not necessarily technical knowledge. Similar artifacts are used in more technical domains such as software engineering and agile development (Holtzblatt et al. 2004) (e.g. essential use cases and user stories), albeit in a different context.

Although several disciplines provide and use notations to describe user needs, these notations are not always suitable to pass information of the user needs to other members of the multidisciplinary team without misconceptions (Haesen et al. 2008). A wide interpretation of tasks and user needs analyses often confuses multidisciplinary team members (Lindgaard et al. 2006). The use of stories combined with sketches in the early stages of user-centered approaches, which is comparable to the use of storyboards, is considered as a powerful technique to reveal errors and to consider temporal and contextual information (Brown et al. 2008).

The professional use of storyboards originates from the film-industry and was introduced in several disciplines, such as advertisement and product design (van der Lelie 2006). For similar visualization purposes, storyboards are used in UCD approaches, where they can have different forms. Storyboards can visually express scenarios of use, or can represent the flow of interaction throughout the application to clarify interactivity in the early stages of UCD (Landay and Myers 1996). We concentrate on the first approach, which considers storyboards as a technique to complement scenarios, resulting in a visual depiction of how a user carries out a task using the system that is to be developed (Kantola and Jokela 2007; Preece et al. 2002).

In UCD, storyboards can be used as powerful artefacts to clarify user needs. In particular, storyboards can depict systems that are used in several contexts of use or on multiple devices. In earlier work, storyboards were used for the design of mobile systems (Sonja and Wally 2005), to provoke empathy in a design team (McQuaid et al. 2003), and to validate conceptual ideas of new interactive systems (Davidoff et al. 2007). In the next sections, we describe COMuICSer storyboards and how they can be specified and used in order to support multidisciplinary teams in UCD.

Definition of a COMuICSer Storyboard

COMuICSer is an acronym for *COllaborative MultIdisciplinary user-Centered Software engineering* and is pronounced as "co-mixer". The name also refers to comics, which have a similar representation as storyboards (Haesen et al. 2010; McCloud 1993). COMuICSer concerns a notation, and accompanying tool support, which are designed specifically to support multidisciplinary teams in UCD. We introduce the COMuICSer notation and COMuICSer tool support separately because a COMuICSer storyboard can be specified by simply using pencil and paper. However, in order to fully benefit from the advantages of COMuICSer storyboards, using the COMuICSer tool is recommended.

The *COMuICSer notation* concerns a storyboard that is defined as *a sequence of sketches of real-life situations, depicting users carrying out several activities by using devices in a certain context.*

Real-life situations, depicting the circumstances in which the future system will be used, are the main component of a COMuICSer storyboard. These situations explain realistic circumstances in which the system is or will be used by means of a scenario. The depictions of real-life situations in a storyboard show the end user needs to all members of a multidisciplinary team and may provoke empathy for the users among the team members.

In UCD, the focus is on the users from the start of a project. Consequently, the *users* have a prominent role in a COMuICSer storyboard. If available, personas can be linked to the storyboard. As stated in ISO 13407 (International Standards Organization 2010), which specifies human-centered design, not only the user should be considered, but also the *activities* carried out by the user, the technology or *devices* that are provided, and the *context* in which a system is used. All these elements can be depicted by the COMuICSer storyboarding notation.

An example of a simple COMuICSer storyboard is presented in the center of Fig. 1. This storyboard depicts a few hours of a journalist's working day. In the first scene, the journalist is working behind his desk. This is how he usually starts his day at work. However, very often he receives a phone call that notifies him about a certain incident in the neighborhood, for instance a car accident. Next, the journalist hurries to the place of the incident, where he takes notes about the incident on his personal device, which is depicted in the second scene. Afterwards, the journalist searches for a park bench and finalizes his article for the newspaper remotely using his laptop, as shown in the third scene.

Bridging the Early Stages of UCD Processes Using COMuICSer Tool Support

The creation of storyboards happens at the early stages of a UCD process, ideally after the observation or analysis of the user needs and the creation of informal design artefacts such as scenarios and personas. A storyboard can lead to user interface designs that carefully take into account the situation in which an interactive system will be used. The interrelationships between a COMuICSer storyboard and other artifacts are shown in Fig. 1. The large light blue arrow shows the general evolution of artifacts in the early stages of a UCD process. The small dark blue arrows indicate that the creation of a storyboard is an iterative process that also considers discussions, evaluations and adjustments within a multidisciplinary team. When informal design artefacts are available in the UCD process and formal artefacts need to be prepared in order to continue the design and prototyping of the interactive system, a COMuICSer storyboard can be used to bridge the gap between both types of artefacts.

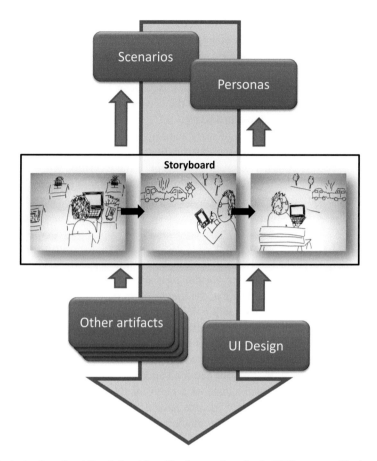

Fig. 1 A storyboard and its relationship with other artefacts in the UCD process. The large light blue arrow presents the general evolution of artifacts in the early stages of UCD, while the small dark blue arrows represent evaluations and iterations of artifacts

Several commercial tools, such as Comic Life,[1] Celtx,[2] ToonDoo,[3] Storyboard That[4] and Indigo Studio[5] support the creation of digital storyboard. These tools mainly focus on comics or storyboards in general, and rarely provide features that consider the use of storyboards in multidisciplinary UCD teams. Ozenc et al. address the need for tools that support refining rough designs and a scenario-

---

[1] http://www.comiclife.com – Comic Life, digital tool for the creation of comics.

[2] https://www.celtx.com – Celtx, digital tool for the creation of screenplay storyboards.

[3] http://www.toondoo.com – ToonDoo, digital tool for the creation of comics.

[4] http://www.storyboardthat.com – Storyboard That, digital tool for the creation of storyboards.

[5] http://www.infragistics.com/products/indigo-studio/storyboards – Indigo Studio, tool for UI prototyping and storyboarding.

driven process (Ozenc et al. 2010). The ActivityDesigner (Li and Landay 2008) tool allows storyboarding at the early stages of user interface design processes. In this tool, designers can extract activities from concrete scenarios, making it possible to include rich contextual information about everyday life as scenes. Based on the scenes, higher level structures and prototypes can be created. However, not all information is visually represented by the scenes and it is not specified which components need to be available in a scene.

Due to the limited availability of storyboarding tools for UCD, the COMuICSer notation is accompanied by tool support to facilitate UCD in multidisciplinary teams (Haesen et al. 2011, 2009). The proof of concept of the COMuICSer tool supports the creation and use of COMuICSer storyboards based on scenarios and personas, and passes contextual information to other artefacts in the UCD process (e.g. UI designs and their interaction sequences).

In practice, a COMuICSer storyboard can be created using pen and paper. However, the COMuICSer tool can be very useful to support the transitions between a COMuICSer storyboard and the other artifacts mentioned in Fig. 1. The following process of creating a COMuICSer storyboard is supported by the proof of concept of the COMuICSer tool (Fig. 2). First, the *scenario* can be written or loaded into the scenario textbox, e.g. by an interaction designer (Fig. 2-1). Next, this scenario can be split into a number of scenes. A sequence in the scenario textbox can be selected, followed by creating a new scene, which appears in the storyboard panel (Fig. 2-2). The selected sequence of the scenario is automatically added as a description to the scene. Now, the interaction designer can load an image and add a title. The image of the scene can be a scanned sketch or a photo of the user observations. As all scenes of the storyboard typically include *personas* and *devices*, this information can be annotated in the scenes. These annotations are made in a similar way to the photo tagging features on Facebook or Flickr. The storyboard in Fig. 1 shows a persona (e.g. Bart, journalist, 43 years old) in the three scenes, which is specified in the properties panel (Fig. 2-3). The device used in the first and third scene is a laptop, while a personal mobile device is used in the second scene. Highlighting or tagging personas and devices enriches the information contained by the storyboard and is useful to make the transition to other artifacts (Haesen et al. 2011, 2009). For instance, automatically constraining a UI design space according to the screen resolution of the selected device decreases the risk of information loss.

By carefully considering the situation of each scene, designers and developers can build an application corresponding to the contextual information, requirements and constraints contained in the storyboard. Interaction designers can use a storyboard to verify that the UI designs take into account all requirements. Using COMuICSer storyboards in UCSE processes increases the visibility of a project's requirements: the accessible notation allows all team members, including end users, to be involved in the UCSE process, and new team members can, for instance, explore the requirements of the project at a glance by looking at the storyboard (Haesen et al. 2011, 2009).

The COMuICSer notation is very accessible and suitable for usage in multidisciplinary teams. The current proof of concept of the COMuICSer tool supports

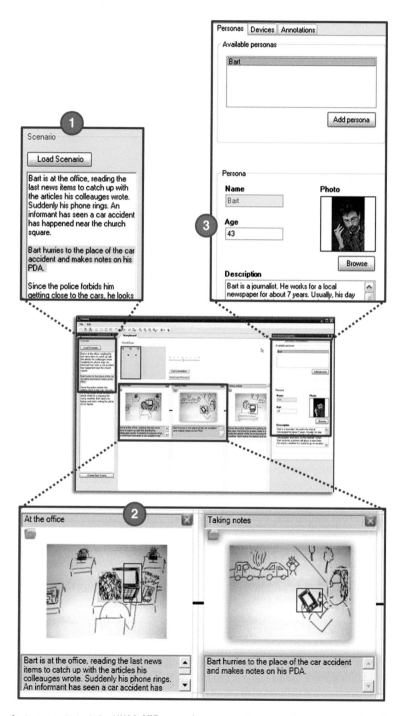

Fig. 2 A screenshot of the COMuICSer storyboarding tool. This tool supports storyboarding by connecting a storyboard with a scenario, personas and other annotations

an individual team member in the creation of a COMuICSer storyboard, which can be shared within a UCD team. To create COMuICSer storyboards during a collaborative co-located storyboarding session, further tool support is needed. Such a tool needs to accommodate co-located collaboration practices, where people gather around a shared surface (e.g. a table or whiteboard) to co-create storyboards in which the considerations of each team member is included. For this purpose, we investigate interactions that occur during collaborative co-located storyboarding.

Collaborative Storyboarding in Multidisciplinary Teams

COMuICSer storyboarding supports the involvement of multidisciplinary teams in the early stages of UCD. A storyboard can be created by one team member, who shares the storyboard with others in a meeting. However, by organizing collaborative storyboarding sessions, all team members and their different skills and perspectives can be included in the storyboard, which is valuable for later stages of UCD.

To get insights into how multidisciplinary teams collaborate during a storyboarding session, we set up a study in an environment where co-located teams can create storyboards using low-fidelity tools such as paper, color pens, scissors, glue, and post-it notes. These tools, although non-digital, provided all necessities to create COMuICSer storyboards.

Multidisciplinary Teams and Study Setup

Three teams of four people participated in the study, all experienced researchers with expertise in the design or development of interactive systems from a HCI and/or user interface design perspective. In order to be able to consider the multidisciplinary aspects of collaborative storyboarding, each participant was instructed to take on a particular role during the study. We used some of the crucial roles considered in UCD (International Standards Organization 2010): HCI specialist, UI designer, systems analyst, and stakeholder (end-user or application domain specialist). These roles were assigned based on the participants' skills and expertise. Results from a post-study questionnaire indicate that most participants felt "comfortable" to "very comfortable" in their role. Three participants felt "neutral" and nobody felt uncomfortable.

Assembling multidisciplinary teams in this way may yield some differences compared to actual teams that have been working together for some time. In real-life settings, one team member can have a combination of skills, which implies that the roles in a team are not always as easy to distinguish as in our three teams. However, in order to observe the influence of the different disciplines involved in a UCD team, we instructed our participants to take on a very specific role. The participants were thoroughly informed about their role and collaborated before in different contexts,

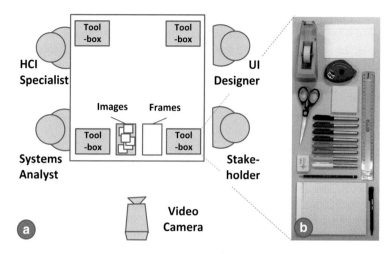

Fig. 3 Setup of the observational study: (**a**) each participant positioned at a different corner; (**b**) contents of the toolbox that was provided to each participant

so there were no social boundaries during the study. We therefore believe that our setup provides a sufficiently similar approach as real-life settings.

The storyboarding sessions were carried out on a regular table (160 by 160 cm), with each team member sitting at a corner (Fig. 3a). We provided a stack of A4 sheets and a box with images representing personas and items. Each participant also had a personal box of non-digital 'tools' (Fig. 3b).

Instructions for the task and a description of the participant's role were provided to each participant. We used personas and a scenario as identical starting points for the three storyboarding sessions, because these documents describe the use of a future software system and can be related to COMuICSer storyboards. First, each participant had 15 min to prepare the storyboarding session individually. They were asked to write down or sketch anything considered to be important, bearing in mind their specific role and goals. Next, the teams were asked to create a storyboard that represents the given scenario. Each team had 60 min.

A video camera recorded the sessions for later analysis, and two observers took notes. Upon completion of the storyboarding task, the participants filled in a questionnaire about their former experiences, and their findings regarding the storyboarding task and collaboration within a multidisciplinary team.

Case: Home Automation

The personas and scenario that were provided to the participants revolved around a home automation system to control the heating and lighting, which can assist a household in saving money on energy consumption. The system can be controlled by different family members (differing in age and technological aptitude), using different devices (e.g. touchscreen, laptop, smartphone). Although not explicitly mentioned, the scenario suggested that the system should take into account settings related to personal profiles and activities, that settings of profiles should be merged in some situations, and that the system should be able to detect people's presence in certain rooms.

Observations and Results

In this section, we present the findings of the study, based on the results of the post-study questionnaire and our observations. To complement the observations, Fig. 4 visualizes the physical activity on the table throughout the sessions. We automatically generated this visualization from the video recordings, based on the movements of participants. Because the participants were seated at a prearranged location around the table, we can roughly associate physical activity with particular participants. This approach obviously has its limits, as it for instance not considers participants reaching across the table, so we rely on our observations to interpret the physical activity correctly.

Individual Preparation

During the individual preparation, several participants highlighted phrases in the provided text. Each participant structured the information in a particular manner: a few used bulleted lists, while others represented it by means of graphical artifacts, ranging from diagrams to sketches.

In terms of content, the roles of the participants were clearly expressed in the artifacts they prepared (HCI specialists focused on relations between personas, devices and tasks, designers on UI designs and requirements, systems analysts on devices and their connections, and stakeholders on general requirements and the personas' needs). In all sessions, participants began to explain their prepared artifacts to the others once the collaboration started, but in two out of three sessions, not all members presented their preparation. Prepared artifacts were rarely included explicitly in the storyboard, but participants did use them during discussions.

Storyboarding Task

The approach to the storyboarding task differed in the three teams. Team A started by shortly discussing their strategy and decided to first depict the equipment and users in the different rooms of the house. Next, they started creating the first scene collaboratively, in a shared space in the middle of the table. Next, the team implicitly split in two to prepare other scenes. Awareness was maintained, since participants

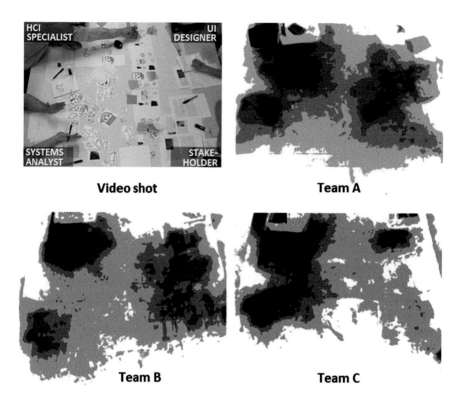

Fig. 4 A snapshot of a video recorded during a storyboarding session and a visualization of the participants' physical activity on the table throughout each of the three sessions, automatically generated from the videos (darker color means more frequent activity)

frequently switched between cooperating with their neighbor and cooperating with the entire team, and a lot of work was done in the middle of the table. The HCI specialist maintained the relationship between the storyboard and the scenario. For team A, Fig. 4 clearly shows the high degree of activity of the HCI specialist, the cooperation between neighbors, and cooperation with the entire team.

Team B first discussed the system based on the requirements mentioned by the stakeholder. After a discussion of approximately 15 min, in which some decisions regarding the system were made, the HCI specialist reminded the team of the storyboarding task and took the lead in creating scenes. The other team members were actively involved in the discussion. Once the HCI specialist started creating a new scene, the stakeholder and designer finalized the previous scene together. Fig. 4 shows the high activity corresponding to the leading role of team B's HCI specialist, as well as the stakeholder and designer collaborating to complete scenes.

Not unlike team B, team C first discussed the system based on the requirements presented by the stakeholder. This discussion lasted nearly 30 min before a first scene was created. While discussing the devices for the system, the available images were put in the middle of the table to debate the different options. Again, it was the

HCI specialist who reminded the team of the storyboard and who started creating scenes. Team C shows the least amount of cooperation in Fig. 4. The seemingly high activity in the systems analyst's region was actually caused by the HCI specialist, who was active in that region while creating scenes.

In the three storyboarding sessions, features that were implicitly described in the scenario led to a lot of discussion. Sometimes it was just one person who noticed a particular requirement, but in many cases, visually representing situations sparked discussions. Participants used the part of the table in front of them as personal workspace, and the available images were scattered across the middle or side of the table to give an overview, somewhat similar to the findings of Scott et al. (2004).

Resulting Storyboards

The resulting storyboards consisted of 7–10 scenes that represent personas and devices, and show the status of particular devices (e.g. a light that is switched on or off). Figure 5 shows how each team structured their storyboard and what materials they used. All teams used some of the available images to depict personas and devices in the storyboards. While teams B and C made extensive use of the images, team A also sketched a lot. Two storyboards contained text to indicate the location or general context of scenes. One team added post-it notes that would remind team members of particular features, difficulties and decisions.

Structuring the scenes of the storyboard was done in different ways. Team A created a visual representation of all rooms and their equipment, and consequently depicted the situation in different scenes for each room. Teams B and C put the scenes in a chronological order, based on the flow of events in the scenario. Extra scenes were inserted into the storyboard sequence when considered necessary. Scenes were labeled with numbers, and in team C, titles were added as well.

Multidisciplinary Team

The participants confirmed in the questionnaire that being part of a multidisciplinary team had a positive impact on the storyboarding session, since it resulted in a combination of different perspectives, ideas and considerations. HCI specialists rated their direct contribution to the storyboard highest on average. Most systems analysts, designers and stakeholders rated their direct contribution considerably lower: analysts and designers prepared artifacts that were used to a lesser extent during the session, while stakeholders were more verbally involved.

We can relate these ratings to the physical activity seen in Fig. 4 (e.g. team B's systems analyst and team C's designer and stakeholder ranked their direct contribution lowest) and to the observation that two of the HCI specialists took the lead for the creation of the storyboard. Furthermore, the HCI specialists of all the teams controlled the link with the scenario by reading it aloud or referring to it. Stakeholders and systems analysts rated their general influence on the storyboard

Fig. 5 Frames from the videos that were recorded during each session, showing the final storyboard of each team. The actual contents of the storyboards is of lesser importance, as we are mainly interested in the general storyboarding approach of the multidisciplinary teams, the materials that were used and the way the storyboards are structured: (**a**) Resulting storyboard of team A; (**b**) Resulting storyboard of team B; (**c**) Resulting storyboard of team C

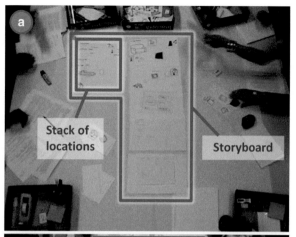

notably higher than their direct contribution. Frequent discussions about feasibility and costs of particular approaches, but less active contributions of these team members to the creation of the storyboard, account for this difference.

Collaborative Storyboarding: Recommendations

The observations of the teams indicate that findings of a study on a creative activity such as storyboarding are not easily generalizable. However, this study shows different aspects that are important for the organization of co-located collaborative storyboarding sessions for multidisciplinary teams in UCD.

Besides the recommendations, we also consider tool support. Several efforts are presented in literature toward the creation of digital tools that focus on individual or collaborative storyboarding as part of a software development process (Atasoy and Martens 2011; Haesen et al. 2011; Truong et al. 2006). These tools concentrate on the (re)use of storyboards, which is useful to support alterations and consistency checks with the system's requirements. Some tools, such as Coeno-Storyboard (Haller et al. 2005) and StoryCrate (Bartindale et al. 2012), assume that one person has the role of a coordinator and organizes artifacts on a timeline. However, none of the tools consider the respective contributions of team members with different skills, perspectives and goals. Because existing tools have demonstrated the benefits of digitizing storyboards for user-centered design and development purposes, we put forward a number of recommendations for a digital tool that supports collaborative creation of storyboards by a co-located multidisciplinary team.

Allow for Differences, Support Agreements

One of the first things to bear in mind is the individual preparation of a storyboarding session. Most designers, for instance, continuously accumulate graphical material, and they frequently use this material as a source of reference and inspiration (Atasoy and Martens 2011). Since paper is still a very ubiquitous medium, team members should not only be able to use digital artifacts, but also tangible artifacts such as paper documents. Prior studies indicate that designers still prefer pencil and paper early in the design process (Bailey et al. 2001).

In our study, the individual preparation resulted in many different artifacts, including device or task descriptions, UI designs, and requirements. Relations between artifacts were also considered during preparation. As the representation style and viewpoints differed greatly and the members of a multidisciplinary team are already accustomed to their specific tools and devices, we should not enforce one particular way of preparing artifacts. The accustomed tools and devices can be supported by allowing the use of personal devices, and by facilitating an easy exchange of data between personal devices and a shared device (e.g. an interactive table or a whiteboard).

Given the differences in the prepared artifacts and viewpoints, the team members had to come to an agreement on several occasions. All teams, for example, had to agree on the devices that would be used in the home automation system. Since involving all team members in the decision making process results in more comprehensive storyboards, active participation should be encouraged. On the other hand, we also observed that it can be beneficial to have someone take the lead, to make sure that sufficient progress is made, focus is maintained, and discussions are called to a halt at appropriate times. A digital tool can incorporate approaches that allow one or more team members to take the lead, while preventing more quiet users from being less involved. Such a tool can facilitate balanced decisions supported by the entire team. For instance, to enforce decisions supported by the entire team, a voting feature (Ryall et al. 2005) may require all users (or a quorum) to agree.

Facilitate Different Approaches in Structuring

Structuring the storyboard happened in two ways: a spatial arrangement, connecting the scenes to a particular location, or a temporal arrangement, organizing the scenes chronologically. Consequently, different approaches in structuring should be facilitated. Mapping scenes to a floor plan can provide insights regarding devices available at a certain location and the use of the system by certain people, while sorting the scenes according to a timeline shows the moments in which particular features of a system are used.

A digital tool should not be restricted to one particular arrangement and should allow teams to create (different alternatives of) scenes and connections between them freely. In Storify (Atasoy and Martens 2011), for instance, a team can add multiple alternatives per storyboard frame, with the purpose of discussing various user experiences. The Anecdote (Harada et al. 1996) tool, on the other hand, allows various design styles by providing several views of the design, including an outline view, timeline view, and scene view. Creating multiple alternatives of a scene and switching between multiple arrangements of scenes can offer different perspectives, but teams will not take advantage of such features if the actions are too time-consuming.

Maintain the Design Rationale

Visually representing the future home automation system stimulated the teams to discuss some unclear and challenging features. Despite the interesting discussions, almost none of those considerations or decisions were included in the storyboard.

Since the design rationale is often valuable for later stages of UCD, it is very important to capture this rationale one way or another.

Wahid et al. (Wahid et al. 2010) state the importance of presenting the rationale in a designer-digestible format based on their study on the relationship between imagery and design rationale. This designer-digestible format depends on factors such as the homogeneity of the team and the familiarity of the team with the problem.

When organizing co-located collaborative storyboarding sessions, it is recommended to capture the design rationale. A digital storyboarding tool can record the rationale and monitor all the artifacts. Furthermore, it can encourage team members to connect those artifacts to the storyboard (e.g. connect a designer's user interface sketches to a particular scene). SILK (Landay and Myers 1995) also enables, for instance, designers to examine, annotate and edit a complete history of the design. Maintaining the design rationale, in combination with the balanced participation of the different roles we mentioned earlier, may lessen the dissatisfaction some participants reported regarding the extent of their contribution, because their artifacts did not end up in the actual storyboard. To keep track of vocal discussions, audio or video annotations can be connected to the storyboard.

To monitor all storyboard artifacts, the tool can track their ownership or origin. Haller et al. (Haller et al. 2005) state the importance of clearly identifying who is manipulating each data object in Coeno-Storyboard. Similarly, Avila-Garcia et al. (Avila-Garcia et al. 2010) use a DiamondTouch (Dietz and Leigh 2001) tabletop to identify the input of up to four different users, because identifying, saving and tracking contributions made by team members can be relevant in a decision making scenario. To support easy logging and audit trail creation, identity-differentiating widgets (Schmidt et al. 2010) or lenses (Ryall et al. 2006) can be incorporated.

Favor Shared Over Personal Space

While preparing, participants each created a personal workspace. Within the boundaries of our observational study, privacy was never an issue when participants shared data. In a real-life setting, however, privacy might come into play from time to time (Shoemaker and Inkpen 2001), although further studies are required to investigate this aspect in the context of storyboarding in multidisciplinary teams.

During the cooperative storyboarding sessions, almost all work was done in the shared space between two participants or toward the middle of the table, even when multiple scenes were being created in parallel. Personal workspaces were still used sporadically, for actions such as writing on a post-it note or consulting the instructions or preparation. The sides of the table were mainly used for storage (e.g. toolboxes, available images, finished scenes). Since space is often at a premium, care has to be taken with personal workspaces or toolboxes taking up lots of space, leaving too little shared space to support a clear overview (as requested by participants in the questionnaire) and effective collaboration.

When asked about the participants' preference for a digital system, all participants favored a shared device such as an interactive tabletop, because it makes collaborating easier and it encourages involvement and discussion. They commented that physical interactions on a shared surface emphasize what is being done and make participants explicitly aware of the progress and contributions. Some participants, however, voiced concerns over the fluency of sketching and text entry on a digital system such as a tabletop. These concerns can be alleviated to some extent by incorporating additional input devices, such as a physical pen and keyboard, or by supporting paper.

Since the use of a large digital tabletop is not always feasible, this approach might not offer adequate space for all team members and their expected tasks. Integrating personal devices can reduce the problem of limited space, since they can act as personal workspaces. An object storage space can be provided to store unused physical objects, and the digital space can be extended by making space-demanding components zoomable.

Another solution is to extend the environment with additional displays. Ryall et al. (Ryall et al. 2004) state that for larger groups it might be necessary to add additional vertical displays for shared information. Avila-Garcia et al. (Avila-Garcia et al. 2010) also suggest the addition of one or more vertical displays. However, their goal is to accommodate passive team members, as one of the displays could show the interactions that are taking place on the tabletop. In our case, we want to avoid members being passive, and adding more displays may have a detrimental effect on balanced participation. About half of the participants also suggested including a personal device to consult preparations or take notes, with the ability to easily share items with others. A point of attention, however, is the possible decrease of mutual awareness and involvement when personal devices are being used extensively.

Conclusion

Design often is a collaborative activity that takes place in multidisciplinary teams. When it comes to UCD of interactive systems, the early design stages, in which informal design artifacts need to be translated into formal design artifacts, often cause difficulties and ambiguity. COMuICSer storyboards and the accompanying tool support can be used for detailing scenarios of use and connecting parts to informal artifacts needed at later stages of design. COMuICSer storyboards support the different backgrounds involved in a multidisciplinary team, and it is important that co-located collaborative storyboarding sessions adequately facilitate the needs of multidisciplinary teams.

We performed an observational study that includes an analysis of group interaction in multidisciplinary teams during co-located storyboarding sessions. Based on this study, we put forward a number of recommendations: allow for differences and support agreements, facilitate different approaches in structuring, maintain the design rationale, and favor shared over personal space. By taking into account these recommendations, multidisciplinary collaboration can be facilitated during co-located storyboarding sessions and digital storyboarding tools, including an extension of the COMuICSer tool, can be designed to engage all team members, while respecting individual contributions and creativity. An extended COMuICSer tool can for instance be provided using a multi-touch tabletop, by considering the aforementioned recommendations in combination with related collaborative tabletop design patterns (Remy et al. 2010; Vanacken 2012).

We strive for a balance in participation among members of a multidisciplinary team, because involvement of all members results in more complete storyboards that carefully take into account different aspects of an interactive system's context of use. The result of a storyboarding session should reflect all opinions and artifacts, also those of more reserved team members. Incorporating viewpoints of multiple disciplines remains a challenge and cannot be entirely delegated to a storyboarding tool. Therefore, the recommendations presented in this chapter should not only be taken into account for the design of digital storyboarding tools, but also when organizing co-located collaborative storyboarding sessions in multidisciplinary teams with the use of non-digital tools.

Acknowledgments We thank the participants of our observational study. The work presented in this chapter is supported by the IWT project AMASS++ (SBO-060051) and the EU FP7 project COnCEPT (610725).

Further Reading

The following sources are recommended for further reading on the topics presented in this chapter.

How talented do you have to be in drawing or sketching in order to create your own storyboards? The following literature explains how you can **easily sketch and draw anything you want in a storyboard**:

- *Understanding Comics: The Invisible Art*, Scott McCloud (1993)
- *The Back of the Napkin*, Dan Roam (2008)
- *See What I Mean: How to Use Comics to Communicate Ideas*, Kevin Cheng (2012)

In order to obtain tips and tricks to create storyboards that **clarify certain aspects of future systems in User Experience design or UCD processes**, you can read the following literature:

- *Sketching User Experiences: Getting the Design Right and the Right Design*, Bill Buxton (2007)
- *Draw Me a Storyboard: Incorporating Principles and Techniques of Comics to Ease Communication and Artefact Creation in User-Centred Design*, Mieke Haesen, Jan Meskens, Kris Luyten, Karin Coninx (2010)

References

Atasoy B, Martens J-B (2011) STORIFY: a tool to assist design teams in envisioning and discussing user experience. In: Proceedings of the 2011 conference extended abstracts on human factors in computing systems, CHI EA'11. ACM, New York, pp 2263–2268

Avila-Garcia MS, Trefethen AE, Brady M, Gleeson F (2010) Using interactive and multi-touch technology to support decision making in multidisciplinary team meetings. In: Proceedings of the 2010 IEEE 23rd international symposium on computer-based medical systems, CBMS'10. IEEE Computer Society, Washington, DC, pp 98–103

Bailey BP, Konstan JA, Carlis JV (2001) DEMAIS: designing multimedia applications with interactive storyboards. In: Proceedings of the ninth ACM international conference on Multimedia, MULTIMEDIA'01. ACM, New York, pp 241–250

Bartindale T, Sheikh A, Taylor N, Wright P, Olivier P (2012) StoryCrate: tabletop storyboarding for live film production. In: Proceedings of the 2012 ACM conference on human factors in computing systems, CHI'12. ACM, New York, pp 169–178

Brown J, Lindgaard G, Biddle R (2008) Stories, sketches, and lists: developers and interaction designers interacting through artefacts. In: Proceedings of the Agile'08. IEEE Computer Society, Washington, DC, pp 39–50

Buxton B (2007) Sketching user experiences: getting the design right and the right design. Morgan Kaufmann, Amsterdam

Carroll JM (2000) Making use: scenario-based design of human-computer interactions. MIT Press, Cambridge

Cheng K (2012) See what I mean: how to use comics to communicate ideas. Rosenfeld Media, Brooklyn

Davidoff S, Lee MK, Dey AK, Zimmerman J (2007) Rapidly exploring application design through speed dating. In: Proceedings of the 9th international conference on ubiquitous computing, UbiComp'07. Springer, Berlin, pp 429–446

Dietz P, Leigh D (2001) DiamondTouch: a multi-user touch technology. In: Proceedings of the 14th ACM symposium on user interface software and technology, UIST'01. ACM, New York, pp 219–226

Haesen M, Coninx K, Van den Bergh J, Luyten K (2008) MuiCSer: a process framework for multi-disciplinary user-centred software engineering processes. In: Proceedings of the second conference on human-centered software engineering, HCSE'08, and 7th international workshop on task models and diagrams, TAMODIA'08. Springer, Berlin, pp 150–165

Haesen M, Luyten K, Coninx K (2009) Get your requirements straight: storyboarding revisited. In: Proceedings of the 12th IFIP TC13 international conference on human-computer interaction, INTERACT'09. Springer, Berlin, pp 546–549

Haesen, M, Meskens J, Luyten K, Coninx K (2010) Draw me a storyboard: incorporating principles and techniques of comics to ease communication and artefact creation in user-centered design. In: 24th BCS conference on human computer interaction, HCI'10, British Computer Society. Swinton, pp 133–142

Haesen M, Van den Bergh J, Meskens J, Luyten K, Degrandsart S, Demeyer S, Coninx K (2011) Using storyboards to integrate models and informal design knowledge. In: Model-driven development of advanced user interfaces. Springer, Berlin, pp 87–106

Haller M, Billinghurst M, Leithinger D, Leitner J, Seifried T (2005) Coeno: enhancing face-to-face collaboration. In: Proceedings of the 2005 international conference on augmented tele-existence, ICAT'05. ACM, New York, pp 40–47

Harada K, Tanaka E, Ogawa R, Hara Y (1996) Anecdote: a multimedia storyboarding system with seamless authoring support. In: Proceedings of the fourth ACM international conference on multimedia, MULTIMEDIA'96. ACM, New York, pp 341–351

Holtzblatt K, Wendell JB, Wood S (2004) Rapid contextual design: a how-to guide to key techniques for user-centered design (interactive technologies). Morgan Kaufmann, San Francisco

International Standards Organization (2010) ISO 9241-210. Ergonomics of human-system interaction – Part 210: human-centred design for interactive systems

Kantola N, Jokela T (2007) SVSb: simple and visual storyboards: developing a visualisation method for depicting user scenarios. In: Proceedings of the 19th Australasian conference on computer-human interaction: entertaining user interfaces, OZCHI'07. ACM, New York, pp 49–56

Kathy R, Alan E, Katherine E, Clifton F, Meredith Ringel M, Chia S, Sam S, FD Vernier (2005) iDwidgets: parameterizing widgets by user identity. In: Proceedings of the 10th IFIP TC13 international conference on human-computer interaction, INTERACT'05. Berlin, Heidelberg, pp 1124–1128

Landay JA, Myers BA (1995) Interactive sketching for the early stages of user interface design. In: Proceedings of the SIGCHI conference on human factors in computing systems, CHI'95. ACM, New York, pp 43–50

Landay JA, Myers BA (1996) Sketching storyboards to illustrate interface behaviors. In: Conference companion on human factors in computing systems: common ground, CHI'96. ACM, New York, pp 193–194

Li Y, Landay JA (2008) Activity-based prototyping of ubicomp applications for long-lived, everyday human activities. In: Proceedings of the SIGCHI conference on human factors in computing systems, CHI'08. ACM, New York, pp 1303–1312

Lindgaard G, Dillon R, Trbovich P, White R, Fernandes G, Lundahl S, Pinnamaneni A (2006) User needs analysis and requirements engineering: theory and practice. Interact Comput 18(1):47–70

McCloud S (1993) Understanding comics: the invisible art. Tundra Publishing Ltd, New York

McQuaid HL, Goel A, McManus M (2003) When you can't talk to customers: using storyboards and narratives to elicit empathy for users. In: Proceedings of the 2003 international conference on designing pleasurable products and interfaces, DPPI'03. ACM, New York, pp 120–125

Ozenc FK, Kim M, Zimmerman J, Oney S, Myers B (2010) How to support designers in getting hold of the immaterial material of software. In: Proceedings of the 28th international conference on human factors in computing systems, CHI'10. ACM, New York, pp 2513–2522

Pedell S, Smith W (2005) Relating context to interface: an evaluation of picture scenarios. In: Proceedings of the 17th Australia conference on computer-human interaction: citizens online: considerations for today and the future, OZCHI'05, Computer-Human Interaction Special Interest Group (CHISIG) of Australia, 2005, pp 1–4

Preece J, Rogers Y, Sharp H (2002) Interaction design. Wiley, New York

Pruitt J, Adlin T (2006) The persona lifecycle: keeping people in mind throughout product design. Morgan Kaufmann, Amsterdam

Redmond-Pyle D, Moore A (1995) Graphical user interface design and evaluation. Prentice Hall, London

Remy C, Weiss M, Ziefle M, Borchers J (2010) A pattern language for interactive tabletops in collaborative workspaces. In: Proceedings of the 15th European conference on pattern languages of programs, EuroPLoP'10. ACM, New York

Roam D (2008) Back of the napkin: solving problems and selling ideas with pictures. Portfolio, New York

Ryall K, Forlines C, Shen C, Morris MR (2004) Exploring the effects of group size and table size on interactions with tabletop shared-display groupware. In: Proceedings of the 2004 ACM conference on computer supported cooperative work, CSCW'04. ACM, New York, pp 284–293

Ryall K, Esenther A, Forlines C, Shen C, Sam S, Morris MR, Everitt K, Vernier FD (2006) Identity-differentiating widgets for multiuser interactive surfaces. IEEE Comput Graph Appl 26(5): 56–64

Schmidt D, Chong MK, Gellersen H (2010) IdLenses: dynamic personal areas on shared surfaces. In: Proceedings of the 2010 ACM international conference on interactive tabletops and surfaces, ITS'10. ACM, New York, pp 131–134

Scott SD, Sheelagh C, Inkpen KM (2004) Territoriality in collaborative tabletop workspaces. In: Proceedings of the 2004 ACM conference on computer supported cooperative work, CSCW'04. ACM, New York, pp 294–303

Shoemaker GBD, Inkpen KM (2001) Single display privacyware: augmenting public displays with private information. In: Proceedings of the SIGCHI conference on human factors in computing systems, CHI'01. ACM, New York, pp 522–529

Truong KN, Hayes GR, Abowd GD (2006) Storyboarding: an empirical determination of best practices and effective guidelines. In: Proceedings of the 6th ACM conference on designing interactive systems, DIS'06. ACM, New York, pp 12–21

van der Lelie C (2006) The value of storyboards in the product design process. Pers Ubiquit Comput 10(2–3):159–162

Vanacken D (2012) Touch-based interaction and collaboration in walk-up-and-use and multi-user environments. PhD thesis, Hasselt University, Diepenbeek, Belgium

Wahid S, Branham SM, Scott McCrickard D, Harrison S (2010) Investigating the relationship between imagery and rationale in design. In: Proceedings of the 8th ACM conference on designing interactive systems, DIS'10. ACM, New York, pp 75–84

Co-Constructing New Concept Stories with Users

Derya Özçelik Buskermolen and Jacques Terken

Abstract One of the challenges for companies when developing concepts for new products, services or applications is whether or not the concepts will make sense to the user. And evidence that a concept will be valuable should preferably become available early in the design process. Involving users in the process of reflecting on new concepts makes sense because they are domain experts. However, in order to judge whether a concept will bring added value, users need to envision future contexts of use. We present the Co-Constructing Stories method, which aims to facilitate this envisioning process for users. In one-to-one sessions of less than an hour, first users are prompted by stories about the current context, helping them recollect relevant real life experiences for sensitization. Next, they are prompted through future scenarios to envision possible future experiences that may be enabled by the concept. In this paper we explain the method and discuss its background and relation to other methods. We introduce a case study in which the method was applied. Based on the insights gathered through this and similar case studies, we provide guidelines for designers who might be interested to use the method in the future.

Introduction

One of the central questions that companies need to answer when generating and selecting ideas for new products, services or applications is whether or not these ideas may ultimately lead to outcomes that will be adopted by the market. To answer this question, the relevant departments (the market or user research department and the design department) need to provide arguments that these products, services or applications will be valuable to people. It is commonly accepted that a sensible way to collect such arguments is by doing user research to understand the users and the contexts of use. From earlier research (Özçelik-Buskermolen et al. 2012) we found that real-life stories from users are considered particularly valuable by design

D. Özçelik Buskermolen (✉) • J. Terken
Department for Industrial Design, Eindhoven University of Technology, Eindhoven,
The Netherlands
e-mail: derya.ozcelik@gmail.com; j.m.b.terken@tue.nl

© Springer International Publishing Switzerland 2016
P. Markopoulos et al. (eds.), *Collaboration in Creative Design*,
DOI 10.1007/978-3-319-29155-0_11

233

teams. While existing methods such as Context-mapping (Sleeswijk-Visser et al. 2005) already provide rich information about the users and the contexts of use, we would like to go one step further and argue that it also makes sense to involve users in co-design or co-reflection sessions to contribute to the concept development or to reflect on the potential value of the ideas. Often, designers are rather sceptical about asking users to reflect on the value of new concepts, as the common assumption is that users are able to think just small steps ahead but are not able to reflect on the value of more disruptive concepts (see Verganti 2009). However, we argue that useful feedback may be collected from users in the early stages of the design process also for novel concepts, provided proper methods are used.

Combining the user stories and the co-reflection lines, we realized that designers are not just designing new products, services or applications. They are also creating a story telling why such a product, service or application would be valuable to people. The Co-Constructing Stories method that we present in this paper aims to bring together the real-life stories from users and the (implicit) story that the designers are creating, leading to an enriched concept story and ultimately a better outcome of the design process, with a higher probability of being successful in the market. The Co-Constructing Stories method is a design research method for early, formative evaluation of (interactive) design concepts mediating user experiences. The evaluation addresses the question of whether or not the new design concept is leading towards the right design. The method helps to get an understanding of whether and why people believe that the new concept will provide added value to their everyday life, and how the concept should be further developed so that the end product is likely to be regarded as valuable by the users. The method consists of two main phases: the first phase aims to elicit relevant real-life stories from people, and the second phase aims to elicit how people envision their future experiences as they will be mediated by the new concept. The method utilizes storytelling and participatory approaches to elicit in-depth user feedback and suggestions specific to the design concept. Stories are used by the designer to set the stage for dialogue and to present the concept, and by the user to communicate his past and anticipated future experiences.

In this chapter, we will first discuss our motivation to develop the method. We will give an overview of existing methods and techniques that utilize storytelling and/or participatory approaches. We will introduce the Co-Constructing Stories method and report a case where the method was applied. We conclude the chapter by providing guidelines for user experience designers who might be interested to use the method in the future.

Motivation

When designing, designers do not only create products and services. They also create a story explaining why this product or service is likely to be useful and valuable for people. The Co-Constructing Stories method is intended to collect information from users, enabling the designer to enrich the story and make it more convincing and credible, therewith providing input to the design process.

The development of the method was motivated by two observations. Firstly, our previous research pointed out that in the early phases of the design process, designers prefer to have user feedback that is contextualized and grounded in real-life situations (Özçelik-Buskermolen et al. 2012). Real-life stories of users are considered valuable by designers by virtue of being trustworthy, informative and inspiring. Secondly, designers need to envision the future context of use to understand how future use situations will be affected by the concept. Methods such as Endowed Props (Howard et al. 2002), Envisioning Use Workshops (Bijl-Brouwer et al. 2011) and Storify (Atasoy and Martens 2011, 2016) help designers to envision future use contexts and establish empathy with users. The Co-Constructing Stories Method offers designers the possibility to involve users in this process of creating such stories. When early concepts are presented to users to elicit feedback and suggestions, users should be helped to imagine themselves in their future contexts and come to a judgment whether and how the concept would bring added value to their life. With the Co-Constructing Stories method we aim to facilitate this envisioning process for the users by helping them revive their past experiences in a related context. We believe that reviving past experiences provides a ground for users upon which they can build their visions about the future and articulate their opinions about novel design concepts. Thus we use this mechanism as the back-bone of the Co-Constructing Stories method.

The Co-Constructing Stories Method

The Co-Constructing Stories Method is designed to be used during formative evaluations of a novel design concept, when the designer is questioning whether it is the right concept and how s/he can further develop it. The method is used during a co-located, one-to-one dialogue between a designer/researcher and the user.

The Set-Up of the Session

A Co-Construction Stories session consists of two phases: sensitization and envisioning (see Fig. 1). The sensitization phase helps participants to revive their past experiences, making the relevant use situations more concrete, so that in the envisioning phase they can better envision the future. The goals of the sensitization phase are similar to for instance the sensitization in the Context-mapping method (Sleeswijk-Visser et al. 2005), the main difference being the duration of the sensitization phase and the approach. In the Co-Constructing Stories method, the sensitization phase is part of the session and starts by a *sensitizing story* presented by the designer. It aims to set the stage for dialogue and introduces the context of

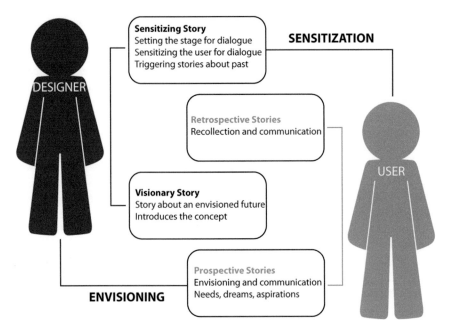

Fig. 1 The Protocol of the co-constructing stories method

interest. After the story ends, the designer asks the user whether s/he recognizes the story, why or why not, and invites him/her to continue the story by telling about his/her past experiences. Non-directive questions should encourage the user to tell a few stories about relevant past experiences. Prompt materials such as sketches of a relevant context of use, pictures and maps are made available to the user to help him/her organize his/her thoughts and communicate them to the designer. The sensitization phase should provide the designer with user stories revealing past experiences, enriching the designer's understanding of the characteristics of the current experience and context of use.

The second phase starts with the *visionary story* told by the designer, introducing the concept in an envisioned context. When the story ends, the designer elicits first impressions of the user about the concept by asking what the user liked and disliked in the story. Then, the designer asks the user to envision him/herself as the user of the concept. The designer invites the user to retell the stories that s/he told in the sensitizing phase by asking: what would this story be like if you would have had the concept back then? What would still be the same and what would be different? How would you feel about it? The user is given prompt materials, such as sketching templates, pictures, maps, etc., which help him/her to communicate the experiences s/he envisions. The designer facilitates this envisioning process with open-ended, non-directive questions. With these questions, the designer encourages the user to supplement the basic story about the concept with contents representing anticipated future experiences, based on the needs, dreams and aspirations of the user. The envisioning phase provides the designer with stories containing envisioned experiences that enable him/her to enrich the story about why the concept will be valuable to people.

Towards the end of the session, participants are invited to compare the current and future situations and to discuss positive and negative points of both situations. Explicitly asking for potentially negative points helps to identify pitfalls. The whole session lasts about 45 min to an hour.

Preparation for the Session

Before starting to prepare materials for the session, the designer should first get knowledgeable about his design space. He should make explicit who the target users are and what benefits the concept is expected to provide to these users. Also, the designer should make explicit what the relevant use situations for the concept are. This results in the initial concept story (or stories). Next, the designer starts preparing the materials needed for the session: two storyboards and associated prompt materials.

Preparing the Storyboards

The Co-Constructing Stories method utilizes two stories presented in storyboards; a method such as Storyply (Atasoy and Martens 2016), which makes a similar distinction in two stories to be created (one consisting of Identifying the Conflict and Envisioning the Consequences and one consisting of Making your Proposal and Envisioning the Improvement) may be used for preparing such storyboards.

The first story is called the sensitizing story and aims (1) to set the stage for dialogue, (2) to introduce the context of interest, and (3) to elicit past experiences of the participant concerning that context. It is an open-ended story which renders realistic characters, situations and experiences that the participant can identify with. When the story ends, the participant is asked whether he has been in such a situation and how the story continued in his case. The participant is encouraged to tell concrete accounts of his past experiences.

The second storyboard represents the visionary story. It is a possible continuation of the first story, including the new concept. It communicates the designer's vision about how the new concept will be used by people. It is important that the participant understands the story and empathizes with the presented situation, but s/he should not be overwhelmed by it and should still feel encouraged to be critical.

Designers may use different tools and methods to create the storyboards, depending on their individual skills and the time they have. Designers may choose making quick sketches, making and editing photos, or creating video prototypes. Whatever medium they use, we recommend designers to keep in mind that the stories should help the participants to empathize with the story. They should avoid any detail that might hinder empathy, both in the story line and while preparing the representations. If they do not have a strong reason to do otherwise, such as using formerly created materials from their archives, we recommend designers to aim for sketchy looking representations; as such representations are known to be effective in enabling empathy and inviting participants to contribute. We recommend designers to present the storyboards on a screen, for example through a simple flipchart animation with voice over, so that the designer and the participant can watch the story together and the participant is not put under pressure while he is reading the storyboard and the designer is waiting for him to finish. Moreover, looking together at a screen puts the participant and the designer in equal positions.

Preparing the Prompt Materials

Prompt materials are artefacts, such as templates for sketching, pictures, maps and mock-ups (Fig. 2). They are used in the sensitization phase to help participants remember certain contexts and their past experiences in these contexts, and in the visionary phase to help participants envision their future contexts and how they would use the new concept in these contexts. Furthermore, the prompt materials help participants to organize their thoughts. Some users find it convenient to use them for clarifying and illustrating their stories. Also, they create a point of attention for gazing, so that the participant is not forced to gaze at the designer all the time. They may also help to trigger the imagination of people. They should be appropriate both for the design case and also the intended user group. Prompt materials should be prepared carefully so that they do not create a threshold for the participant to interact with and they should support him/her in recollecting past experiences and envisioning future experiences.

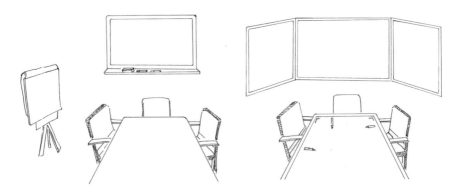

Fig. 2 Templates used as prompt materials while investigating the value of an interactive meeting room (having a smart screen and multi-touch table) for collaborative design sessions. *Left* for sensitizing phase (current situation). *Right* for envisioning phase (future situation)

Choosing the Setting

Before conducting the session the designer should also decide where he will meet with the participants. He should create a relaxing atmosphere, so that the participant feels comfortable. The designer should also decide how he will capture the sessions. We recommend recording the sessions with video camera or microphone so that the conversation is not interrupted by the need to take notes. If a camera is used, the visual and gestural information are also captured besides the verbal information.

Facilitation of the Session

The protocol of the method and the stories and prompt materials assist the designer in facilitating the sessions. For example, participants are invited to envision the future after they have been helped to explore their past experiences in a related context. Initiating each phase by showing a story to the participants instead of starting questioning immediately not only helps participants to understand the concept and the context of interest but also works as an ice-breaker. However, to sustain the dialogue and obtain relevant insights and user feedback, further facilitation is needed.

In the sensitization phase the aim of the designer is to elicit past experiences of the participant in a context that is relevant to the design case. In this phase, the designer should first help the participant to situate him/herself in a particular situation that s/he has experienced. The designer can help the participant to concentrate on specific situations by asking about a particular happening, such as the first or the last time s/he has experienced the situation, or about the time s/he felt the most frustrated or happy. For example, if the focus of the sensitization phase is on collaborative design meetings (because the designer is working on a design concept that can support collaborative design meetings), the designer may ask the participant to tell him/her about the most recent collaborative design meeting. Then the designer should help the participant to explain the situation vividly, asking questions such as where s/he was, what the context was, with whom s/he was, what s/he was doing, whether s/he enjoyed it, why s/he was so frustrated or happy, etc.

In the envisioning phase the aim of the designer is to obtain feedback of the participant on the design concept with an underlying reasoning of the feedback. In this phase the designer should help the participant imagine how s/he would use the new concept in his/her everyday life. The designer should start the dialogue by asking first impressions of the participant about the concept. This helps the participant to think and talk about the concept. Then the designer asks the participant to think about the situations that s/he told in the sensitization phase and re-tell those stories by thinking how these situations would be different (for good or bad) if s/he would be using the concept. The designer can help the participant by reminding him/her about the details of the stories that s/he told in the sensitization phase. The designer should guide the dialogue with non-directive questioning and open-ended questions and pave the way for the participant to talk more. He should also listen to the participant attentively without prejudice and without judging and/or framing the opinions of the participant.

Towards the end of the session, the participant is asked to compare his/her lived experiences (the stories s/he told in the sensitization phase) with his/her envisioned future experiences (the stories s/he told in the envisioning phase). The designer asks the participant what s/he likes and/or is concerned about in each situation. The designer may help the participant by reminding him/her of the things s/he said like "you also talked about [x] while you were telling this story to me". To be able to do this, the designer should be listening to the participants attentively all through the session. At the end of the session the designer debriefs the participant and thanks him for his contributions.

We recommend designers to carry out these sessions themselves as our previous experience showed that designers learn more from users if they are the ones who are in dialogue with users. Carrying out sessions themselves, designers can steer the dialogue to the direction of their interest and interpret the results more easily. However, carrying out such sessions require some skills and practice. Pilot sessions may help the designers to practice their interviewing skills. Designers may also choose to prepare the sessions in collaboration with a researcher. Such a researcher can facilitate the sessions while the designer observes.

Analyzing and Using the Results

The Co-Constructing Stories method elicits stories of past and anticipated future experiences. These stories can be used in different ways depending on the case and the needs and interests of the designer. One possibility is to use the raw materials for inspiration during the further design process (Sanders and Stappers 2013). In this case the designer immerses himself in the stories told by the users to gain empathy and get inspired. A second possibility is to use the concrete feedback and suggestions of the users to give direction to the design decisions. A third possibility is to analyze the stories and use the re-occurring elements, such as people, places, objects, activities and motivations, as an inspiration to create overarching user stories. The storytelling game of Doug Lipman (Quesenbery and Brooks 2011) and the Storyply tool by Atasoy and Martens (2011, 2016) may help the designers in creating overarching stories. Finally, the designer can use the stories told by the users to learn what matters to them: as the stories are about past (real) and future (envisioned) experiences, they typically provide information about how the concept might give rise to valuable experiences. A structured method to extract this information is to apply qualitative data analysis such as thematic analysis. Guidelines for conducting thematic analysis can be found in Boyatzis 1998; Braun and Clarke 2006; Richards 2005; and Taylor-Powell and Renner 2003. Thematic analysis requires considerable time, and not all designers may want/need to conduct such a thorough analysis. In all cases, the stories told by the users should enable the designer to enrich the concept story.

Relation to Other Tools and Methods

The Co-Constructing Stories method builds on previous work on scenarios, story-telling and participatory design. Therefore, it has some elements that also appear in existing tools and techniques; however, why and how these elements are brought together is unique to the method.

Like Fictional Inquiry (Dindler and Iversen 2007) and Storytelling Group (Kankainen et al. 2012), the method aims to elicit visions of people about the future. However, different from these methods the Co-Constructing Stories method does not inquire any dream about the future but the anticipated future of the user him/herself as it will be mediated by the design concept. The Co-Constructing Stories Method helps people to focus on their own lives and imagine how they would experience the new concept. That is why, the Co-Constructing Stories Method is used in individual sessions, in contrast to the above mentioned methods which are used in group settings.

Similar to Generative Techniques (Sanders 2000), Contextmapping (Sleeswijk-Visser et al. 2005) and Co-reflection (Tomico and Garcia 2011) the Co-Constructing Stories method also uses past experiences of users to trigger them to envision the future. However, in the Co-Constructing Stories method the past experiences and future visions of the user are strongly connected. The user relies on concrete accounts of his/her past experiences as a ground to build his visions about what his everyday life would look like if he would use the design concept. While the above mentioned methods generate information about the context of users and their latent needs, the Co-Contructing Stories Method elicits feedback on the design concept that is supported by contextual information.

In the sensitization phase the method elicits accounts of users' past experiences and while doing so it uses a dialogue which is similar to Explicitation Interviewing (Light 2006); however, in addition to this technique the method also uses scenarios and prompt materials to facilitate the dialogue. We believe that asking users to make comparisons between different views may lead to fruitful elaborations, like in Focus Troupe (Salvador and Howells 1998) and Contravision (Mancini et al. 2010). Unlike these methods, we ask participants to compare their past experiences with their envisioned experiences and to elaborate which aspects they would favour in each experience. Finally, in comparison to many other methods listed above, the co-constructing method is expected to take less time and effort, i.e., up to 45 min to an hour for a single session. Finding participants for qualitative research is usually challenging, and asking for an hour of their time, instead of half a day or more, might make it easier to convince people to participate.

Co-Constructing Stories Case Study: Enhancing
the Experience of Waiting for a Train on a Railroad Platform

The Co-Constructing Stories method was used in a design case in which the designer
was developing design concepts that aim to enhance the travelling experience of
railroad passengers while waiting for a train on a railroad platform. The designer
received the assignment from a railroad company in the Netherlands. The company
observed that people do not spread evenly on the platform but tend to wait in certain
areas (usually closer to the entry to the platform), which causes certain carriages
of the train to be overcrowded while other train carriages are rather empty. The
company wanted to guide people to spread more evenly over the platform while
waiting for an upcoming train, so that they would distribute more evenly over the
train carriages, leading to a more pleasant train travel. The designer came up with
the idea of informing the people waiting on the platform about how full the carriages
of the upcoming train will be by means of light panels placed next to information
boards. A dark panel means the carriage stopping at that part of the railroad platform
will be rather full, a half lit panel means the carriage will be half full and a totally
lit panel indicates an empty carriage (Fig. 3).

Fig. 3 The idea of informing people about how full/ empty the train carriage is through the use of
light panels

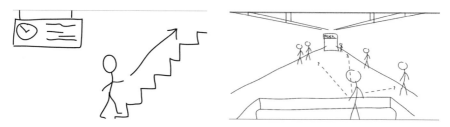

Fig. 4 Stills from the sensitizing story

The designer conducted Co-Constructing Stories sessions with train passengers in order to get to know the current behaviour of people in train stations and how people perceive and might use the new system that was proposed. The designer talked with 13 participants in one-to-one meetings.

The meetings started with the sensitization phase. After being welcomed, the participants were invited to watch the sensitizing story with the designer. The sensitizing story was about a train passenger who is walking to the train station after a tiring day. He was hoping to be able to sit in the train and have a rest during travel. The story ended while the main character was climbing the stairs towards the platform. The story was shown to the participants through a video prepared as a flipchart animation with a voice-over (Fig. 4).

After the story ended, the participants were asked whether they recognized the situation and what might happen when the main character in the story is waiting for a train on the platform. They were asked to share the accounts of their real-life experiences with waiting for a train on a railroad platform. The designer prepared a wooden mock-up of a railroad platform. On one side of the mock-up some light bulbs were installed to represent the light panels. The other side did not have any light bulbs and represented a generic railroad platform. The designer used the latter part of the mock-up for the sensitization phase and asked the participants to use the mock-up while they were telling their experiences. The designer elicited two to three distinct stories per participant within 15–20 min.

In the envisioning phase, the participants were shown the visionary story, which presented the concept, in the same manner. In this story, the main character enters the railroad platform and notices the light panels next to the information boards. He remembers that new light panels have been installed in the station informing about how crowded the carriage of the upcoming train will be. He realizes that the light panels near to the stairs are dark, meaning that the carriages here will be full. Then he realizes that the light panels at the end of the station are lit up totally and he starts walking to the end of the platform (Fig. 5). After the story ended, the participants were asked what they liked and disliked about the story. Then, they were asked to pick one of the stories they told in the sensitization phase and to imagine how the story would unfold if the light panels had been available back then. They were told that the light panels could provide any information they desired and could behave any way they want. The designer turned the mock-up so that the side with light

Fig. 5 Stills from the visionary story

Fig. 6 A participant using the mock-up while he is telling his story

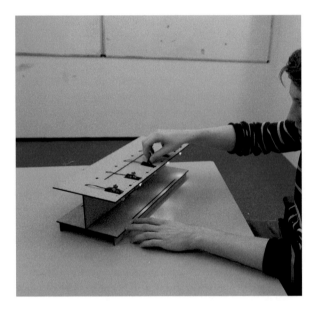

bulbs pointed towards the participants, and asked the participants to use the mock-up while they were telling their stories (Fig. 6).

The designer videotaped the sessions. He transcribed the conversations with participants and carried out extensive qualitative analysis.

After he carried out the study, we asked the designer to reflect upon his experiences with the method: how he experienced the method regarding different aspects such as getting prepared for the sessions, the sessions themselves and the results he obtained from the sessions.

The designer told that the preparation for the sessions did not take much of his time. What took most time was building the mock-up and putting light bulbs in the mock-up. Once the mock-up was ready, he quickly built the stories. He used hand-made rough sketches in his storyboards and in the visionary story he used photos of the mock-up. He advised designers not to spend much time on visualizing the stories. He said: "It is mainly about communicating your ideas and what you want to talk about and not about how it looks. Actually it would be better if it is not such a good drawing so that it gives room for participants to make their own interpretation of what you are presenting".

The designer said that the protocol of the method – knowing that the first phase is about past experiences and the second phase is about envisioning the use of the concept – helped him and also the participants to keep the conversations focused. He said: "First presenting the sensitizing story to the participants helped to get them going, get them into the story, talking about themselves and how they behave. If I had introduced the concept right away they might have been more hesitant about it". The protocol also helped the designer to repeat the same structure with all the participants, making results comparable. Furthermore, he pointed out that different elements he used in the sessions – stories and the wooden mock-up – raised the interest of the participants towards the sessions. He thought that participants enjoyed the sessions as some sessions lasted longer than planned while participants did not seem to mind.

The designer said that the feedback of the participants pointed out things that he either was not aware of or he was aware of but underestimated their importance. He found the feedback and suggestions of the participants very inspirational and he also took some design decisions based on the feedback and suggestions. For example, the feedback pointed out that people are quite positive about being informed about how full or empty the carriages of the train are. However, knowing only whether the carriage is rather full or empty is not enough for people. They would like to know how many seats are available and whether they are first or second class seats. Furthermore, they also expected that the information should not only take into account how full the carriages of the upcoming train are but should also take into account how many people are waiting for the train on the platform. The designer added that the stories that people told were also useful in creating awareness about the user group and the use context among other stakeholders like the railroad company. He used the stories of participants during his discussions with other stakeholders. The stories helped to create a common vision that was shared by all stakeholders.

Practical Guidelines for Applying the Co-Constructing Stories Method

Based on our experiences with the method and also the experiences of other designers we prepared the following guidelines for designers who may want to use the method in the future.

Before the Preparation

- Before starting to prepare the materials for the method, get knowledgeable about your user group and the use context you are designing for.
- Make your aim for the study and what you expect to learn from users explicit.

The Preparation

- While preparing the scenarios, keep in mind that users may comment on any detail you put in the scenarios, thus avoid details unless user feedback on these details is welcome.
- Prepare the sensitizing story such that users can empathize with the story and be drawn into it. Incorporating the known traits and attitudes of the user group and the general emotions associated with the context helps users to empathize with the stories.
- Make sure that the sensitizing story does not depict your view on the current context of use but raise an interest towards it so that when the story ends you can ask the participant to tell you his past experiences in that context. For example, if you are interested in the context of a design meeting, the sensitizing story can depict how the main character gets prepared for the meeting and ends when the main character walks into the meeting room.
- Prepare the visionary story such that the users can understand the concept. Pay attention that the story is not overwhelming and does not try to convince the users that the proposed concept is necessarily the optimal one.
- Choose a proper medium for presenting your stories. We recommend preparing low-fidelity storyboards and presenting them as a flip chart animation with voice over so that the users can comfortably watch the stories instead of trying to read accompanying text.
- Prepare prompt materials such that it is not hard for users to work with them. Starting from available materials can be easier for people than sketching their view on past or imagined situations from scratch.

The Sensitization Phase

- After your participant reads/watches the first storyboard you created, ask him if he recognizes the situation and which aspects in the story make the situation recognizable for him.
- Elicit concrete real-life experiences. Make him concentrate on specific situations by asking about the last time he experienced such a situation or the first time, or about when he felt the most frustrated or the happiest.
- Help your participants explain the situation to you vividly, by asking questions such as where he was, what the context was, with whom he was, what he was doing, why he was so frustrated, why he was happy (and some other details that you might be interested to learn), etc.
- Elicit more than one experience. The first experience your participant remembers may not be the most interesting one, as he is also getting used to the process. In addition, talking about one situation may make him remember further situations which might be more interesting for you.

- Ask your participants the things he liked and disliked regarding each situation. Elicit his emotions and the underlying reasons for the emotions.
- Note the experiences that your participants told you about and also the things he said. You may need this information to refer to in the envisioning phase and at the end of the session while comparing the past experiences with envisioned ones. Writing down keywords as mnemonics for the experiences may offload your memory, but avoid interrupting the conversation by taking extensive notes.

The Envisioning Phase

- After your participant watched the second storyboard, ask him what he thinks about it. Is the story recognizable to him? What does he think about the concept? What does he like about the concept and what not?
- Ask your participants to imagine what the situation would look like if he had had the concept in the situations he told about in the sensitization phase. Ask how things would be different (for good or bad).
- If needed, help the user to remember the details of the story that he told in the sensitization phase.
- Listen to the user attentively, with minimum preconception, without prejudice and without judging and/or framing the opinions of the participant.
- Repeat the situation for every single situation he told you about during the first phase.
- Note the situations that your participants told you and also the things he said. You will need this information while comparing the past and envisioned experiences. If needed, again write down keywords as mnemonics for the experiences, but avoid interrupting the conversation by taking extensive notes.
- Ask your participant to compare his past and envisioned experiences. Ask him what he appreciates in each situation. What are the points he is concerned about or does not like in each situation? What would be the added value of either situation over the other? What are the down sides of each situation if compared to the other? Overall which situation he would prefer and why? Or maybe in which kind of situations would he prefer to have the concept and in which situations he would see no value?
- If the user produced sketches, you can put the past and envisioned situations next to each other to facilitate the discussion of the past and envisioned situations, as they are placeholders for the stories that your participants told you. If no such materials were produced, you may use your notes to help your participants. You may remind him of the things he said like "you also talked about [x] while you were telling this story to me."
- End the session by thanking your participant.

Conclusions

In this chapter, we presented the Co-Constructing Stories method, which aims to elicit visions of users about the added value of design concepts in their future lives. We shared our motivation for developing such a method and described the proposed procedure. We discussed the method in relation to other design research methods and techniques and shared a design case where the method was applied. According to our experience there are certain advantages of the method in comparison to existing methods. First of all, the method elicits concrete past experiences of users. These experiences are valuable information for the design process in their own right; however, the method uses these experiences to help the user envision how the design concept would fit in his life in the future, by connecting the envisioning activity to concrete contexts of use. Such a view helps the user to evaluate the value of the concept in the concrete, specific settings of his personal life and provide deep, personal, grounded feedback instead of having to make generalizations. Secondly, the method elicits stories from people. Stories have two main advantages. They help the designer to gain empathy with the user. Moreover, they are easily remembered and communicated, thus they establish a shared vision among the members of the design team. Finally, the proposed method takes relatively little time, while at the same time providing deep understanding about the users. Especially the sensitization phase of the method is rather efficient as participants can talk about two to three concrete experiences in 20 min and reveal several anecdotes.

The method also embodies a challenge. It has been argued that the outcomes of design research methods that are based on dialogue depend to a considerable extent on the skills of the facilitator. This argument is also relevant to the current method. However, skilled facilitators cannot generate good results if they are not supported by proper methods. We argue that the procedure of the Co-Constructing Stories method has the potential to assist designers in facilitating the sessions with users. For example, the sensitizing story triggers participants to talk about their past experiences, and making these experiences explicit helps participants to envision the future. In conclusion, based on our experience with the method we believe that it helps to elicit in-depth feedback that is specific to the concept, in relatively little time.

References

Atasoy B, Martens J-B (2011) Crafting user experiences by incorporating dramaturgical techniques of storytelling. In: Procedings of the second conference on creativity and innovation in design. ACM, New York, pp 91–102. doi:10.1145/2079216.2079230
Atasoy B, Martens JBOS (2016) STORYPLY: designing for user experiences using storycraft. In: Markopoulos P, Martens JB, Malins J, Coninx K, Liapis A (eds) Collaboration in creative design. Methods and tools. Springer, New York

Bijl-Brouwer M, Boess S, Harkema C (2011). What do we know about product use? A technique to share use-related knowledge in design teams. Presented at the IASDR 2011 4th international congress of international association of societies of design research, Delft, The Netherlands. Retrieved from http://doc.utwente.nl/82017/

Boyatzis RE (1998) Transforming qualitative information: thematic analysis and code development. Sage, Thousand Oaks

Braun V, Clarke V (2006) Using thematic analysis in psychology. Qual Res Psychol 3(2):77–101. doi:10.1191/1478088706qp063oa

Dindler C, Iversen OS (2007) Fictional inquiry—design collaboration in a shared narrative space. CoDesign 3(4):213–234. doi:10.1080/15710880701500187

Howard S, Carroll J, Murphy J, Peck J (2002). Using "endowed props" in scenario-based design. In: Proceedings of the second Nordic conference on human-computer interaction. ACM, New York, pp 1–10. doi:10.1145/572020.572022

Kankainen A, Vaajakallio K, Kantola V, Mattelmäki T (2012) Storytelling group – a co-design method for service design. Behav Inform Technol 31(3):221–230. doi:10.1080/0144929X.2011.563794

Light A (2006) Adding method to meaning: a technique for exploring peoples' experience with technology. Behav Inform Technol 25(2):175–187. doi:10.1080/01449290500331172

Mancini C, Rogers Y, Bandara AK, Coe T, Jedrzejczyk L, Joinson AN, Nuseibeh B (2010) Contravision: exploring users' reactions to futuristic technology. In Proceedings of the SIGCHI conference on human factors in computing systems. ACM, New York, pp 153–162. doi:10.1145/1753326.1753350

Özçelik-Buskermolen D, Terken J, Eggen B (2012) Informing user experience design about users: insights from practice. In: CHI'12 extended abstracts on human factors in computing systems (CHI EA'12). ACM, New York, pp 1757–1762

Quesenbery W, Brooks K (2011) Storytelling for user experience. Rosenfeld Media, New York

Richards L (2005) Handling qualitative data: a practical guide. Sage, London/Thousand Oaks

Salvador T, Howells K (1998) Focus troupe: using drama to create common context for new product concept end-user evaluations. In: CHI 98 conference summary on human factors in computing systems. ACM, New York, pp 251–252. doi:10.1145/286498.286734

Sanders EBN (2000) Generative tools for codesigning. In: Scrivener SAR, Ball LJ, Woodcock A (eds) Collaborative design. Springer, London, pp 3–12

Sanders EBN, Stappers PJ (2013) Convivial toolbox: generative research for the front end of design. BIS Publishers, Amsterdam

Sleeswijk-Visser F, Stappers PJ, van der Lugt R, Sanders EB-N (2005) Contextmapping: experiences from practice. CoDesign 1(2):119–149. doi:10.1080/15710880500135987

Taylor-Powell E, Renner M (2003) Analyzing qualitative data. University of Wisconsin – extension, cooperative extension, Madison

Tomico O, Garcia I (2011). Designers and stakeholders defining design opportunities "in situ" through co-reflection. In Participatory innovation conference. Presented at the participatory innovation conference, Sønderborg, Denmark, pp 58–64

Verganti R (2009) Design-driven innovation. Changing the rules of competition by radically innovating what things mean. Harvard Business Press, Boston

idAnimate – Supporting Conceptual Design with Animation-Sketching

Javier Quevedo-Fernández and Jean-Bernard Martens

Abstract Creating animations is a complex activity that often requires an expert, especially if results need to be obtained under time pressure. As animations are potentially relevant in many different contexts, it is interesting to allow more people to use them for communicating ideas about time-varying phenomena. Multi-touch devices create opportunities to redesign existing applications and user interfaces, and new classes of animation authoring tools that use gestural interaction have therefore started to appear. Most of them focus on specific applications, such as cartoon and puppet animation. This chapter presents idAnimate, a low-fidelity general-purpose animation authoring system for sketching animations on multi-touch devices. With idAnimate, the user can manipulate objects with natural hand gestures while the system records the trajectories and transformations, using them to build animations.

Introduction

During the early stages of the design process designers need to make their ideas explicit so that they can be visualized, shared and discussed, both within the design team and with external stakeholders. Creating adequate visualizations can be cumbersome and time-consuming, though, especially when the design concepts contain dynamic elements such as user interactions, or time-varying behavior. Existing prototyping tools such as Axure (2015) or InVision (2015) require a substantial amount of effort and level of commitment from their user, while static sketches can be too limiting in the information they can convey. idAnimate helps designers bridge this gap by allowing them to rapidly create animated sketches in a simple and intuitive way, enabling them to quickly show their concepts, even those that include dynamic aspects. Thanks to idAnimate, designers can reduce the amount of effort required to iterate in the early stages of design by quickly externalizing, sharing and discussing their ideas without the need for expensive

J. Quevedo-Fernández • J.-B. Martens (✉)
Department of Industrial Design, Eindhoven University of Technology, Eindhoven, Noord-Brabant, The Netherlands
e-mail: j.b.o.s.martens@tue.nl

© Springer International Publishing Switzerland 2016
P. Markopoulos et al. (eds.), *Collaboration in Creative Design*,
DOI 10.1007/978-3-319-29155-0_12

prototyping. This chapter presents idAnimate, discusses it in relation to existing design tools, and guides the reader through an example to illustrate how it may be utilized in a design scenario.

Overview

idAnimate is an application for iPad® devices that helps designers sketch interactive products and services (idAnimate – An animation sketching tool for iPad; Quevedo-Fernández and Martens 2013). Designers can describe the behavior of products and services, as well as how users interact with them, through animations.

Figure 1 shows how idAnimate can be used to complement existing design tools in the early stages of the design process. The effort required to build design artifacts grows as the complexity of the artifacts and the tools used to create them increase in successive phases of the design process. idAnimate introduces a design activity in between sketching on paper and building wireframes and mockups, where the designer can describe with little effort the concept, its behavior and its dynamics by creating animated sketches. Simplifying such a phase can possibly trigger designers to carry out more iterations in the available time before committing to a particular solution.

In a nutshell, idAnimate is designed to complement paper sketches by providing designers with the ability to introduce time-varying events. Since these animated visualizations convey additional information, it makes it easier for stakeholders to be involved effectively (Tversky and Bauer Morrison 2002), discussing the dynamics of the concepts, collecting feedback and rapidly modifying the ideas. Thus, idAnimate is especially intended for:

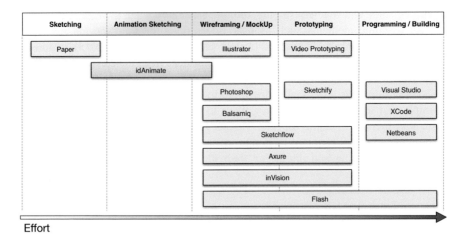

Fig. 1 idAnimate in relation to existing design tools and the phase of the design process

- increasing the information conveyed in the design artifacts exchanged at a conceptual stage;
- increasing the awareness and shared understanding of design proposals;
- supporting participatory design in workshops and brainstorms;
- allowing designers to more easily communicate and discuss early-stage concepts with a broad range of stakeholders such as marketing people and end users;
- collecting insights and feedback already at a stage where no high-fidelity prototypes are available yet;
- involving end-users early in the design process by providing them with representations of the design proposals that are easily accessible and interpretable.

Background

In the early stages of the design process when ideas are vague and imprecise, designers usually work with paper sketches to explore their imagination and articulate their ideas (Purcell and Gero 1998). The reason behind this is that sketches are easy, fast and cheap to create, while they provide a very flexible medium for expression (Buxton 2007). In general, sketches are used for:

- Exploring and expanding the space of alternative solutions.
- Communicating design concepts.

Discussing and Refining the Core Ideas Behind the Concepts with a Team of Stakeholders

Not surprisingly, sketches are not ideal for describing highly dynamic concepts. This is due to the fact that most of the behavior and time-related aspects that are to be conveyed are either left implicit, or roughly described through arrows and annotations (See Fig. 2). This implies that the understanding of the information transported within the sketch heavily relies on the imagination of the interpreter (Stacey et al. 1999). Thus, sketches can give rise to misunderstandings and misconceptions, especially when they are used to express highly dynamic concepts.

Theories about creativity and design support the idea that the thoughts evoked by reflecting on the visual artifacts that are created during the creative process determine for a large part the quality of the outcome of a design process (Schon 1986). In essence, the materials that are used during the creative activity enable, but also limit, the creative capabilities of the practitioner. Consequently, working with static visualizations may not always help the designer to foresee unexpected events, or more in general, enable the designer to adequately explore the space of

Fig. 2 Description of a
web-based registration form
using a static sketch

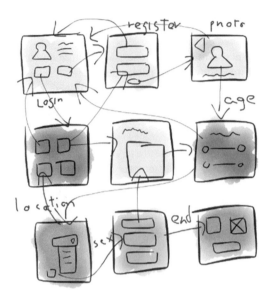

solutions. Thus, using static sketches to explore the space of solutions of highly dynamic concepts may lead to incomplete design solutions, while using them to communicate behavior and time-dependent ideas may lead to misunderstandings. As a result sketches are normally only used for a limited time during the design process, generally only at early stages. Using high-fidelity prototypes and mockup tools can resolve most of the aforementioned issues. However, creating such artifacts is generally time consuming and expensive (Buxton 2007). As a result, high-fidelity prototypes are mostly used at later stages of the design process, when a commitment to a particular solution has already been made.

Creating or modifying high-fidelity prototypes usually requires a set of skills that is present only in a subset of the members in the design team, which limits the possibilities to provide input by other members of the team, as they do not possess the technical skills needed to modify or alter the prototype. As a result, design teams increasingly find themselves in a situation where there is a need for tools that allow them to go beyond the expressive capabilities of paper sketches, i.e., by augmenting them. The cost and time involved in creating such new forms of visualizations can however not increase significantly, as the threshold for using them within early stages of the design process would otherwise be exceeded.

We propose that creating animations in a way that resembles sketching could be an interesting approach towards more effectively exploring and communicating the dynamic aspects of concepts. In order to investigate and test this idea more concretely, idAnimate was designed and built.

Related Work: Computer-Supported Animation Authoring Systems

Animation authoring is a tedious and complex activity that requires extensive time to learn and master. The goal of idAnimate is to make this process easier and more accessible. In a similar fashion, multiple computer-supported animation systems have been developed over the years. In order to understand the current state-of-the art of computer-supported animation systems, we hereby present a selection of related systems.

Over the course of the years there have been many different approaches to supporting the creation of animations with the help of computers. In 1969 Ronald M. Baecker presented Genesys, the first computerized animation system (Baecker 1969). Genesys allowed the animator to draw an object, and dynamically change the position, orientation and shape of the visual objects with the help of a stylus. Since Genesys, many additional animation techniques have been introduced.

Using the flipbook metaphor, GIF animations grew popular during the 1990s, partly because they could be easily distributed digitally. In the mid 1990s Macromedia presented Flash (Curtis 2000) and Director, which popularized the key-frame technique, in which the animator creates two frames and the system interpolates in between. Flash is probably the most widely used animation-authoring system today. However, while being obviously very useful in the hands of skilled animators, Flash does not explicitly address novice (or infrequent) users. Learning Flash is not simple, creating animations takes a substantial amount of time, and generally speaking, the quality of the outcome is unnecessarily high for the (early) phase of the design process that idAnimate is meant to support(Ozcelik-Buskermolen and Terken 2012).

Some newer systems have focused on developing techniques that feel more natural and intuitive to non-experienced users. Two specific examples are: (a) articulating 3D figure animations based on 2D figure sketches (Davis et al. 2003) and (b) sketching the motion of a character (Thorne et al. 2004). While non-expert animators can use such tools, they offer interaction mechanisms that are tuned to a specific application, i.e., articulating the motion of a character, and are therefore not necessarily suitable for general-purpose animation.

More interesting in view of general-purpose animating is the motion-by-example technique described by Moscovich (2004). This technique, in which the user can drag an object around the screen while the system records the location trajectory as a function of time, has been widely adopted in animation tools that target novice animators, such as K-Sketch (Davis et al. 2008), Sketch-n-Stretch (Sohn and Choy 2012), or Sketchify (Obrenovic Ž and Martens 2011). Sketchify is primarily intended for quickly creating prototypes with real-time input from a diversity of sensors, and therefore its interface is simply too sophisticated and complex for rapidly sketching animations.

The idAnimate tool shares similarities with K-Sketch and Sketch-n-Stretch, as it extends the motion-by-example technique, which tracks an object's position, to also include orientation and size. The extension to multi-touch devices (specifically, the iPad) can potentially speed up an animation process, as the user can transform multiple properties at a time instead of consecutively (See details in The Animation Technique section). An additional advantage of multi-touch devices is that multiple users can potentially cooperate on a single animation, although the full potential of this idea is only likely to emerge on multi-touch surfaces that are larger in size than the iPad that we used in the current case study.

Ceylant and Capin developed a multi-touch interface for animating 3D Meshes (Duygu and Tolga 2009), that produced a sense of movement in objects by deforming and reshaping visual objects through the fingers, which is similar to the multi-touch technique by Takayama and Irigarashi for manipulating the shapes of 2D characters (Takayama and Igarashi 2007). While both techniques share some features with our own prototype, mostly due to the common use of a multi-touch interface, they are mainly intended for character animation, and are hence fairly specialized. To the best of our knowledge, the most similar existing tools to idAnimate are Toontastic (Russell 2010) and Photopuppet HD (Cooke 2012), which use multi-touch gestures for creating animations. The latter tool is mostly intended for cartoon animation in a professional studio, as it offers many sophisticated and dedicated features, while Toontastic has a very specific application domain, i.e., storytelling for children.

idAnimate

Design Goals and Non-functional Requirements

idAnimate is an animation authoring tool designed to support creativity, discovery and innovation, and its design was inspired by the design guidelines formulated by Shneiderman and Resnicks (Shneiderman 2000; Resnicks et al. 2005). These guidelines assisted us in mapping the general objective of sketch-like animations into more specific design goals that can also be verified in subsequent validation studies.

Low Threshold

In order to make the process of creating animations easily accessible to non-experienced animators, the threshold for using the tool should be extremely low. This means that novice users, including those using the tool for the first time, should be able to create simple animations with minimal instruction.

Speed

The tool should allow users to rapidly create animations. Ideally, the cost of creating an animation should be similar to that of creating static sketches, so that they can just as easily be discarded and replaced by alternatives.

Support Exploration

The tool should allow users to try out different alternatives and examples, and should make it easy to roll back to past situations.

Flexibility and Wide Walls

The tool should not limit its users to a predefined set of patterns or prescribed scenarios. The tool should ideally allow the majority of users to describe most ideas that they can think of, within the reasonable boundaries of what animations can describe.

Accessibility

The tool should be available for as many people as possible, and in as many situations as possible.

Simplicity

Besides having a low threshold, the tool should feel simple and intuitive to use after a prolonged period of time. Instead of providing an extensive set of highly configurable features, the tool should provide only those that are strictly necessary to achieve the desired flexible outcome.

Requirements

This section describes the functional and non-functional requirements of the idAnimate tool. These requirements were collected through interviews and workshops with active designers.

Functional Requirements

- 1. Visual Object Creation – The user should be able to create visual objects, which are images that can be used in an animation, and that are either created specifically for the purpose or retrieved from an existing repository. These can include:

  - 1.1. Sketches – The user should be able to make simple sketches within the animation tool itself.
  - 1.2. Camera/Pictures – Photographs of real objects or paper sketches are an efficient and natural way to enter visual material. This can be achieved by making pictures with the device camera or retrieving existing pictures from the camera roll.
  - 1.3. Internet Search – It should be possible to search image archives on the Internet, and to easily import selected images into the application.
  - 1.4. Library of Objects – The tool should provide access to libraries (or catalogues) of saved and reusable objects. The tool should include a collection of preinstalled library objects, while the user should be able to download and install additional libraries of objects, called Library Packs. The user should be able to create custom libraries of objects, for instance starting from collections of images available in his/her desktop computer (custom Packs).
  - 1.5. Import from external tools – It should be possible to exchange visual objects with external tools such as Papers (Paper by FiftyThree). More specifically, idAnimate should provide mechanisms to seamlessly integrate external sketching tools for creating visual objects.

- 2. Animations. The user should be able to animate the visual objects that are part of an animation.

  - 2.1. Object transformations – Changes in Position, Size and Orientation over time of visual objects should be easy to create, store and modify.
  - 2.2. Object Appearances – Visual objects may have one or more visual Appearances. A visual appearance is how an object looks at a particular time. For instance, an object that represents a light bulb may have two visual appearances, one for when turned off, and one for when turned on.

- 3. Storyboards – The user should be able to create animated Storyboards. Animated Storyboards are composed of multiple animated sketches, each of which includes a representative image, a textual explanation, and a sequence number.
- 4. Sketchy look and feel – The animations produced by idAnimate should have a sketchy look-and-feel, i.e., express the fact that they are temporary in nature and purposely unpolished. This should for instance be reflected in the sketching tool incorporated within idAnimate.

- 5. Sharing – The user should be able to share projects created within idAnimate via multiple mechanisms. This includes sharing by email and social networks such as Facebook.

  – 5.1. Project Files – User should be able to share project files (in idAnimate format), which can in turn be opened for viewing and/or editing on other iPad devices (similarly to sharing a word document).
  – 5.2. Movies – idAnimate should be able to convert its animations into movies with a standard, well-recognized format. Users should be able to export and share such movies. Movies, as opposed to project files, cannot be modified, but they can be included in a Website, PowerPoint Presentation, etc.
  – 5.3. idAnimate.net Public Gallery – Users might be interested in sharing selected projects with other idAnimate users (e.g., on the idAnimate.net online gallery), supporting the development of a community of users.

- 6. Learning – idAnimate should possibly include within-app functionality to assist the user in learning how to use the tool most advantageously.

  – 6.1. Examples – The tool should provide access to a collection of examples that illustrate different features.
  – 6.2. Video Tutorials – The tool should provide access to video tutorials that show users how to use available features.

Design Rationale

Tablet devices were selected as the platform to host the animation tool.

Sketching and collecting images is at the heart of animating, which led us to consider alternative devices to the Personal Computer. In particular, we focused on those devices that had stylus or finger input methods, and with an embedded camera to take photographs of the environment.

Personal computers, including laptops, can compromise accessibility due to their size and form factor, while tablet devices on the other hand are lighter and more portable. Furthermore, tablet devices can be very personal and intimate, while they can also easily be shared, and can even support simultaneous interactions. Multi-touch also offers interesting opportunities for reducing the complexity and improving usability of the task of animating.

Tool Organization

The documents produced with the tool are called Animated Sketches. Animated Sketches contain one or more visual elements called Animated Objects.

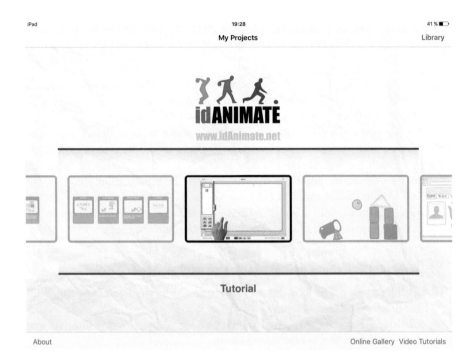

Fig. 3 idAnimate project selection window

Fig. 4 On the *left* the idAnimate sketch editor. On the *right*, the animated storyboard editor

IdAnimate supports Animated Sketches as well as Animated Storyboards (see Fig. 3). Animated Storyboards are composed of a number of Animated Sketches. The tool includes a project browser (see Fig. 4) where the user can create, import, export, or share the animations (as either video or project file).

It is possible to import visual elements from different sources into the application, including attachments in emails, images resulting from internet searches (Bing, Google or Flickr images), photographs taken with the built-in camera (if the device

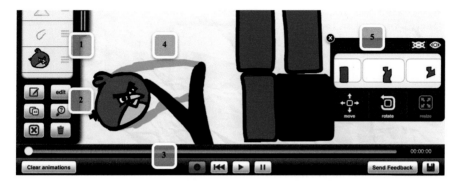

Fig. 5 idAnimate being used to animate a game concept (Angry Birds). *Mark 1* shows the list of objects, *mark 2* shows the possible actions on an object, *mark 3* the record and playback controls, *mark 4* the canvas and *mark 5* the object inspector

has one), or images already available in the users library. The tool also includes a simple sketching application that can be used to create new visual objects, or to modify existing ones.

The Animation Editor is show in in Fig. 5. Mark 1 shows the list of objects that are contained in the current animation. By modifying the order of objects in the list, they move in front or behind each other. Mark 2 shows the controls for adding a new object, editing the images associated with an object, duplicating, inspecting, deleting or clearing the animations of the selected object. Mark 3 shows the animation controls including the timeline and the play, pause, record and rewind buttons. Mark 4 shows the canvas where the objects are displayed and where the user interacts to define the animations. Mark 5 shows the inspector, which allows users: (1) to change between the different images that an object can display, (2) to manage the geometrical properties of an object that can be transformed, and (3) to specify when an object is visible or invisible.

Transformation by Example, the Animation Technique

The transformation-by-example technique is an extension of Moscovich's motion-by-example technique for multi-touch devices. In essence, it allows users to freely manipulate objects using a multi-touch surface while recording the changes in position, size and orientation over time.

Animated Sketches are composed of one or more visual elements called Animated Objects. An Animated Object is a region of varying position, size and orientation that can at any instance display one of a set of possible images, where most often the image reflects the state of the object concerned. This way an object can be moved, rotated and scaled while its visual appearance changes.

Fig. 6 Transforming by example a visual object. The user rotates, pinches and pans the object while the system records the changes, and is able to replay them at a later stage

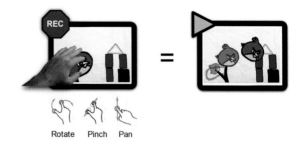

Rotate   Pinch   Pan

Transforming Through Gestures

Multi-touch gestures are the key to intuitively defining the transformations on Animated Objects. Commonly accepted multi-touch gestures are generally perceived as highly intuitive thanks to the direct manipulation of objects through the tip of the fingers, which effectively maps the finger movements to the geometric parameters that are being controlled by the system.

Since our technique complies with common practice, it should be perceived as easy to learn and easy to use. As stated before, next to being engaging and inviting, multi-touch gestures are potentially faster, as three different transformations can be applied simultaneously.

In order to animate an object, the user has to select it from the list of available objects in a scene. Using two fingers, a user can translate (move), rotate and scale simultaneously the selected object in a single action (see Fig. 6). While recording, all of the transformations performed on the device are registered and time-stamped relatively to the animation start time. When the user initiates the record action, a 3 s countdown is displayed to give him/her time to prepare the position of the hand and fingers.

In order to provide control over small objects, due to the limited size of the tablet display (9.7 inches in diagonal), the user does not necessarily need to touch the actual (selected) object, but instead he/she can interact by touching anywhere on the device's screen. Thus, when an object is selected, the touch interactions on the entire screen are mapped to the properties of the selected object. This way the user avoids blocking his view on the selected object, and is able to more precisely define the animations. However, the consequence of this design choice is that it is currently not possible to animate two or more objects simultaneously, which is something that will probably need to be reconsidered when migrating to large-sized multi-touch displays that are operated by several users at a time.

Synchronization and Concatenation

Complex animations are accomplished by adding new object transformations to existing animations, effectively concatenating transformations. In order to coordinate the animations of different objects, we use the record-while-playback technique. The user can replay a previously created animation, and meanwhile new transformations can be applied on other objects. These additional transformations are recorded, time-stamped and synchronized with the existing animation.

If a user animates an object that contained previous animations, all existing object transformations from the moment that he/she starts to interact are overwritten. Therefore the user can easily and intuitively refine or redo part of the animation of an object without compromising earlier parts.

States

The user can dynamically change the image associated with a visual object. In order to change states, the animation can be replayed, and during the playback, the user can select the desired state from a menu, and consequently the image to be displayed. The changes in the state are recorded and time-stamped with the rest of the animation. An alternative way is to change the states not while recording, but while the animation is stopped, which allows for more time and precision. Specifically, users can navigate to a particular point on the timeline and select the desired state to be displayed from that time on.

Switching states allows to change the visual appearance of a region, for instance changing the color of a light from red to green, the text displayed on a screen, or the different expressions of a character. Flipbook animations can easily be constructed in this way.

Supporting the Early Stages of the Design Process with idAnimate

To illustrate how *idAnimate* can be used for the aforementioned purposes, we will follow an example utilizing it to generate design solutions for an exemplary design brief.

A Practical Example: A Smartphone Payment System for Gas Stations

Let us assume that a concept about a gas station payment system for mobile devices has to be communicated, and suppose the concept includes the following features:

- payments are conducted using a mobile device (such as a smartphone);
- payment initialization relies on proximity, i.e., bringing the mobile device close to the payment artifact;

- the selection of the type of product (type of gas), amount to fill (volume or money) is accomplished on the smartphone;
- payments need to be confirmed on both the gas pumping device, and the smartphone.

Preparation for Illustrating with idAnimate

Collecting Material

Animations and storyboards are generally composed of three core elements: (1) the Place or setting where the situation occurs, (2) the Object(s) or props involved, and (3) the Actor(s) who carry out the interactions (usually, as some form of dialogue with technical systems).

idAnimate users are recommended to start by creating a collection of images related to the concept to be illustrated. This can be done in multiple ways: by sketching on the built-in sketchpad, by preparing collections of PNG images on a separate computer, by taking pictures with the iPads® built-in camera, or by pulling images from internet sources.

In our particular example we combine a specific sketching application for the iPad® (Paper by FiftyThree) with idAnimate in order to create the visual elements.

Places

Places constitute the setting and context for the product, the user and the interactions. While it is an optional element in an animation, situating the interaction in a specific place usually helps to better understand how and why things happen (see Fig. 7).

Object(s)

Props are the objects that have relevance in the story; mainly those that the actor will interact with.

Fig. 7 A sketchy looking gas station

Fig. 8 On the *left*, a Hand (actor) interacting with a Mobile Device (prop). On the *right*, a Gas Station Pump Screen

For our particular example the selected props are a smartphone, the gas hoax and the screen of the gas pump, as well as a collection of user interface elements intended for the application on the mobile device (see Fig. 8).

Actor(s)

The actor interacts with the environment and the props, providing them with an essential role in the animation. The actors will carry out the interactions with the objects, triggering the product behaviors and responses. In our example the actor is a hand of the user of the smartphone device (see Fig. 8).

Setting up the Scene

Casting the Objects

The first step to set up a scene is to cast the elements previously created. Since the images of our example were sketched with a different application, we will bring them into idAnimate by exporting and importing them to/from the device's gallery of images, which is shared across the two applications.

In order to do so, we select the "create a new object" inside idAnimate, and import the desired sketch into it by selecting the import button.

Placing the Objects

On the left side of the screen we can find the object selector (See left side of Fig. 9), which allows the user to choose which object to move, scale or rotate at any specific time on the digital paper. Doing this repeatedly for every object allows the user to define the initial placement of all the elements in the animation. Additionally, the user can define the layered order of the objects by swapping their position inside the object selector.

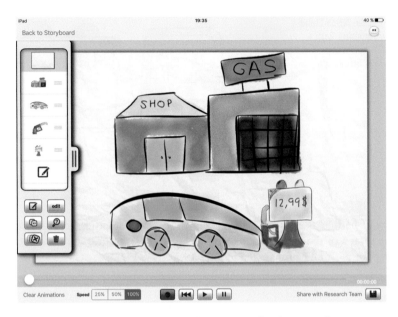

Fig. 9 Animation editor of idAnimate. Exploring concepts for the gas station payment system within idAnimate

Defining the Motions

Everything is prepared to start defining how the action develops, which is achieved by animating the objects. The approach to follow is simple: First we select the desired object to animate from the object selector, and then we tap on the Record button. A countdown will be displayed, giving the user time to prepare for acting out the motion. Whatever movements and transformation are carried out on the selected object will be recorded as part of the animation until the user decides to stop recording. This procedure can be repeated for each of the objects in the animation, making it possible to record new motions while others are being replayed, allowing the user to synchronize the movement of different objects.

In our example, the interaction starts by bringing the mobile phone close to the gas pump screen. When both elements are sufficiently close, the screen of the phone shows the interface for selecting the product and amount to refill, while the gas pump turns on an orange light (see the defining visual appearances of objects ahead for more details). Once we have recorded the motion of the device, we can start defining how the actor interacts with the user interface, selecting the type of gas and the product. Having done this, we can act out the gas hoax to show what happens while the gas is being pumped into the car's deposit, until completion.

Fig. 10 Example of
sketching multiple visual
appearances of a character

Fig. 11 The object inspector
allows the user to select the
visual appearance to be
displayed from a particular
moment in time onwards

Defining Multiple Visual Appearances

Objects may have multiple visual appearances (see Fig. 10), i.e. multiple images
that can represent their visual state. Think of visual appearances as different outfits,
which can be changed for instance to show two stages of a light bulb (On or off), or
two different facial expressions of a character. As shown in Fig. 9, the sketch editor
helps the user create these distinct appearances in a way that resembles using an
onionskin notebook.

   In the animation editor, the user can proceed to a particular moment in time and
select the desired visual appearance to display from that moment on. This selection
is done using the object inspector as shown in Fig. 11. Users may hide and show
objects during the course of an animation by switching between visual appearances
with content and visual appearances that are empty.

Fig. 12 A storyboard showing the different steps of the gas pump scenario in detail

Creating Alternative Scenarios

Once we have an initial animation it is easy to make small variations to show alternative scenarios or use cases. What happens when the user pulls the hoax before the gas has been fully loaded? What happens if the tank is full earlier than expected? How does the system display errors or react to different circumstances? To show this, users can duplicate a project and then rapidly make the appropriate changes to it.

Storyboards

idAnimate's storyboards (see Fig. 12) are sequences of animations with textual captions. Storyboards can be used to illustrate a story with multiple scenarios, or to show a particular element in more detail. In our specific example, the screen can become too cluttered when introducing all the elements at the same time. We can improve this by separating the story into four different animations. The first one shows the car arriving at the gas station, and the user placing the gas hoax inside the car's deposit, and the gas pump screen changing state. The second and third storyboards show the interaction between the user and the displays of the smartphone and the gas pump to select the amount of gas to fill, and to confirm the payment. The last animation shows the car leaving the gas station.

Similarly to animations, storyboards can be duplicated to create modified scenarios.

Sharing and Discussing Animations

Once the animations or storyboards have been created, they can be shared with team members. Other members cannot only watch them, but can also propose modifications to the ideas, quickly creating and sharing alternatives of the concept or scenario.

In addition, it is possible to export movie clips to embed them in a Powerpoint presentation, or share them on Facebook.

Conclusion

Tools for sketching animations can help designers narrow the gap between sketching and prototyping. We have shown how the idAnimate tool can be utilized at an early phase to support designers to be creative, and enable them to communicate dynamic ideas to other designers or external stakeholders (such as end-users) to collect feedback and input for the next design iteration. In order to do so, designers can use idAnimate to rapidly sketch user interfaces, scenarios and storyboards to generate design concepts and discuss in an iterative way its implications in the resulting user experience.

Where to Find Additional Information

A research version of *idAnimate* can be downloaded for free from its website (http://www.idanimate.net), and from the Apple® App Store™. To do so, you can use the link on the site, or simply search for *"idAnimate"* in the Apple® App Store™ with your iPad® device. The idanimate.net website also includes a series of video tutorials to help users learn the basic features of the tool, as well as the more advanced functionalities.

References

Axure (2015) Axure prototyping tool. http://www.axure.com. Accessed July 2015

Baecker RM (1969) Picture-driven animation. In: Proceedings of the May 14–16, 1969, spring joint computer conference – AFIPS'69, ACM, New York, 273–288. doi:10.1145/1476793.1476838

Buxton W (2007) Sketching user experiences: getting the design right and the right design, 1st edn. Morgan Kaufmann, Amsterdam/Boston. ISBN 000–0123740371

Cooke M (2012) PhotoPuppet HD iOS application. In: Apple App Store. https://itunes.apple.com/us/app/photopuppet-hd/id421738553?mt=8. Accessed June 2013

Curtis H (2000) Flash Web Design: the art of motion graphics. New Riders Publishing, Indianapolis. ISBN 978–0735708969

Davis J, Agrawala M, Chuang E et al (2003) A sketching interface for articulated figure animation. In: Proceedings of the 2003 ACM SIGGRAPH/Eurographics symposium on computer animation. Eurographics Association Aire-la-Ville, Switzerland, Switzerland © 2003, pp 320–328

Davis RC, Colwell B, Landay JA (2008) K-Sketch: a "Kinetic" sketch pad for novice animators. In: Conference on human factors in computing systems, 2008. Copyright 2008 ACM 978-1-60558-011-1/08/04

Duygu C, Tolga Ç (2009) A multi-touch interface for 3D mesh animation. Bilkent University, http://citeseerx.ist.psu.edu/viewdoc/summary?doi=10.1.1.369.9740

idAnimate – An animation sketching tool for iPad. http://idanimate.net. Accessed July 2015

InVision (2015) InVision App. http://www.invisionapp.com. Accessed July 2015

Moscovich J, Hughes JF (2004) Animation sketching: an approach to accessible animation. Technical Report CS 04–03. Computer Science Department, Brown University, Providence

Obrenovic Ž, Martens J-B (2011) Sketching interactive systems with sketchify. ACM Trans Comput Hum Interact18(1), Article 4 (May 2011):38. doi:10.1145/1959022.1959026http://doi.acm.org/10.1145/1959022.1959026

Ozcelik-Buskermolen D, Terken J (2012) Co-constructing stories. In: Proceedings of the 12th participatory design conference: exploratory papers, workshop descriptions, industry cases – vol 2 – PDC'12. ACM Press, New York, p 33

Paper by FiftyThree. https://www.fiftythree.com/paper. Accessed 28 Sept 2014

Purcell AT, Gero JS (1998) Drawings and the design process. Des Stud 19(4):389–430. doi:10.1016/S0142-694X(98)00015-5

Quevedo-Fernández J, Martens J-B (2013) idAnimate: a general-purpose animation sketching tool for multi-touch devices. In: CONTENT 2013, 5th international conference on creative content technologies, IARIA, 38–47

Resnick M, Myers B, Nakakoji K et al (2005) Design principles for tools to support creative thinking. National Science Foundation workshop on Creativity Support Tools. Washington DC. http://repository.cmu.edu/isr/816/

Russell A (2010) ToonTastic: a global storytelling network for kids, by kids. In: Proceedings of the 4th international conference on Tangible, embedded, and embodied interaction (TEI'10). ACM, New York, pp 271–274. doi:10.1145/1709886.1709942

Schon D (1986) The reflective practitioner – How proffesionals think in action. Basic Books, New York. ISBN 978–0465068784

Shneiderman B (2000) Creating creativity: user interfaces for supporting innovation. ACM Trans Comput Hum Interact 7(1):114–138. ACM, New York. doi:10.1145/344949.345077

Sohn E, Choy YC (2012) Sketch-n-stretch: sketching animations using cutouts. IEEE Comput Graph Appl 32(3):59, 69. doi:10.1109/MCG.2010.106

Stacey, Claudia ME, McFadzean J (1999) Sketch interpretation in design communication. In: Proceedings of the 12th international conference on Engineering Design. University of Munich, Munich, 923–928

Takayama K, Igarashi T (2007) 2d animation authoring system with FTIR multi-touch table. Interactive Tokyo. http://www-ui.is.s.u-tokyo.ac.jp/en/projects/

Thorne M, Burke D, van de Panne M (2004) Motion doodles. ACM transactions on graphics (TOG). In: Proceedings of ACM SIGGRAPH 2004, vol 23(3), August 2004. ACM, New York, pp 424–431

Tversky B, Bauer Morrison. (2002) Animation: can it facilitate? Int J Hum Comput Stud:247–262. doi:10.1006/ijhc.1017. Elsevier Science

Using Video for Early Interaction Design

Panos Markopoulos

Abstract This chapter discusses how to use video for prototyping interactivity during early phases of design. Advantages and limitations of video prototyping are discussed and related to other ways of representing early design concepts. The chapter traces the introduction and development of this method in the field of human computer interaction (HCI), moving on to discuss how video can help involve stakeholders in the design process and especially users. A range of techniques, methodological choices, and practical advice for future video prototype creators are discussed, and illustrated with examples.

Introduction

Since the very first steps of cinema at the start of the twentieth century cine-matographers experimented with special effects to represent envisioned futuristic or imagined technologies. An early French cinematographer called George Méliès pioneered several special effects in his 1902 film "*Le Voyage dans la Lune*" (A trip to the moon), including stop motion, overlaying images, time-lapse photography. Special effect techniques developed rapidly, often contributing to great box office success, as in the famous 1933 production '*King Kong*', where 3D models were animated. Video technology and computer graphics have made these techniques widely available and accessible outside the film industry, so that they can now be applied fruitfully in the context of technological design without even the need for specialized training in film-making.

Video has been used successfully to represent envisioned interactive technologies for some time. In the early eighties Robert Spence of Imperial College used video inventively to introduce the concept of bifocal displays (Spence and Apperley 1982), a new and at the time difficult to implement information visualization technique, using a simple scroll of paper. More elaborate and expensive productions of that era included the seminal Knowledge Navigator by Apple (Dubberly and Mitsch

P. Markopoulos (✉)
Department of Industrial Design, Eindhoven University of Technology, Eindhoven, Noord-Brabant, The Netherlands
e-mail: P.Markopoulos@tue.nl

© Springer International Publishing Switzerland 2016
P. Markopoulos et al. (eds.), *Collaboration in Creative Design*,
DOI 10.1007/978-3-319-29155-0_13

271

1992) and the Starfire by Sun Microsystems (Tognazzini 1994), each representing corporate visions for interactive computing that have been studied since by several generations of interaction designers and HCI researchers, as testimony both of visionary thinking but also as paradigmatic uses of video to represent future interactive technologies.

By now video prototyping has become established as a popular medium for representing design concepts mainly because of its ability to visualize interactivity while circumventing technological realization challenges. This chapter discusses how video can be used for this purpose; it builds on some of the pioneering writings in the HCI field, which have contributed to the maturing and popularization of the technique. Two key authors are Laura Vertelney who introduced video prototyping in a brief article in a professional bulletin for interaction designers (Vertelney 1989), and Wendy Mackay who codified methodological knowledge for video prototyping in her research articles and tutorials presented at major international conferences (Mackay and Fayard 1999; Mackay et al. 2000; McCurdy et al. 2006).

This chapter is an introduction to video prototyping for designers yet unfamiliar with this technique; it provides guidance on how to create and use video prototypes for interaction design, based on practical experience and methodological research. The next sections introduce video prototyping and motivate its use as a design representation, laying out its advantages and disadvantages. A simple guide of techniques for video prototyping is provided as a resource for designers and researchers wishing to apply this method. Finally methodological issues are examined and pointers to related research and more extensive texts are provided.

What is a Video Prototype?

A video prototype is a short movie, typically not longer than a few minutes, that represents a usage scenario, illustrating how one or more users interact with an envisioned system. Usage scenarios are traditionally and more typically rendered in text rather than video; they are short stories that describe an envisioned and idealized usage of a system to achieve a goal or experience a benefit from the system.

Text scenarios are very extensively used to capture user requirements. They are a very flexible and accessible design representation medium and so they are used widely for interaction design (Cooper et al. 2014; Rosson and Carroll 2002), and product design (Suri and Marsh 2000). They can be created by design teams or even in collaboration with users and can be illustrated with sketches or even storyboards.

Showing scenarios in video allows us to visualize vividly an envisioned system that does not yet exist or has not yet been implemented. The video prototype (sometimes also called video scenario) may illustrate system use by using film techniques such as animations, special effects, as will be elaborated below. Actors may be shown to interact with very realistic looking artifacts or even with paper mock-ups or rough props that bear only a rudimentary similarity to the technologies designed. Video shows user activity and interactions with the system unfold over time, something that is harder and sometimes tedious with static media such as sketches and storyboards. By filming videos on location one can easily show how actors are expected to use technology in its intended physical and social context.

The level of detail shown, the quality and cost of producing a video scenario can vary substantially. On the low end designers may film a simple role play or some manipulations of low tech props to represent their intended design. On the other extreme a video prototype can be a carefully crafted expensive production with an elaborate plot to show several aspects of the intended design in a way that audiences may find engaging.

Most typically video prototypes are short (typically less than 10 min) video clips that are used at early stage design as representations of interactivity and the intended user experience before the envisioned system is built. This chapter is not about filming demonstrations of existing technologies or about creating polished productions that visualize a future vision, though both these uses of video have several similarities to video prototyping. Importantly, the creator of video prototypes is assumed to be a designer rather than a film-maker, so film-making skills are not required or assumed.

Why and Why Not Use Video for Representing Design Concepts?

Designers have a wide range of media at their disposal to represent early design concepts. The main contenders are text scenarios, drawings, paper mock ups and storyboards which can be cheap to produce and flexible requiring simple and very accessible materials. When done well, these early representations direct the attention of the designer and the audience to the most essential aspects of a concept or design challenge, suspending consideration of engineering and less relevant details. Still it is often difficult to represent dynamic aspects of interactivity and system reactive behavior in static media. Video helps overcome this challenge as it is able to capture dynamic events, portraying fluently user actions and (ostensible) system reactions in a very fine grain, it can highlight social dynamics succinctly and subtly by means of dialogues and non-verbal behaviour (e.g., a facial expression, proxemics, glances and brief utterances between actors), and it can capture and communicate physical and social context with much less effort than a textual description.

Compared to technological prototypes such as wireframes and software prototypes, video prototypes are typically faster and cheaper to create and, crucially, they can be created without special training. Apart from designers and engineers, this makes it possible even for end users to get involved in creating video prototypes in a participatory design fashion. Indeed authors such as Mackay have been advocating this approach (Mackay and Fayard 1999). Another advantage of video prototyping is that it can be applied flexibly during different moments of the design process. During very early stages of design it can be combined with very rough paper prototypes or low tech mock ups, while later in the design process the video could include footage of functional prototypes.

Vertelney (1989) enumerated a few reasons why video should be useful; all of which still hold today:

- Video can provide an entire range of visual expression; from rough video "sketches" to highly refined and believable video productions.
- Video sketches enable designers, clients, and end users to visualize interface ideas and get feedback early in the design process reducing the costs incurred by late design changes.
- Interface designers can produce many design alternatives in the same amount of time that it would take to implement just one design.
- Interface designers don't have to write software in order to prototype user interfaces.
- Video is useful for specifying user interfaces for technologies, which do not yet exist and can emulate the mechanics of real systems without actually having to build them.

Vertleney also identified a few disadvantages of video that one has to pay attention to. First she argued that videos are hard to change or manipulate and that productions need to be carefully planned ahead of time. A quarter of a century later video is much easier to create and edit, so to some extent this problem is has been overcome; still good planning and preparation will make a better and more effective video prototype. Another argument that is even more true today than in the eighties is that:

- Sometimes it is better to prototype directly in the target environment, e.g., using software prototyping tools.

While the specific software tools one would use are different today than at the time of her writing, software prototyping has become a lot easier and democratized since. Where the specific look and feel are important to show, when an information architecture or a visual design need to be conveyed, a software prototype may be indeed preferable and rather easy to make. On the other hand, video may be better at representing the envisioned context of use. In many cases, it may even be productive to combine software and video prototyping in order to combine the strengths of both.
  Further disadvantages noted in (Vertelney 1989) are that:

- Video can mislead people into believing they see the finished prototype, making it difficult to distinguish a visualization of a concept from a working computer system.
- Users cannot test video prototypes.

The last two points are important caveats to keep in mind. On the one hand, expectations can be created among stakeholders that are hard to meet or audiences may even be deceived unintentionally. On the other, while users cannot have the intended user experience first hand from a video prototype to allow actual testing, video prototypes can be valuable for soliciting feedback from viewers on several aspects of the design: functionality, fit to context, etc. This issue is revisited below in the section on video prototyping methodology.

Further to the above, one should add that video is fun to make and watch which can help liven up a design session, injecting some energy and direction to a design teams, it can be a fun and creative team activity, it is very useful for communicating persuasively to managers and stakeholders, and very importantly it has proven its value in industrial contexts.

Creating Video Prototypes

Getting Started

Creating video prototypes does not require extensive infrastructure. Nowadays, consumer level digital video cameras offer very satisfactory quality; even entry level smartphones provide sufficient video capturing functionality to support basic video prototyping. Arguably, the simpler the equipment the more focus will be drawn to minimal representations of design concepts that can be created fast, viewed and discussed instantly, and thrown away as soon as they are not needed anymore.

Recording and joining simple film segments can help visualize vividly an envisioned interaction sequence. Objects that are moved, annotated, or extended between captured frames can be filmed with stop-motion to give the illusion of interactivity and continuity in time, effortlessly communicating action-reaction or cause and effect relations between different events.

A simple video prototype can be made by video recording of mock-ups, e.g., a paper or foam prototype that a designer animates according to a predefined script. In this way, the prototype comes to life, manipulated or modified by the designer between shooting frames. The added advantage is that the whole interaction visualized can be rendered at the intended pace and even in the intended context making it much more effective as a communication tool than animating the paper prototype 'live' in front of the user. Similar to paper prototypes, drawings, storyboards, or still pictures can be displayed on video in succession. All these materials can be brought to life with narration or explained with subtitles that can be inserted digitally through specialized tools, or when one wants a less polished and easier to make result, they can be written out on notes and filmed together with the relevant interactions they refer to.

Vertelney (1989) enumerates seven possible ways to create a video prototype:

- Drawings successively framed turn a storyboard to a video.
- Paper cut outs are manipulated in front of the camera to provide the illusion of interactivity.
- Physical props or objects are manipulated in front of a camera.
- Non interactive on screen renderings whith which an actor ostensibly interacts, providing the illusion of a fully interactive system.
- On screen renderings can be enhanced with paper to simulate interactive components.

Fig. 1 The Immersion at a Distance project, explored how holographic imagery could be exploited to support several futuristic functionalities could be combined to support fluent and serendipitous interactions among co-workers, such as listening into and joining a conversation, attracting attention with proxemics, peripheral awareness, eye contact, etc. The video served as a future vision to orient research and development work at the Alcatel Lucent labs in Antwerp, Belgium. (Credits: Herjan de Heuvel)

- Projections on physical objects, even on cardboard, can give the illusion of a fully interactive embedded display.

These very simple techniques are still as valid and valuable today as they were in the mid-eighties. The availability of digital video editing facilities allows also to create such effects by editing video and inserting still images directly into the footage, allowing high resolution renderings of supposedly interactive applications to be shown in a context provided by the film. (see Fig. 1).

The list above focuses on rendering graphical user interfaces which were the dominant interaction paradigm of that era. However, the advantages of video as a prototyping medium are even more compelling when considering contemporary mobile and ubiquitous applications, where further to the screen contents, the context of use and other multi-modal means of interaction are of importance. Consider for example, what could be perhaps the simplest video prototypes to create: smart light switches. One can film the lights switched on or off while an actor acts out or narrates the interaction by which the lights are ostensibly controlled, e.g., "Ceiling based sensors track her movements, turing the lights on as she enters the room and turning them off in the corridor".

Two more ways interesting ways to create video prototypes can be added to the list above:

- Role-play, users or designers can be actors for the video, acting out interaction scenarios using props and mock-up's created in advance, e.g., see (Laurel 2003; Y\lirisku and Buur 2007).
- Blue or green screen effect: actors are filmed in front of a blue or green screen, which can then be replaced by a video footage which illustrates envisioned system operations. This can also be done to replace parts of objects (e.g., a screen) with footage showing the intended interaction (Halskov and Nielsen 2006).

Exercise Create a simple video prototype with your camera phone. Show how speech input can work on a conventional device/appliance in your surroundings.

Showing Action in Context

Often, it helps to film in the intended context of use or to simulate this context in a location where filming is easier or cheaper. A useful practice for doing this is to start with initial 'establishing shot', e.g., showing the physical location with a wide angle shot or a panning movement of the camera, before drawing attention to the actors and their interactions with the system. For example, in the Starfire video (Tognazzini 1994) the time of the intended action is indicated subtly, by showing the actor walk under a conference banner dated 10 years later.

Apart from representing the context visually, one can use subtitles or narration to inform viewers about when the action is supposed to be taking place, where, what has happened before, and what the persona is aiming to achieve. Particularly for projects envisioning future technologies, it can be crucial to communicate the time horizon that the designer targets, e.g., "it is 2025, and Jenny starts a day at the office". Even if the video prototype is shot at the designer's studio or the office next door, narration or subtitles can provide a frame of reference that guides audiences, or even helps suspend disbelief to tune into what the designer wishes to emphasize with a video prototype.

Exercise Consider the case of sharing applications between devices using the "cloud" – explain how the two devices share data using narration and embedding it in a dialogue. What works best in your opinion?

Dialogues, Monologues and Narration

The sound track is one of the most powerful tools for the video prototype maker:

- Actors can ostensibly verbalize their thoughts spontaneously, and in doing so describe and explain actions that are not visible or clear in the video, e.g., "let's select the weather report option".

- Actors can speak out voice commands, which are followed by showing their intended effect to simulate speech input.
- A narrator can describe the context of the filmed interaction, or can explain what the actor supposedly perceives or aims to do.
- Commentaries accompanying interaction can explain the supposed inner workings of a system, guide viewers in interpreting system output or even adding information that is not communicated by the video the footage alone.

The latter commentaries can be recorded live by one of the team members standing behind the camera, or they can be a post edited as a 'voice over' of someone reading out a text. Text to speech software can also help here when native speakers are not easily available.

Short messages and subtitles can provide a very useful alternative for all the purposes listed above. These though have to be used parsimoniously: long explanations written in text can be tiring and distract audience's attention from the interaction shown.

A more subtle approach than narration and voice over, but one that requires a bit more preparation, is to embed explanations of a design concept in what the actors say. This makes for a more natural and pleasant result. Tognazzini (1994) warns against actor monologues that explain the interaction: they are easy to make but tedious to watch. Crafting such information into the scripted dialogues between actors can be more natural (see for example the Vista prototype (Wichary et al. 2005) discussed below); where necessary dialogues can be supported by subtitles and narration.

Props

Physical props of all kinds are useful. One of the most versatile ones are sticky notes, that can be placed on top of objects to visualize notifications presented to the user, they can be un-stacked one by one to visualize successive states of an interactive dialogue, or they can be placed directly on available physical objects and props to show explicitly what kind of interactions an object is intended to support. For example, a sticky note on a cardboard box saying 'touch sensitive surface' appeals to viewers to suspend disbelief and 'relabel' the artefact accordingly when viewing the action.

The attention one puts in crafting the physical props, can vary. They can be very carefully crafted, showing mechanical properties of a product (see for example the cardboard modeling chapter by Frens in this volume). On the other extreme, the physical props for a role play can be improvised by grasping any object at arm's reach and 'relabeling' them implicitly through action and dialogue, or very explicitly attaching sticky notes to them, or by providing subtitles and comments that declare what these objects stand for.

The attention one should spend on crafting props, depends on the intended audience and the purpose of the video prototype. During brainstorming a role play with available props that only remotely resemble the intended physical form and interactivity may be enough. For communicating to a wider audience or to obtain feedback from users, some more attention is needed to ensure the prop is understood as intended by users and the video scenario is more believable.

Acting

It is not necessary to have acting talent or training to act in a video prototype. In the author's experience, most designers easily perform the role-play required and enjoy doing so. On some occasions when users are very different than the designers, e.g., when designing for children, it may be necessary to recruit representatives of the target user group as actors, or to use video prototyping techniques that do not require actors, e.g., filming paper cut-out animations or physical models. For example, Jlisirsku and Buur (Ylirisku and Buur 2007), illustrate the use of toy-people as puppets to act out a video scenario for future kitchen appliances. Paper cut outs of sketches or stick figures or clay models could equally well be used for this purpose (see Fig. 3).

Plot and Humor

(Tognazzini 1994) referring to the Starfire video advocates that the video should show things going wrong. An idealized situation as one can read in typical text scenarios for scenario based design, can result in a very bland video that will fail to convey its message (viewers may even lose focus). A plot with some twists and even bad characters can keep the audiences attention. In the Starfire video the lead character is an executive competing with a colleague who 'plays foul' to steal the deal from her. By using the envisioned system she fights back and gains her deserved success, while the whole interactivity visualized is shown seamlessly as part of the story with added humorous elements that make the video pleasant to watch and the enacted social interactions less staged.

Video can be fun, but humor and plot should be used sparingly to avoid distracting attention from the real message that is being communicated. A good example of using humour is the Vista video prototype (Wichary et al. 2005) featured below. A typical beginner's mistake is often to draw excessive attention to the plot, and overelaborate it, at the expense of the design concept shown. In the worst cases such an approach can result in a dubious quality 'soap' or comedy. Until one has some experience with video prototyping it is better to use plot and humor conservatively.

Showing Designed Interactions

The very process of video prototyping tends to reveal to designers aspects of the design they have not yet considered. For example, when a mechanism for sharing data is being filmed, one may realize that they have neglected to consider which feedback should be provided to users or that secondary tasks such as access control have not been thought through.

Storyboarding can help as a first version of the design, as is suggested by Mackay (Mackay and Fayard 1999), but designers may still brush over important interaction steps that they only become aware of during filming.

Reflecting on the production of the Starfire prototype discussed above, Tognazzini (1994) puts forward the guideline that the interaction shown on video should be designed and tested just like any other interface would prior to shooting the video: "*Most of the interactions seen in the film were built and tested in isolation to ensure that they would work*". This guideline is particularly applicable for higher budget productions, having the purpose to communicate and persuade rather than to externalize and share early design ideas among designers. As with any other medium video prototypes improve with iteration, so one could start making very rough prototypes intended to externalize and discuss design ideas, before proceeding to design and film a more polished and detailed interaction design. This guideline is less relevant when using videos to encourage participation of end-users or to support idea generation sessions, where speed and free association are key.

Filming the Impossible?

As noted by Tognazzini, the video medium draws towards filming interactions that cannot be realized within the time horizon of a project (Tognazzini 1994). Artificial Intelligence can work perfectly on film, and actors can achieve effortlessly complex interaction goals. This can be very liberating for divergent phases of a design process where many ideas, playful, far fetching and even impossible ones are desirable. Representing those on video can be usefully combined with brainstorming (Mackay and Fayard 1999).

The power of video to show the impossible, as if it were real, is a strength but can also be detrimental. As the film industry has shown us on several occasions, even time travel and instant teleportation are possible to visualize convincingly in moving pictures. To avoid misleading audiences or neglecting crucial feasibility issues, video prototype creators should scrutinize their videos or their storyboards before filming as to whether the functionality shown is feasible in their time horizon and within the scope of their project.

Conversely technology limitations, or even unresolved design issues can also be useful to show. In the Starfire video the system is shown to try to interpret erroneous user input (a misplaced sandwich that the computer vision software tries to recognize). The potential misuses of the system are also hinted at when private moments are accidentally observed by another (see (Tognazzini 1994) for a more extensive discussion). Showing things 'going wrong' with the designed system and the limits of technology can help convey clearly the nature of the technology envisaged, the design choices made but also to trigger debate about quite subtle issues, such as privacy, control, appropriation, etc.

Some Easy Mistakes to Make and Avoid

With video prototyping designers take the role of film directors and actors. This can be fun and rewarding, but mistakes are unavoidable from a film-maker's perspective. Poor use of the camera, e.g., moving the camera too much and too fast, zooming in and out excessively, shooting against the light or in poor lighting conditions are typical examples that are immediately noticeable and can be corrected on the spot.

Sound quality is another aspect of video prototypes that easily goes wrong. Typically designers rely on the microphone built-into digital cameras for capturing audio. This can result in inaudible voices of actors who are far from the camera or capturing sounds that were not easily noticed during the recording, e.g., the sound of wind hitting the microphone. Investing in an external microphone can help, but in many cases, this can be overcome simply by rethinking the positions of actors and the camera.

A bit less obvious are errors that refer to how the video prototype is 'directed'. For example, actors often start lecturing to the camera explaining how the envisioned system is supposed to work, making the footage resemble a cooking program on television or a filmed laboratory demonstration.

It is only too easy to forget to design and film important interactive steps; if the prototype is meant to collect feedback on low level interaction then this interaction must be designed and pilot tested prior to filming, at least with a thorough walk through inspection. In this way, the very process of creating the prototype draws the attention of designers to interaction issues that would otherwise be brushed over when using a more static prototyping medium such as text, sketches, and storyboards.

Editing

Editing is an essential part of video prototyping. It is only for very rough prototypes that playing small video segments pasted together, or showing a simple stop motion animation will be sufficient to illustrate the intended concept. Inevitably, some sequences have to be removed and still photos can be inserted in a video sequence to show clearly some specific detail, or an alternative angle of shooting.

Editing and adding visual effects can absorb an enormous amount of time but can be instrumental in visualizing interactivity effectively; using widely available video editing software can give a professional look to a video prototype, and it can increase the pace of the action.

There are plenty of tools for editing video available and learning to use a serious video editing software can be a very useful investment for a designer (e.g., popular ones are Adobe Premiere Pro and Adobe After Effects). Simple video prototypes can be produced with freely available and simple software that also presents a very low threshold for beginners (e.g., Microsoft Movie Maker). For designers starting to make video prototypes, it can even be advantageous to start with rudimentary video editing software that allows them to focus on designing and producing the video, rather than investing effort in learning complicated software and in elaborate video editing.

Invisible Design

An alternative approach to using video to represent design concepts and trigger feedback from stakeholders is the 'Invisible Design' or 'Obstructed Theater' approach. In this case, the actual system designed is never shown in the video clip, which focuses on how humans interact with and experience a system. Its purpose can be to draw viewer's attention to more general considerations about a class of technology rather than the specific of any particular instantiation. The videos made by (Briggs et al. 2012) were also humorous making them memorable and triggering discussions. This technique has also been used effectively with children to solicit design ideas from them during participatory design sessions (Read et al. 2010). A variation of this approach requires actors to perform the movements related to using a product without actually manipulating any visible physical object (Buur et al. 2004); this draws designer's attention to what users have to do with their hands, and thus encourages the design of related motions and gestures.

Narrated Stills

A 'discount' approach to video prototyping is to take some carefully framed or even directed still pictures. These can be put in sequence in a slide show, adding narration or even voice over for actors. The result can be very similar in purpose to a video prototype and can be obtained at a fraction of the cost. For cases where the prototype is meant to trigger debate on the role of the designed artifact, this can be sufficient.

The Role of Video Prototyping in the Design Process

The diversity of purposes that video can serve during the design process should already be clear: It can be the final output of a process aiming to envision a future technology, or to communicate a vision, or it can be an intermediate product that is created to increase understanding of the problem domain and user needs.

The ease with which video prototypes can be created democratizes the design process enabling users and other stakeholders to get actively involved in prototyping. Accordingly video prototyping has been seen as instrumental in enabling participatory design processes (Mackay and Fayard 1999). Users can act out their ideas and even do so in the actual context of work or every day life which the design project targets; for example (Ylirisku and Buur 2007) illustrate an example of a maintenance worker who improvises some design solutions in context.

Videos are by their nature 'throw away' prototypes: they cannot evolve into the final system as one might do with refining a software prototype this can encourage experimentation and prevent premature commitments. If necessary they can be produced with relatively little effort and cost, which means that they do not constrain the design process. One can decide to explore different design issues at any moment during the design process; of course they are most useful early on where few design decisions have been made and a broader exploration of the design space is most needed.

As a powerful communication medium video can help members of the design team to form and share a clear understanding of what exactly they are designing. This can be contrasted to more formal models or specification documents that are either difficult or tedious to read and as a result may be ignored or misunderstood by some audiences. On the other hand, viewers of video prototypes have to invest very little effort to acquire a shared understanding of the design concept they are working towards.

The power of video to communicate and convince means also that it can focus the attention of a design team to those issues highlighted on video and to neglect those which are brushed over or not shown. For example (Batalas et al. 2012) explain how several information management issues which were not key to the essence of the initial design idea captured with a video prototype were ignored by the interaction designers until decisions that had to be made during software development brought them to the foreground. (see Fig. 2).

Using Video for Evaluation

One of the most common and valuable uses of video prototypes is to obtain feedback from users and other stakeholders. The video can be used as a stimulus for an interview, a group discussion, or even a focus group. There can be many variations on how one runs such an evaluation session; one can even solicit user feedback to design concepts by remote viewing online. A simple set up would be to gather a

Fig. 2 The Siren video prototype visualized the viral transmission of information through opportunistic encounters of actors. The video drew attention away from issues such as initializing and maintaining one's document collection, which were thus neglected until software development started

number of stakeholders, introduce the video and what kind of feedback the designers are interested in and play it, allowing viewers to stop the video to discuss emerging issues.

One could wonder about how well users are able to grasp design concepts shown to them on video since they cannot experience the intended interaction directly. By extension one could doubt how valid their feedback might be about a design concept they have never used either in real life or even as a prototype. These questions motivated two studies that compared the feedback obtained from viewers of a video prototype versus that of test participants in usability evaluations. Viewers could play the video as many times as needed and were surveyed regarding the potential usefulness of the design concept shown and its ease of use. The study of (Zwinderman et al. 2013) did this with a computer vision based mobile application (Google Goggles) which was at the time of the study new and unknown to users, with a reverse engineered video prototype of this same application. Here, reverse engineered means that the video prototype was constructed post-hoc to represent the main uses of the actual system, by filming drawings of screen content manipulated by a designer in front of the camera to simulate the actual interaction with Google Goggles (see Fig. 3).

Another study (Bajracharya et al. 2013) compared the findings from a laboratory based user evaluation of a context sensing system for supporting awareness of remote family members with feedback pertaining to a reverse engineered video

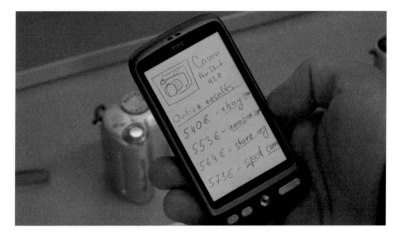

Fig. 3 A screenshot from the reverse engineered video-prototype of the vision based application. The prototype combined an actual phone with hand-drawn sketches of the user-interface to give the impression that the application was not actually developed yet

prototype. In both these studies, no major differences or inconsistencies could be found, showing that video prototype based evaluation results are reliable. The video seemed to draw more attention to contextual aspects of the user experience and the role of the technology, while the usability test seemed to draw attention more to the learnability and ease of use aspects.

Prototype Fidelity and Video

A choice that designers face in creating a video prototype pertains to its fidelity as an interaction prototype. Prototype fidelity is concept extensively studied and debated, that refers to the degree to which a design prototype resembles the intended final system or product (Virzi 1989). Low fidelity prototypes (e.g., paper prototypes) can look sketchy or incomplete and are often used in early stages of design (Rudd et al. 1996). High fidelity prototypes simulate some aspect of the intended product or system more realistically and typically offer a more elaborate visual design. The notion of fidelity can pertain to different aspects of a prototype, such as its appearance, its functionality, its interactivity, or even the data model used (McCurdy et al. 2006). For example, hand-drawn sketches and wireframes usually belong to the lower end of the visual fidelity spectrum they only give a rudimentary indication of system function, and may use fake data rather than actual domain data, scoring low on fidelity on the latter two aspects.

An obvious advantage of low fidelity prototyping is that it demands less time and resources than high fidelity prototyping. This efficiency facilitates exploration and may make it easier for designers to abandon unsuccessful ideas and explore new ones. A widely held belief is that low fidelity prototypes will also help solicit user feedback regarding the essence of a design concept rather than its superficial details (Virzi et al. 1996). On the other hand high fidelity prototypes may be more successful in simulating complex systems or systems requiring physical manipulations, and may be perceived as more "professional" or "attractive" than low fidelity ones. Such an impression can be especially important when promoting a product or an idea to management or customers.

Specifically for video prototyping, this notion of fidelity can be applied to two levels: one concerns the representation of the system or product that actors use in the film, for which the general discussion on fidelity applies directly. The second level applies to the filming technique and quality: Is the video shot on location? Are real users or professional actors involved? What is the quality of the production?

The choices facing designers are numerous and one might expect the general trade-offs regarding prototype fidelity to apply equally well to video prototypes. There has been very little research on this topic; some related investigations are discussed briefly below.

A recent study (Dhillon et al. 2011) compared two video prototypes of an envisioned persuasive system to encourage stair use; the two prototypes differed in terms of the realism of the filming. Using RFID-technology, the Stepper's Club System (SCS) would be able to recognize company employees and dynamically adjust the ambience (lighting and music) of the staircase according to their preferences. By climbing stairs, employees would earn points, which could be exchanged for snacks and coffee. SCS also consisted of public displays on every floor and a web application. The public display would show the ranking of a user in relation to other employees and the web application would provide a more analytical, detailed display of statistics, scores and ranking.

The low fidelity video was filmed on a tabletop using the same equipment as for the high fidelity video. The sketches depicted the scenes that were filmed in the high fidelity video. Two researchers were swapping and manipulating the sketches and their hands were visible in the video, even giving a playful mood to the presentation that emphasized the sketchy nature of the design representation. Such an approach is faster and easier than stop-motion animation or attempting to film at a higher quality. (See Fig. 4)

A team of two trainees spent approximately thirty-two person hours to prepare, film and edit the high fidelity video prototype and about twelve hours for the low fidelity video prototype.

The feedback obtained from people who viewed the two videos online were compared to each other but no difference could be found regarding the volume of comments one can get from viewers from one or the other, the number of 'problems' that the two would help identify with the design concept, or even the suggestions and ideas the two videos would trigger with viewers. Viewers appreciated visually the paper cut out prototype more than the enacted video prototype.

Overall, these studies seem to suggest that very few and small differences should be expected between the feedback users give from actual usability testing of working systems and from viewers of video prototypes. While initially flying

Fig. 4 Acted out video versus cut-out animation for Steppers Club (Dhillon et al. 2011)

against expectations that differences should be expected in user feedback both in number and content, there can be a very simple explanation about it. All these media, when used by skillful designers, succeed in communicating the designer's intent and to draw the attention of the user to the appropriate and most interesting design issues. It could be a sign of a poor prototype to divert viewer's attention to issues outside what the designer aims to communicate. If indeed this explanation holds, the conclusion of fielty comparison studies for video prototypes is that cheaper and more efficient productions of video prototypes are preferable for supporting iteration and exploration during early design.

Two Instructive Cases

This section discusses how video prototyping was used in two design projects carried out by graduate students of the Eindhoven University of Technology, as part of their training in interaction design. The first, concerns the design of Vista a system to support social interaction at the work place, described in (Wichary et al. 2005). The second one called Behand (Caballero et al. 2010) set out to invent novel gesture based interaction techniques for mobile devices.

Both projects have been published and readers can find more about them in related publications (Caballero et al. 2010; Wichary et al. 2005). The discussion here is limited to the role of video prototyping and to the inventive uses of this medium by the two design teams. The two projects followed very different design processes; Vista followed a complete design cycle that started with user studies, identifying needs, and ended with the field deployment of a working system. Behand was an exploration of novel interaction techniques that relied exclusively on video as a prototyping medium.

The Vista Video Prototype

The Vista video prototype was developed as part of a twelve week long project by a team of five designers, who examined how technology could support informal interactions between office workers. The video prototype was the main deliverable for the conceptual design phase of the project, which was followed by software implementation, lab-based evaluation, and field tests. The prototype itself took about a week to make for the team.

Its aim was:

• To assess whether intended users would understand and appreciate the concept.
• To support communication with stakeholders of the project, including intended users and engineers.

The screenplay for the movie was based on two persona scenarios prepared in the user requirements phase. Three people interact with Vista, thanks to which they end up interacting with each other at the coffee corner at their work place. The designers deliberately focused on showing the overall concept of the Vista, its role, and not the specific interaction details. Vista can be described as a 'walk up and use', interactive public display which helps start informal conversations at the workplace and 'engineer' serendipity. Even though the emphasis was not on low level interaction, this had been thought through, and is featured during in the video, e.g., color-coding different categories of information, animated transitions, different modes of operation: serendipitous browsing and a directed search for information on colleagues.

For the filming, Vista was simulated by a standalone (traditional) whiteboard placed next to the coffee and vending machines in the hall of an office building. Scripted animations from a software prototype were edited in the area of the whiteboard using video effects software. The movie was edited using Adobe Premiere Pro.

The actors were all designers acting out different user personas; the dialogues and subtitles provided a context for the action and the short clip keeps the interest of the viewer with a simple plot. Rather than just filming a demonstration of the intended system explaining its features blow by blow, it features the episode of a 'green' new employee being recruited into the project of some more seasoned colleagues. Humoristic elements are added but these are carefully balanced to be subtle and to avoid distracting attention from the design concept.

Fig. 5 The new employee shown to interact with the system (Wichary et al. 2005) to find out information about his colleagues; during filming the designer acted out his interactions in front of a whiteboard and the interface animation was post-edited

The video was successful in its purpose to convince the client to take this concept further and to support it materially; it also remains years after the implementation of a working prototype an excellent description of the overall concept that abstracts away from implementation and low level interaction details (Fig. 5).

The BeHand Video Prototype

A very different use of a video prototype can be seen in the BeHand video (Caballero et al. 2010), a video prototype that aimed to explore and illustrate futuristic gesture based interaction techniques for smartphones. BeHand was the product of a team of three postgraduate designers who worked for 10 weeks dreaming up and exploring novel interaction techniques. The project explored possibilities and challenges from an interaction design perspective, suspending any consideration of technical feasibility and implementation.

The envisioned interaction technique is that the smartphone camera records the movements of the hands in the space behind the smartphone and superimposes an image of the gesturing hand in a virtual world shown on the screen. This enables a range of typical interaction tasks such as sketching, 3D manipulation, etc. (see Fig. 2).

From a video prototyping perspective the BeHand video features a very interesting approach. The actual interaction was first designed, and then animations were created of the supposed screen content. The video was filmed by having actors reproduce the shown gestures behind the smartphone device that was of operating on a play back (non-interactive) mode (Fig. 6).

Fig. 6 Image from the
Behand video (Caballero et
al. 2010). The actor's gestures
behind the smartphone appear
on the video as gesture based
manipulations of virtual
objects

Conclusion

There can be endless variations and combinations of the basic techniques discussed
in this section to create interesting to watch prototypes that are effective communication and exploration tools for design. Making such prototypes is becoming one
of the standard skills for interaction designers, who find themselves often in the
roles of actor, operator, or director. Though affinity with film making and editing
can contribute to a high standard of video prototypes, interaction designers should
be able to create at least simple video prototypes effective for their purposes.

As a medium to be used during the design, video prototypes can be relatively
cheap to produce, and can be an invaluable tool for communicating design ideas to
stakeholders, obtaining feedback, or simply externalizing/representing design ideas
to help develop them further.

Although lots of things can be shown on video here this chapter is concerned
with capturing the essence of a design idea or concept, adhering to the fundamental
prototyping principle (Lim et al. 2008): *"Finding the manifestation of a concept
meeting some requirements that in its simplest form filters the qualities in which
designers are interested, without distorting the understanding of the whole"*. This
principle runs throughout the use of video prototypes: the whole production and use
of video prototypes is driven by the needs of a design process that requires iteration,
involvement of users and other stakeholders, and avoiding premature commitment
to technological details.

There are no hard and fast rules for guiding video prototype creators – some
methodological choices have been discussed above, but there is a lot of room for
creativity and for exploring new uses of the medium during the interaction design
process. Related research is gradually providing us some methodological guidance,
particularly with the impact of different video prototyping approaches on viewers
and on the feedback that they give. As technology to process video is becoming
cheaper and more effective, interaction designers should be using video even more,
finding their own ways to embed its use in the design process.

Acknowledgments Many thanks to Jacob Buur for providing critical feedback on an early version of this chapter. The following postgraduate designers trainees of the USI program are thanked for their video creations featured in this article: Herjan van de Heuvel for the Immersion at a Distance video, Nikos Batalas, Hesther Bruikman, Annemiek Van Drunen, Elaine Huang, Dominika Turzynska, Vanessa Vakili, and Natalia Voynarovskaya for the Siren video prototype. Luz Caballero, Maria Menéndez, Valentina Occhialini for the Behand video. Marcin Wichary, Lucy Gunawan, Nele Van den Ende, Qarin Hjortzberg-Nordlund for the Vista video prototype. Matthijs Zwinderman, Rinze Leenheer, Azadeh Shirzad, Nik Chupriyanov, Glenn Veugen, Biyong Zhang for the Goggles prototype.

Further Reading

Designers wishing to learn about video prototyping could benefit by reading the very beautifully written book by Ylirisku and Buur on Using Video in Interaction Design (Ylirisku and Buur 2007). The book covers both the use of video during design research and the use of video as a design representation. It covers practical and theoretical aspects, introducing readers to epistemological questions regarding qualitative research and the use of video in fieldwork, supported by a number of short videos that illustrate the techniques explained in the book.

Capturing human activities on video can help designers understand their temporal structure or social interactions between people. The Interaction Analysis Laboratory (Jordan and Henderson 1995) is a structured collaborative observation process that design teams can apply to analyze such footage. A game like variant of this method was introduced is action scrabble (Buur et al. 2014) where designers note their observations on cards, which they organize in a scrabble like game to represent physically temporal aspects of the activity shown on the video.

Tognazzini's account and reflection on the making of the Starfire video (Tognazzini 1994) is engaging and interesting to read along with viewing the seminal video production it refers to, which is available freely online.

Video as a medium raises several ethical challenges. The discussion by Mackay on ethics aspects of video in design (Mackay 1995) extends beyond video prototyping, and should be definitely on the reading list of designers starting to work with video.

References

Bajracharya P, Mamagkaki T, Pozdnyakova A, Da FS Pereira MV, Zavialova T, de Zeeuw T, Dadlani P, Markopoulos P (2013) How does user feedback to video prototypes compare to that obtained in a home simulation laboratory? In: Distributed ambient pervasive interaction. Springer, Berlin/New York, pp 195–204

Batalas N, Bruikman H, Van Drunen A, Huang H, Turzynska D, Vakili V, Voynarovskaya N, Markopoulos P (2012) On the use of video prototyping in designing ambient user experiences. In: Ambient intelligence. Springer, Berlin/New York, pp 403–408

Briggs P, Blythe M, Vines J, Lindsay S, Dunphy P, Nicholson J, Green D, Kitson J, Monk A, Olivier P (2012) Invisible design: exploring insights and ideas through ambiguous film scenarios. In: Proceedings of the designing interactive system conference. ACM, New York, pp 534–543

Buur J, Jensen MV, Djajadiningrat T (2004) Hands-only scenarios and video action walls: novel methods for tangible user interaction design. In: Proceedings of the 5th conference on designing interactive system processes, practice, methods and techniques. ACM, New York, pp 185–192

Buur J, Caglio A, Jensen LC (2014) Human actions made tangible: analysing the temporal organization of activities. In: Proceedings of the 2014 conference on designing interactive system. ACM, New York, pp 1065–1073

Caballero ML, Chang T-R, Menéndez M, Occhialini V (2010) Behand: augmented virtuality gestural interaction for mobile phones. In: Proceedings of the 12th international conference on human computer interaction with mobile devices and services. ACM, New York, pp 451–454

Cooper A, Reimann R, Cronin D, Noessel C (2014) About face: the essentials of interaction design. Wiley, Indianapolis

Dhillon B, Banach P, Kocielnik R, Emparanza JP, Politis I, Pączewska A, Markopoulos P (2011) Visual fidelity of video prototypes and user feedback: a case study. In: Proceedings of the 25th BCS conference human-computer Interaction. British Computer Society, pp 139–144

Dubberly H, Mitsch D (1992) Knowledge navigator. In: ACM conference in human factors in computing systems CHI'92 special video program: conference on human factors in computing systems

Halskov K, Nielsen R (2006) Virtual video prototyping. Hum-Comput Interact 21:199–233

Jordan B, Henderson A (1995) Interaction analysis: foundations and practice. J Learn Sci 4:39–103

Laurel B (2003) Design research: methods and perspectives. MIT Press, Cambridge, MA

Lim Y-K, Stolterman E, Tenenberg J (2008) The anatomy of prototypes: prototypes as filters, prototypes as manifestations of design ideas. ACM Trans Comput-Hum Interact TOCHI 15:7

Mackay WE (1995) Ethics, lies and videotape . . . . In: Proceedings of the SIGCHI conference on human factors in computer system. ACM Press/Addison-Wesley Publishing Co, New York, pp 138–145

Mackay WE, Fayard AL (1999) Video brainstorming and prototyping: techniques for participatory design. In: CHI99 extended abstract on human factors in computer system. ACM, New York, pp 118–119

Mackay WE, Ratzer AV, Janecek P (2000) Video artifacts for design: bridging the gap between abstraction and detail. In: Proceedings 3rd conferece on designing interactive system: processes, practices, methods, techniques. ACM, New York, pp 72–82

McCurdy M, Connors C, Pyrzak G, Kanefsky B, Vera A (2006) Breaking the fidelity barrier: an examination of our current characterization of prototypes and an example of a mixed-fidelity success. In: Proceedings of the SIGCHI conference on human factors in computer system. ACM, New York, pp 1233–1242

Read JC, Fitton D, Mazzone E (2010) Using obstructed theatre with child designers to convey requirements. In: CHI10 extended abstract on human factors in computer system. ACM, New York, pp 4063–4068

Rosson MB, Carroll JM (2002) Usability engineering: scenario-based development of human-computer interaction. Morgan Kaufmann, San Francisco

Rudd J, Stern K, Isensee S (1996) Low vs. high-fidelity prototyping debate. Interactions 3:76–85

Spence R, Apperley M (1982) Data base navigation: an office environment for the professional. Behav Inf Technol 1:43–54

Suri JF, Marsh M (2000) Scenario building as an ergonomics method in consumer product design. Appl Ergon 31:151–157

Tognazzini B (1994) The "Starfire" video prototype project: a case history. In: Proceedings of the SIGCHI conference on human factors in computer system. ACM, New York, pp 99–105

Vertelney L (1989) Using video to prototype user interfaces. ACM SIGCHI Bull 21:57–61

Virzi RA (1989) What can you learn from a low-fidelity prototype? In: Proceedings of the human factors and ergonomics society annual meeting. Sage, London, pp 224–228

Virzi RA, Sokolov JL, Karis D (1996) Usability problem identification using both low-and high-fidelity prototypes. In: Proceedings of the SIGCHI conference on human factors in computer system. ACM, New York, pp 236–243

Wichary M, Gunawan L, Van den Ende N, Hjortzberg-Nordlund Q, Matysiak A, Janssen R, Sun X (2005) Vista: interactive coffee-corner display. In: CHI05 extended abstract on human factors in computer system. ACM, New York, pp 1062–1077

Ylirisku SP, Buur J (2007) Designing with video: focusing the user-centred design process. Springer Science & Business Media, London

Zwinderman M, Leenheer R, Shirzad A, Chupriyanov N, Veugen G, Zhang B, Markopoulos P (2013) Using video prototypes for evaluating design concepts with users: a comparison to usability testing. In: Human-computer interaction–INTERACT 2013. Springer, Berlin/New York, pp 774–781

Part IV
Tools for Creativity and Collaboration in Early Design

Early Stage Creative Design Collaboration: A Survey of Current Practice

Paul Bermudez and Sara Jones

Abstract In an effort to gain a deeper understanding of how to better facilitate creative collaboration during early stage design activities, a survey and a number of interviews were undertaken to determine current practice in this field. A key aim of this research was to review how tools and technologies are currently used to support collaborative creativity and problem solving in design. The outcome is an overview of early stage activities, the environments in which they are currently carried out, and the role of tools and technologies within them.

Background

Much of our current understanding of creativity comes from the study of the different professional practitioners of creativity.

Many of these studies have also attempted to define a model of creativity within specific disciplines, industries or environments (Brophy 2001; Coughlan and Johnson 2008; Howard et al. 2008; Zhu 2011). These models of creativity have ranged from describing creativity in very specific situations and environments, to more general and broad ranging. Other variations of these creativity models have been used to describe the cognitive and organisational aspects of creative processes.

One approach to modelling creativity that is of particular relevance here has been to focus on the different types of interactions during creative processes. This might be an interaction with a tool (or external "artefact") or between people. Here, "interactions" are described as the link between mental creativity and physical representation (Coughlan and Johnson 2009). This particular research has yielded categories of interactions, such as "Productive Interactions", "Structural

P. Bermudez
Centre for Human-Computer Interaction Design, City University London, Northampton Square, London, EC1V 0HB UK

S. Jones (✉)
Centre for Creativity in Professional Practice, Cass Business School, 106 Bunhill Row, London, EC1Y 8TZ UK
e-mail: s.v.jones@city.ac.uk

© Springer International Publishing Switzerland 2016
P. Markopoulos et al. (eds.), *Collaboration in Creative Design*,
DOI 10.1007/978-3-319-29155-0_14

297

Interactions" and "Longitudinal Interactions", which may or may not occur in a linear or chronological way (Coughlan and Johnson 2009).

"Productive Interactions" could be considered those types of interactions that occur with the purpose of generating a new idea as well as possibly a visual, or physical, representation of that idea (Coughlan and Johnson 2009). These types of interactions are frequently dependent on the rapid and spontaneous representation of ideas. Particularly within the visual arts, sketching is frequently used as a mode of idea generation and exploration (Sedivy and Johnson 1999). This type of sketching during early stages of idea generation tends to be fast, spontaneous and flexible. These attributes support a process through which early ideas can be identified, explored, and refined, all possibly in a short period of time (Sedivy and Johnson 1999).

As key elements of this idea generation and exploration are speed and spontaneity, the tools used must be accessible and easy to use. Research has suggested that immediately accessible, or "ubiquitous" tools are key to the representation of ideas during early stages of creative processes (Coughlan and Johnson 2008), and recent research has also suggested that analogue tools, such as pen and paper, are generally preferred because of their ease and immediacy of use (Coughlan and Johnson 2008). However, studies have shown that while practitioners prefer the use of analogue tools for the early representation of ideas, they recognise the potential "organizational advantages" of supporting technology (Coughlan and Johnson 2008).

Whether it is through analogue or digital means, the capture (storage), organisation and reuse of materials and/or creative ideas used or generated in early stages of design can be crucial (Coughlan and Johnson 2009). The tendency of creative practitioners to continually save ideas and resources is well documented as a pervasive and constant aspect of their lives (Coughlan and Johnson 2009). Research suggests that access to physical artefacts used during idea generation is extremely valuable, even if the ideas have been digitised and are digitally available (Geyer et al. 2011). However, one benefit of digital tools over analogue that is referred to in the literature is the ability to facilitate the sharing of ideas across distance, with larger numbers of people, and also to do so asynchronously (Geyer et al. 2011; Gumienny et al. 2013).

Exploration and refinement of ideas are aspects of creativity that are supported by another key element of any creative process, reflection. Creative reflection has been described as an internal mental "conversation", whereby ideas and goals are reviewed and possibly iterated on (Casakin and Kreitler 2011). Reflection can also take place in a more physical actionable way. A designer might redraw a sketch multiple times in order to explore variations and subtle changes (Coughlan and Johnson 2008). This process frequently includes a degree of internal dialogue whereby the designer compares the resulting changes to previous versions, original goals, or a mental model of a desired outcome.

Acts of creative reflection can also be found within "Structural Interaction." In the context of a review of current practices of creative collaboration, the category of

"Structural Interaction" is particularly relevant. "Structural Interaction" is defined as having a focus on the "self-reflective component" of creativity and is integral to the formation, evaluation, and evolution of creative processes (Coughlan and Johnson 2009). Iterating on existing creative process can theoretically lead to new original processes (and tools), which may in turn produce new innovative ideas (Coughlan and Johnson 2009).

Another common feature of creative processes is the sharing of ideas between peers. Collaboration amongst peers can take many different forms and can involve participants in a variety of ways. Spontaneous informal personal interactions have repeatedly been found to be integral to co-located collaborative workflow. Research has shown that organisational creativity and productivity is often dependent on informal collaboration, and suggested that digital tools developed for creativity in collaborative environments should be designed with this in mind (Bellotti and Bly 1996). Other aspects of the environment that facilitate or hinder informal collaboration, such as workspace proximity, noise, available meeting space and social dynamics can also be important (Bellotti and Bly 1996).

A key area that has been found to be integral to facilitating informal collaboration is the concept of awareness of other participants or potential collaborators (Bellotti and Bly 1996; Gutwin et al. 2008). For example, this could be "awareness" of availability of a co-worker, or recognition of the status of a project or task. This can be further extended to the awareness of the availability of tools, physical artefacts, or even environments (e.g. an empty meeting room). Greater awareness of these different elements potentially facilitates informal collaboration and communication by making it easier for practitioners to initiate spontaneous and informal collaboration.

In the rest of this chapter, we describe the methods and then the results from our study, in which we sought to investigate the extent to which issues such as the above were still important in the work of practicing designers.

Research Questions and Methods

The aim of the study reported here was to understand the current state of practice in collaborative, early stage creative design, in order to determine whether current practice as observed when the study was conducted (in the autumn of 2013) corresponded to what had been reported in the existing literature. The study sought to ascertain whether, as a result of changes in the fast moving world of technology, practice had moved on since the earlier studies, reported above, had been conducted.

The study began with an online survey to which 37 practicing designers responded. Of those 37, 23 were based in the UK, 7 in the US, and 7 in other locations including Canada, France, India, Peru and Malaysia. 7 of the respondents were under 30, 14 were aged 30–39, 14 were 40–49 and 2 were 50–59. In terms of experience, while 7 were relatively new to design with 2 years' experience or less, 11 had between 2 and 5 years, 5 had 6–10 years, and the remaining 14 each had over

Table 1 Interview participants

Participant	Job title	Design type	Organisation type
P1	Associate Design Director	Product design	Corporate
P2	Researcher	User experience	Corporate
P3	Masters Student	Applied creativity	University
P4	Interaction Designer	User experience	Corporate
P5	PhD Research Student	Research	University
P6	UX Practitioner	User experience	Corporate
P7	Interaction Designer	User experience	NGO
P8	Web Designer	Web design	Corporate

10 years' experience as a designer. Respondents defined their professions in terms of user experience design, web design, industrial design, marketing, merchandising, video production, motion graphics, graphic design, architecture and hairdressing, with two defining themselves as inventor and artist. All participants took part in some form of creative collaboration, with 97 % of respondents described themselves as participating in creative collaboration at work, and 57 % describing themselves as collaborating on personal creative projects.

Following the survey, eight face-to-face interviews, each lasting approximately an hour, were conducted with respondents who had agreed to participate. Of these participants, four were male and four female. An indication of the type of design these interviewees participated in, as well as their job title and the type of organization in which they were working, is shown in Table 1.

The interviews were conducted as semi-structured conversations. As the self-reflective analysis of process and methods has already been identified as being a common activity (and existing skill) amongst creative practitioners, the study sought to engage them in further analysis of their current processes. The participants were asked to reflect on their creative process and their experience in collaborative situations and environments. The basic outline of the interviews was centered on exploring the specifics of the early stages of each participant's creative process. Key areas of interest during these interviews were: when, where, how, and even why, collaboration took place. Participants were also queried on the types of tools they used throughout their creative process, as well as which tools they wished they could use. Regarding tools, a key point of interest was to gain insight into why certain tools were being used at certain points, and which related functions they supported (e.g. collaboration). Participants were also asked to elaborate on the process of choosing tools, and the key factors in deciding which tools to use, or not use to use.

The final area of focus for the interviews was on the activity of reflection, within the context of a creative process. As the reviewed research literature consistently identified this activity as a key element of creative processes, the interviews were structured in order to gain insight into its relevance to current creative practice. The interviews were intended to explore when, where, and how this reflection occurred and what impact it had on creativity. In addition, participants were queried about possible instances of collaborative reflection, or review of ideas. Finally, participants were again asked to indicate which tools and technologies were used to support these activities of reflection.

Interviews were transcribed, and a thematic analysis of the transcripts was conducted. The remainder of this chapter presents the results obtained from our survey and interviews. Much of this is written using the words of the interview participants, with the aim of providing a more immediate sense of how work in this area feels to those who are doing it.

Early Stage Design Activities

In this section, we present our findings in relation to the various forms of collaboration that take place during early stage creative design, as well as the different types of activity and processes that designers engage in.

Collaboration

As stated earlier, all of the respondents to the survey participated in some form of creative collaboration, either as part of their work or personal projects. The value of collaboration in design was also acknowledged by many of the interview participants. Collaboration was frequently felt to be an integral aspect of the creative process, which potentially generated a greater range of novel ideas, or facilitated more effective problem solving.

Designers in our survey typically worked with teams of different sizes: 13 people said they collaborate on average with just 1–2 people; 18 said they worked with 3–5 people; 4 with 6–10 and 2 with more than 10. Interview participants frequently described collaboration as occurring in many different forms and a variety of environments. Collaborative activities were described as both formal structured events as well as impromptu, unstructured events. The proportion of survey respondents taking part in more structured collaboration processes as well as informal and spontaneous collaboration is shown in Fig. 1.

In this section, we first look at how the communication necessary for collaboration takes place, and then at the different ways in which first informal and then more formal or structured collaboration takes place.

Direct Communication

Participants frequently described their collaboration with others as occurring in the same space, at the same time, talking to each other while sketching things on a shared surface (whiteboard or paper), in an effort to either come up with ideas or

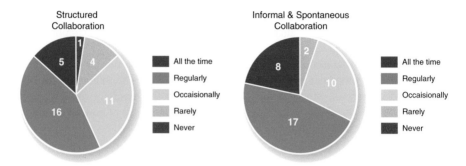

Fig. 1 Numbers of survey respondents engaged in structured and informal collaboration

to solve a problem. This act of collaboration, with designers working in the same place at the same time, with open and spontaneous communication, was recognised by a number of interviewees as being a key element of their creative collaboration, especially during the early stages of a project.

Many of the participants described this type of collaboration as being integral to the way they worked. Participant P1 went as far as suggesting this was something that his business depended on. A key aspect of this type of collaboration was to allow each individual to contribute by developing their own ideas simultaneously, while also being able to communicate with others directly. This type of simultaneous idea development and communication was described within the context of a wide range of circumstances and environments.

P2 described the differences between remote and co-located synchronous collaboration: 'If they're not in the same room, you don't have all the other things that maybe contribute to that environment. So like you draw something, I draw something, you do something, I do something. Rather than everyone's like, you're in this corner and I'm on the floor, and we're like talking, and you're like 'Oh that's interesting'. It's just lack of a shared space.' The ability to immediately interact and respond to other individuals directly was consistently highlighted as one of the most valuable aspects of co-located collaboration, during the interviews. Further more, interviewees suggested that the lack of immediacy and direct interaction was one of the key weaknesses of remote collaboration, and the technologies that support it.

Informal and Spontaneous Collaboration

Many of the designers interviewed referred to the spontaneous nature of informal collaboration and the regularity with which it occurs. Being able to share ideas and get direct feedback from peers was frequently commented on as being an important part of their creative process. As P6 put it, 'There's lots of spontaneous gatherings and ideation.'

The types of collaboration that could be categorized as 'informal' were instances where there was minimal planning, and the collaborative activities were able to take place in a wide range of environments, and frequently with minimal infrastructure or technological requirements. The main requirement described was the co-located sharing of ideas.

The informal collaboration was frequently described as being triggered by the introduction of a problem, or an initial idea. As one participant described it: 'Whenever we had an idea we'd just go and do it. We'd get the brief ... and figure out what the problem was. And then we'd try to talk through it ourselves. I'd say 'Lets go get some paper and go and do it.' My colleague was sitting right beside me so it was really easy' (P7). This further confirms that this type of collaboration is less dependent on structured events and environments, and can occur in an unpredictable manner.

There were a number of different factors that were indicated to influence the accessibility of informal collaboration. The physical environment was consistently mentioned as a factor in either facilitating or inhibiting this type of collaboration. For example, interviewees frequently commented about how physical proximity to peers can facilitate the sharing of ideas and a variety of collaboration. P7 commented: 'we sit next to each other. It's kind of 'what do you think of this?'; and another participant confirmed, 'I sit in the center of a design team. .... So we have lots of side conversations all the time.' (P2).

Physical proximity was also important in enabling the visibility or awareness of the availability of co-workers. This again allowed collaboration to happen in a more spontaneous, unplanned manner. For example, P8 described how: 'We sit at one long desk. I just stand up and keep an eye on if they're working, or if they're not 100 % busy, and then I'll just ... tap them on the shoulder and say 'Martin, can I have a minute?'

Structured Collaboration

Formal or more structured collaborative events were frequently described as being necessary in order to frame a project and set goals. While the details of this type of collaborative event differed amongst interviewees, some more formal collaboration was often felt to be a necessary part of the early stages of design and problem solving.

Some of the interviewees talked about the transitions between informal and more structured collaboration, and the role that technical tools can play in facilitating that. In particular, P4 talked about gathering together the whole of a project team for a kick-off meeting, even when some members of the project are in different locations, by using web conferencing tools: 'That kickoff meeting can be massive. It can even have people in on [web conferencing tool]. ... So typically I'd say between 15 and 20 people. ... It's just so later on people can't go 'I wasn't included, so I'm not going to contribute'. It's all about bringing people on, ... It's just so everyone knows what's going on.'

This comment also illustrates how these formal collaborative events can potentially remove barriers to further collaboration. The inclusion of people at this stage enables further collaboration amongst project participants by giving those included a common reference point to work from.

Creative Activities and Process

During the study, as well as with much of the research reviewed, the activities that were seen to be most effective, and return the most creative results, frequently involved the simplest and easiest tools to use, in easily accessible environments with few technical components. Participants continually described early stages of idea generation as being most effective when organizational, environmental, and tool or equipment barriers are removed.

Idea Generation and Representation

Idea representation usually consists of the act of putting pen to paper and beginning to describe an idea, either visually or with words. The process of representing early ideas is often referred to as an "explorative" activity, where the details of an idea are gradually described and built upon. The activities related to representing an idea are often felt to be the best way to develop and evolve initial ideas, as well as to trigger the generation of new ideas.

The representation of the idea in essence frees the individual (or group) from the cognitive load of remembering, or mentally visualizing the idea, and allows the practitioners to turn their creative energies to investigating other ideas, or developing further details of the initial idea. This in turn can facilitate greater divergent thinking, as the practitioners are better able to develop completely unrelated and different paths of thinking.

Additionally, initial representations of ideas are especially important in collaborative environments where developing shared understanding and common mental visualisations is at the very least difficult. In these situations, shared representations of ideas are often used as the building blocks for further idea generation, or idea evolution. In our study, the initial act of representing potential solutions was frequently felt to stimulate creativity and possibly lead to more innovative ideas. As participant P6 explained: 'It's not necessarily ideation, it's just thinking together in a group. "Here's a problem, let's go to the boards, let's go to the wall, let's scribble." Because it's always helpful, to put things down. ... Because it sparks other people's ideas'. As this quote illustrates a key function of these activities at this early stage is not to develop a fully formed solution or design, but to rapidly explore possibilities. A similar approach was outlined by P7, who talked about the use of collaborative sketching to jointly develop an idea: 'Then we'd spend a few days to a week, trying to abstract out the idea of what we were being asked to do. And we'd come up with some sketches. We'd sit down for a couple of hours. One of us would start, just by drawing whatever. And then the other would take that idea and change it.'

The key goal of this act of sketching and representing ideas is to have something tangible result from this exercise that can be used and referred to at later stages. It should also be noted, that while "sketches" are frequently referred to as a desirable outcome of early stage design, other representations of ideas such as lists, or groups of post-its were also often referred to.

The most common attribute of all of these early "sketches" of ideas was that the starting point was more often than not created without digital tools, but with "analogue" tools, such as pen and paper, markers or whiteboards. Analogue tools are a natural fit for this as they produce physical artefacts that are generally easy to interact with. This is particularly important in co-located collaborative environments, where ideas need to be rapidly represented and shared.

Idea Capture and Reuse

The capture and reuse of previously generated ideas can take many different forms, and can include something as simple as creating and referring to a physical artefact, or generating and reviewing digital documentation of something that emerged from an early ideation activity.

The question of how to capture the outputs from collaborative sketching sessions, such as those described previously, was mentioned by many designers, both in the survey and in the interviews. P1 explained that: 'A lot of the time if we're having a discussion and creating things there's someone who's sort of notetaking or doing diagrams and things that just summarize the areas that we're looking at. And that's often really handy . . . you're having a conversation and somebody is just writing on a flipchart, writing the key things and drawing sketches, and then puts it up on a wall.'

This sort of "note taking" (or documenting) is a common occurrence. However, a theme that emerged from the interviews was that while analogue tools (such as pen, paper and post-its) are great for capturing ideas in the moment, as they are generated, there is still a necessary digital element. As P6 described: 'How we use post-its is [to] get ideas out, but then you have to still digitise them.'

One issue that came up during the study was that referring to physical artefacts outside of the context of the situation or environment that they were created in is frequently physically problematic, particularly if that initial context was collaborative. Physical artefacts in many cases may not lend themselves to group reviews, depending on size and format (e.g. a sketch in the margins of a sketchbook is not easy to submit to group review). Even in the case of an individual review of physical artefacts, if the object has not been in some way digitized, the individual must have access to the object and be physically present in the same location in order to review it. Even for a portable object, such as a large sheet of paper with ideas written on it, the circumstances in which it can be reviewed may be limited, due to its size and format.

The context in which the ideas were generated was often an important aspect of understanding what was being reviewed. Participants frequently commented on how they wished to capture, not just the ideas, but also some aspects of the context in which they had been conceived. 'For us, for me personally going back into that space, and seeing that work on the wall is a positive thing. Because it takes you back to that thought process.' (P6) P5 also expressed a similar sentiment in relation to attempting to create digital records of ideas generated in workshop settings: ' . . . you try to photograph things, but you're not necessarily seeing the process that was around it at the time.'

The need to digitise the ideas (and in some cases the environment) was a consistent aspect of their creative process. Several of the interviewees described capturing such outputs simply by photographing them: 'We're using smart phone cameras. That's how we digitise and it works great . . . We're very low key. Low tech.' (P1); and one explained that although photos are not perfect, they are better than text-based transcriptions of ideas, as they capture at least some of the contextual information: 'When you're capturing these ideas, and you take a picture. You capture so much more when you're doing that, then typing something up and sending someone an email about what happened. You get the context as well a little bit.' (P6).

The use of smart phone cameras was consistent amongst all of the interviewees. Many of them indicated that the simplicity and ubiquitous nature of camera phones was a key factor in their use. It frequently wasn't so much that they were chosen over other methods of digital capture, but rather that they were common devices that most people involved in the collaboration would have immediately at hand and be able to use without any difficulty.

Reflection

During the interview phase of the study, participants frequently described activities as part of their own creative process that could be classified as reflection. There was a wide range of activities described, occurring in a variety of circumstances and environments, that all served the purpose of reconsidering previous ideas. This range of activities and environments included planned formal review; spontaneous and informal, collaborative review; and background, low focus thinking which all served as a form of reflection.

One of the most common forms of reflection discussed was a focused review of previous sketches or notes. This was often described as an informal personal reviewing exercise, primarily dependent on access to notes, via a sketchbook, or digital device (e.g. smart phone) and occurring in a wide range of situations (e.g. commuting on public transport).

A number of the designers interviewed described how reflection could be a background activity that happens as a natural part of their creative process. As P5 explained: 'If you're working on a project it's hard to control when you think about it, when you reflect on it. There'll be times when some random element will trigger something that sits in the back of your mind. And then you'll just sit there and think about it for a while.' In a similar way, P7 explained how: 'you can't just kind of turn it on and turn it off. So you'd come home and you'd be sort of thinking in your head "Oh what about this, or what about that" and you would still be thinking about it.'

This type of low focus reflection is a common aspect of creativity and can often lead to new ideas being generated. The background processing of information and ideas was continually cited as something that the interviewees were aware of doing, but not necessarily able to control or influence. They also generally placed great value on this background processing of ideas, based on past experiences and the impact it had on generating further creative results.

A common response to the desire to be able to reflect on current work at any time is to carry a physical notebook. Most of the participants mentioned using notebooks, or in some cases for example, P7 described how: 'Since I started working in UX I carry a small notebook always. I always did before, but now I actually scribble in it. Because a lot of times I find myself coming up with a really good idea and then forgetting about it. Which is really annoying. So now you know, if I'm thinking about something in particular, and something comes up out of nowhere, or out of somewhere, then I'll go 'Oh that's a good idea', and I'll sketch it down.'

The use of notebooks was seen as an effective way of documenting new ideas and also as tool to support reflection. Browsing through a notebook filled with related ideas can be a valuable form of reflection, where the sequence of ideas and thoughts can also help the practitioner recreate the context within which the ideas were created.

Interviewees described the use of both analogue and digital tools for these types of reflective activities. In describing her preference for reflecting with non-digital tools, P3 also offers some insight as to how reflection helps in the further development of ideas: 'But these days, because I use my computer for so many different things, for research, and making my project happen, for [using a video conferencing product], for email, for everything, I find it a relief to be able to just sit down with nothing digital around me, and to write down ideas. In terms of reflection it helps me clarify some thoughts.' Thus, as well as describing the reasons for choosing an analogue tool over a digital one for reflection, another key point that P3 appears to be making is how reflection can aid the clarity and development of early seeds of ideas.

Finally, one of the interviewees (P1) mentioned a process of collaborative reflection, in the sense of working with a small group on jointly reviewing and iterating on ideas and goals in a physical way when working with physical tools: 'So if you're having a conversation and somebody is just writing on a flipchart, writing the key things and drawing sketches, and then puts it up on a wall. The thing is that as you move along, you might have one wall, you can then refer back to what you've been discussing, and start making links, and you might have a pause half way through … And we'll kind of look back through and maybe star the things that we think are really valuable and where we don't like to go. And so it's this running dialogue, and it's a way of being able to review what you've already done.' This is a particularly good example of how reflection can occur in a collaborative environment as a group, and also in a more formal structured setting.

Supporting Early Stage Design

This section covers the various elements that when combined together appear to be particularly effective in supporting the early stages of design, or creative problem solving. These supporting elements range from interaction qualities, or types of interactions with tools, to attributes of the environment, or Press (Rhodes 1961), in which the designers in our study are working.

Interaction Qualities

Two general themes that emerged in relation to the nature of interactions between designers and their tools were as follows.

Immediacy

Immediacy was a commonly occurring theme during the interviews, and was described as having an impact on a range of issues relating to supporting creativity and collaboration in early stage design. The importance of immediacy in early creative stages manifests itself in a number of different ways. Ease of use, familiarity, skill requirements, accessibility, and responsiveness are all qualities that can be linked to immediacy. These qualities were frequently described as key enablers of creativity with the tools used and also to some degree within the environments in which they are used.

'Ease of use' has previously been identified as an important factor here, for example by Sedivy and Johnson (1999), who report how an interview subject commented that an advantage of a multi-modal sketching tool under consideration was that it allowed for performing actions without having to think about how to access them. Participants in our study also commented on how familiarity with tools (i.e. pen and paper) made a tool easier to use in the sense that little thought was required while using it. Tools, which were immediately usable without any thought, were felt to be especially beneficial to creativity and allowed for greater productivity and efficiency.

This ease of use was also connected to the issue of skill requirements. The more a tool requires prior knowledge, expertise, or training, to use, the greater the barrier to immediate use. This was frequently a comment of interviewees that the training requirements of new digital tools could be a major drawback.

Another aspect of immediacy is the need for tools to be available whenever they are required to record ideas that arise unexpectedly, perhaps during periods of reflection or informal collaboration as described previously. An example of a context in which a designer might have an immediate need for a tool of some kind to record an idea was given by P4: 'Bringing ... a pad of paper feels unnatural sometimes because it's like we're at the copy machine and it's like 'ahh no, we've got to quickly sort this out'. So when it's not planned, you know.'

In this case the immediacy is directly tied to availability and accessibility. Having a tool immediately accessible in instances where there is an unplanned need is critical to spontaneous creativity. This is even more important for creative collaboration, where an impromptu gathering might be more difficult to re-create and coordinate, if the desired tools were not initially available.

A related issue is that whatever tools are employed should be quick to use and should not interrupt the designers' creative flow. P1 described the disadvantages of the latency inherent in many digital sketching tools: 'If I take it down to the level of being a designer and doing a sketch, think about that kind of thing. There's issues of latency. So the speed between your input and the reaction of the device. You know [tablets] and [smartphones] are great, because that's a media thing. But if you start to do note-taking and stuff, ... there's always a bit of a lag, that just kills it.'

This element of responsiveness is frequently considered a major issue with current technologies. Instances where a tool suffers from any sort of delay in application can bring the user's focus to the device they are using as opposed to the activity they are trying to accomplish.

Flexibility

A desirable characteristic of creative design tools that is closely related to immediacy, is flexibility. Flexibility, which was proposed by Guilford as one of three key factors (along with fluency and originality) that can characterize divergent thinking abilities (Guilford 1957), is a key theme in the literature on creativity. It has been suggested that tools developed with a high degree of flexibility of use are less likely to 'disrupt the flow of thinking and action' (Edmonds et al. 2005). This is very similar to the study's findings regarding immediacy, as described previously.

Recent research has also suggested that creative practitioners frequently work in an improvisational manner and that the tools they use should support this (Gumienny et al. 2013; Hoeben and Stappers 2005). This is supported and explained in a statement from one of our interviewees: 'You've got to be flexible. Because as a designer your brain needs to be really flexible.... because you then start making connections with unexpected things and that's where new and really valuable things come from' (P1).

This concept of flexible thinking allows for practitioners to be more effective in their divergent thinking and to develop ideas without being locked into a linear progression of similar ideas. Flexible thinking raises the potential of having creative jumps from one idea, related to one path of thinking, to another path of thinking that might only tangentially be related to the first idea, thus resulting in a possible novel idea.

Previous work has linked creative design to flexible work environments (Bellotti and Bly 1996), and this was evident in our study when interviewees talked about moving things around to configure their work spaces. Environments with fixed infrastructure such as interactive tables or wall-mounted devices can be problematic in this respect. Such a lack of flexibility can manifest itself in various negative ways. For example the use of a digital table-top, with an interactive surface, might result in a more rigid environment where people might tend to be stationary in seats around the table. This could potentially inhibit the types of personal interactions and collaboration (Fernaeus and Tholander 2006). Additionally, the physical dimensions of the tabletop might enable only a very limited number of users (Geyer et al. 2011).

Related to this is a theme that arose a number of times in both the interviews and survey in relation to flexibility: that of portability, or being able to move things around. If a device or tool is not portable and can only be used in a special room, then it is not suitable for supporting the kind of spontaneous and informal collaboration described above, and any use also requires additional administrative work to plan and coordinate its application.

Several of the participants explained how useful it is to be able to move post-it notes around during creative design activity. P5 explained: 'It's kind of the flexibility of just being able to pick something up, move it, stick it down, shift it, that's what makes it work well.'

P1 further elaborated on how the portability or mobile nature of small analogue design artefacts can positively impact informal collaboration: 'So its the real, this is what I mean, they've printed stuff off, often it can be lots of post it notes, lots of paper, and its about . . . These things are really mobile, and you can be sitting at your desk where you'll prop these against the wall referring to these as you work and other teams might come over to you and say 'what was that idea that you were talking about?' '.

Other issues relating to flexibility that were raised by survey respondents included the restrictiveness of some digital tools due to their form factor (for example providing only small screens), the need for specialized equipment, and compatibility with other tools. These were mainly mentioned as potential disadvantages of digital tools.

Environment

Characteristics of the physical environment that emerged as being important in supporting early stage design were as follows.

Individual Spaces

Within the collaborative context, direct communication, as previously described, is particularly important in the development of ideas. However, even within collaboration the act of representing specific ideas is still very much an individual act, performed by individuals. When designers were asked about where and how they were creative (or generated creative ideas), individual workspace was emphasised as an integral part of their creativity.

This concept of individual workspace seemed for some to include the idea of a personal information space, such as a personal notebook for sketching when alone on the train, for example: 'I really use my notebook and have it with me all the time.' (P3). This personal space allows for designers to be creative and develop ideas without fear of their ideas being evaluated by peers, before they are prepared to share them.

Others talked about how initial creative work might happen within an individual designer's head before being shared and developed with others. For example, P3 explained how: 'When it's a personal project, I think the creativity does happen at the very early stages. But it's happening within myself. ... But then I may need to bring in people to help me figure out what it is that I'm doing.'

This is perhaps the more crucial element of allowing for individual workspace within a collaborative environment. While an individual will usually have the personal mental creative space to come up with ideas and to develop them, it is also crucial to facilitate a personal space where idea representation and reflection can happen on an individual level. Allowing for this space can remove some of the evaluation apprehension from the situation and provide a safer, risk free space, to explore and be more creative.

Communal Spaces

As has already been mentioned above in relation to informal collaboration, communal spaces were frequently referred to in our interviews as environments where both informal and facilitated collaborations could take place. The basic requirements for such spaces appeared to be quite minimal, for example P1 described how: 'We go outside and sit on the bench and just talk about it.' P4 explained how: 'if it's a real big problem, ... then we'll go in some room, like a meeting room, or something where we've got loads of whiteboards and loads of marker pens and all that stuff. And we'll just draw things out.'

Three of the interviewees mentioned that spaces for collaboration should ideally be comfortable. P3 described one of her favourite spaces for collaboration as being both flexible and comfortable: 'there's also just couches, and cushions, and you can move them around. And it's a brilliant place to collaborate. And we did use that space quite a lot, and meet there, and it was sort of comfortable, and a great place to discuss ideas.'

Additionally, P2 described the use of both public spaces, such as coffee shops, for collaboration, as well as more dedicated spaces within her own company's offices: 'A lot of people come here [coffee shop] actually. We also have, we have that open space where we all sit. We also have smaller little rooms. Comfy chair, a white wall that we can draw on. So that we can have more of a closed space, that's not distracting.'

There was great variety in the descriptions of the communal spaces that were preferred and used. To some extent this variety also seemed to be a valuable aspect. One participant described comfortable collaborative spaces with the appropriate infrastructure, but also highlighted that busy public spaces were frequently used for collaboration. Where these public spaces might be "distracting", they also appeared to offer variety of environment, which could stimulate creativity.

P2 described her working environment as follows: 'Open plan, we all sit together. Like in small groups. We have whiteboards behind us to sketch. We have a glass wall. In some places we draw on the glass. We get post it notes up there. Yeah, its pretty cool.' P1 also described how writing things down in communal spaces is useful for sharing ideas: 'Yeah, you want to be able to see it. You want to be able to get it in front of people and share it. ... We're doing that in an analogue way. You know, being able to use this whole wall as a way of mapping out an idea and a thought process is incredible.'

Both of these quotes highlight one of the most important attributes of communal spaces, which is public visibility. This public visibility allows for the results from the creative collaboration to be easily shareable both during and after the collaboration. Persistent visibility of ideas can in turn stimulate further creativity and also facilitate reflection.

Persistent Cues

The study found that one of the possible ways to support creative collaboration, was through awareness of earlier work, and the provision of persistent cues that could enable shared understanding and reflection on ongoing projects. While the issues related to informal collaboration and communal spaces have already been discussed, this section focuses more on the advantages of persistent cues enabling awareness of ideas previously generated during collaborative design.

The informal collaboration and communal spaces are key elements of the potential of persistent cues in that they enable immediate access (sometimes in a passive low focus manner) to the earlier ideas. These earlier ideas can either be shared or reflected on in either a direct manner or in a passive indirect way where indirect sharing, or reflection, could be the result of the persistent display of artefacts (digital or analogue) generated from an earlier creative collaboration.

In a number of cases interviewees expressed the desire to go back and view the scene of informal collaboration, where ideas had been generated. This was thought of as a way to possibly stimulate new creativity. In describing this P1 spoke about how the boards used to capture ideas as analogue drawings and artefacts were very useful to have in plain sight while continuing to work on a project. Talking about outputs from a design meeting, he explained: 'We'll just have them there facing out. That people can sort of glance at them, ... Yeah, ideally you want to have them out all the time. Because there's something, your brain is always working and that's just a reference to go back to. But it's a very ad hoc way of doing things, but it seems to work.' In this case the physical artefacts "facing out" and displaying the outputs from earlier ideation enable the persistence of the ideas and both focused and passive reflection.

Finally P7 described how he used a persistent display of competing designs to help shape their development and facilitate collaboration: 'I just draw each thing. So I put it up. So I have 4 designs. Each separated by like a line of white masking tape. ... And I can have people come over and go 'Uh I don't like that' or 'That's a good idea.''

This indicates that the persistence and accessibility of earlier creative results can also further stimulate ongoing collaboration. The facility of co-workers to be able to engage in impromptu conversation (informal collaboration), can be a very valuable quality to a creative enterprise.

Tools for Early Stage Design

Throughout the interviews many of the participants described how they use certain tools to help them in their current creative collaborations. The themes previously identified as being crucial to activities in the early stages of creative collaboration, and the environments in which they take place can also be linked to these tools and the roles they perform. In this section, we provide an overview of the strengths and weaknesses of current digital and non-digital (analogue) tools being used by designers in our study to support those activities in those environments.

In reviewing the benefits and drawbacks of the various tools used, it is clear that analogue tools have a very important role during creative collaboration. For example, the proportion of survey respondents using analogue and digital tools 'when initially coming up with ideas' in early stages of design is shown in Fig. 2. From both the reports of interviews above and the figure below, we can see that the use of analogue tools at this stage is currently more prevalent than that of digital tools. However, the limitations of analogue tools such as pen and paper in terms of capturing such outputs for longer-term reuse and sharing were mentioned by around a half of survey respondents.

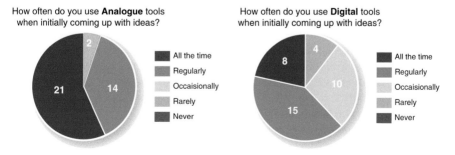

Fig. 2 Numbers of survey respondents using analogue and digital tools when initially coming up with ideas

Almost all of the interviewees talked about the importance of writing things down in shared spaces, and it is interesting to note that this is still being done using simple analogue tools such as whiteboards and pens. For example, P4 described how 'we've got a room in the office where it's just whiteboards. And the world's your oyster. You can just scribble over everything' and P6 how 'We've got these big whiteboards. Massive things on wheels. Because it gives us the space to just scribble and try ideas.'

The flexibility offered by analogue tools was cited as an advantage of analogue over digital by around a quarter of survey respondents, and explained by one of the designers we spoke to as follows: 'Paper, or analogue of any kind, is very flexible for [workshop] participants. There are … many fewer constraints and restrictions on what they can do, and I think fewer pre-conceived ideas. I think asking people to work immediately in a digital format can be a little bit more prescriptive of what they are going to do' (P5).

The comments above offer insight as to how analogue tools support flexibility, and immediacy. The ability to "scribble over everything" for example implies a high degree of both qualities. Due to the freedom inherent in an environment filled with analogue tools, there is a wide range of ways that ideas can be represented (flexibility) and the facility to do so is immediate and accessible. Furthermore, the physical affordances of analogue tools, in combination with the environment, can in turn support direct communication, where communication between multiple people can happen simultaneously, in parallel with sketching.

Additionally, the physical and 'haptic' nature of analogue tools has previously been seen as a key ingredient to enabling both individual expression and collaborative interactions (Geyer et al. 2011). This was mentioned by around a third of survey respondents as being an advantage of analogue tools. Previous work has suggested that the physical aspect of analogue tools and their corresponding affordances allow for interactions that are not always possible with digital tools. For example the haptic nature of physical tools allows for unsighted inaudible feedback, which is frequently absent from digital tools (Treadaway 2007), an example being where the sense of pressure exerted on paper using a pencil, or marker, might give the user feedback as to how thick or dark a drawn line is, without actually seeing the line.

Currently one of the key differences between analogue and digital devices relates to the theme previously discussed, 'immediacy.' Almost half the survey respondents referred to speed of use as being an advantage of analogue tools, and slowness as a disadvantage of digital. Survey respondents also cited loss of flow as a disadvantage of some digital tools. This was also reflected during the interviews. P4 reported how the need for everyone on the team to be able to use the same tools without expending additional effort on learning how to do so meant that tools that were perhaps technically less suited to a task were sometimes used anyway, just because everyone was able to use them.

Six of the survey respondents specifically mentioned familiarity and ease of use as advantages of analogue tools, and difficulty of use or an associated learning curve as a disadvantage of digital tools. Additionally, two of the interviewees commented specifically on the learning curve associated with some digital tools being off-putting. In relation to sketching and note-taking on tablets and smart phones, P1 commented: 'And there's a big learning curve to doing all that kind of thing.' and P7 complained that sometimes 'I'm spending more of my time learning tools than actually working.'

Immediacy was also cited as an issue with remote collaboration. While the ability to support remote communication and collaboration through the use of digital technologies was cited by around a third of the survey respondents as an advantage of digital tools, the difficulties of attempting to substitute same time, same place collaboration with distributed collaboration were described as significant drawbacks. As P1 described: '... the video you've got isn't normally of everybody in the room getting this eye-to-eye contact, the resolution isn't that great, and again there's that issue of latency'.

Similarly, P4 commented: 'We've found that it's really difficult to do Agile or all that kind of stuff over [web conferencing product]. Communication is really difficult. Even though you can see everything, people talk over each other, people don't feel like they're a part of it, or there's a bad line, there's always something that just doesn't work.' P4's comment also points to the inherent difficulties associated with attempting to use digital tools to support direct communication.

P2 also described how digital tools restrict this type of activity: 'But also, you're doing one thing at a time when you're using technology to do something, rather than multiple sketches quickly ...' While this issue of only being able to do "one thing" relates to how technology can restrict both immediacy and flexibility, it also further illustrates the inherent difficulties with using digital tools in a collaborative environment, where multiple activities between multiple people are crucial to the fluency of creativity.

However, one of the key benefits of digital tools is the capture and digitisation of the results of collaborative activities. The possibilities of capturing ideas in a way that they can be stored, organised, archived, shared, and used for documentation, in which changes to ideas during the course of a project can be tracked, were cited by around a third of the survey respondents as some of the most significant benefits of using digital tools in a creative collaborative process. Conversely this was also described as a drawback of analogue tools, which are not easily shareable, awkward to store and can get lost.

There can still be difficulties in the digital collection of ideas in that there is also a dependence on the organsiational systems. Where as analogue tools are dependent on the organizational efforts of the users (i.e. filing systems and preferences), digital tools require an added element of system design to enable reuse of the ideas captured.

However, digital tools can, through the effective capture of ideas, facilitate acts of creative reflection. One of the interviewees (P8) adopted a digital approach to her own reflection: 'I've got this note thing on the phone. Yeah it's kind of random thoughts. ... Sometimes when I was in conversation, ... when I get home later I'll be like 'Oh there was something really interesting'. ... and then I would type it down on the notepad. And then once in a while I'll look through it', and described how reflection was facilitated for her by the constant availability of a digital device such as a phone: 'I would take out the phone in a situation like on the train, or after a meeting, where I don't have any paper. I'm just going to jot down whatever I'm thinking. Little thoughts. I actually tried to do audio once. It did not work.'

This also points to one of the potential advantages of digital devices supporting reflection (and other creative activities), in that the ubiquitous nature of a device can allow for a wider range of activities. A device that is already carried for other purposes (e.g. a phone) which supports a variety of activities can also support a greater degree of flexibility and immediacy in relation to creative activities. However, the collaborative, creative work that designers are doing is complex, and as P2 explained: 'The problem with technology and collaboration tools is that there are so many and not one thing does everything you need them to . . . . usually these tools are designed in isolation for one problem, rather than a whole situation.'

Summary

This study has confirmed that current practice is dependent on a number of interconnected elements that potentially increase the effectiveness and efficiency of creative collaboration.

Participants continually identified collaboration, in a variety of forms, as being integral to their current creative process. Additionally, whether it was a quality of the tools they were using (e.g. immediacy) or an element of the environment (e.g. individual spaces), it was clear that there were a number of fundamental attributes of designers' tools and working environments that were felt to positively influence creative outcomes. Where previous research has identified many of the themes present in this study, the interconnectedness, and possible mutual dependencies between these themes has perhaps not previously been explored to any great extent. An example from our study that illustrates the dependency between some of the themes discussed is the fact that while informal collaboration has been identified as a key element of creative collaboration, in idea representation, capture and reuse, both flexibility and immediacy are required of the tools within the environment for this to occur. This is illustrated by P2 describing how: *" . . . the most creative collaboration that happens with people in my proximity, where you have informal idea generation from a side conversation and you start to talk about it, and if its interesting you start to sketch ideas."* In this example the collaboration is dependent on communal space, which supports immediate, flexible, informal collaboration, resulting in idea representation (sketching ideas), which is in turn also dependent on the immediacy of tools (in terms of accessibility, ease of use, and immediacy of output). This example also demonstrates how immediacy and flexibility are underlying themes that support multiple facets of creative collaboration.

In addition, as past research has highlighted, collaboration amongst creative practitioners is 'supported by very simple low-tech tools, such as Post-It notes, color pencils, sketch papers, tapes and so on' (Zhu 2011). Our study confirms these findings, but also reveals how designers are increasingly finding a role for digital tools, such as smart phone cameras, video conferencing software and email, to support their work. One of the strengths of some of the more frequently used digital tools, such as smart phones, is their ubiquitousness, which mirrors the availability and accessibility of analogue tools.

The combination of different types of tools in early stage creative collaboration is indicative of the effectiveness of using the distinct strengths of analogue and digital tools to complement each other. For example, analogue tools are well suited to fast, spontaneous idea representation in a collaborative environment. Participants commented on using sketchbooks to develop ideas and then scanning or photographing the pages of the sketchbook in order to share the ideas generated. Digital tools, such as smart phones, are then frequently used to capture the outcomes. In this case the digital tool potentially facilitates sharing, reflecting on, or reusing the original ideas in more flexible environments (e.g. reviewing images on a smart phone in a café).

Sharing of these digital records seems mainly to be done using basic technology solutions such as email. At present, the organization of the resulting information and digital assets remains a challenge and can potentially limit the effective review and reuse of initial representations of ideas. Participants commented on the difficulty of finding and reusing specific digital assets quickly in large and possibly complex storage systems. Existing digital systems (software and digital file storage systems) designed to manage creative outcomes, such as the ideas generated from early stages of creative collaboration, could be greatly improved. The powerful capabilities of digital systems to store and archive information is frequently not matched by the usability of those systems and the accessibility of the stored information.

Another real strength of digital tools is their ability to support remote collaboration, although awareness of the activities of designers on remote sites is still not as good as it would be for same time same place collaborations. However, risks associated with digital tools include their lack of immediacy and flexibility, and associated possibility of breaking the creative flow of both individuals and collaborative teams, as well as the lack of support they currently provide for the kind of informal spontaneous collaboration that is obviously so common. Additionally, the technical overhead (i.e. setup, maintenance, training) involved in digital tool use is a barrier to use in early stage creative collaboration.

In conclusion, although many of the key themes identified in the existing literature are obviously still very relevant to the work of current designers, and analogue tools still appear to be preferred for a number of activities, such as fast and spontaneous collaboration and idea representation, it appears that there may also be a subtle but ongoing shift amongst designers towards the increasing adoption of digital tools, especially simple and ubiquitous tools including smart phone cameras, video conferencing software and email, that begin to mirror some of the key interaction qualities, such as immediacy and flexibility, of the analogue tools that have so long been favoured.

References

Bellotti V, Bly S (1996) Walking away from the desktop computer: distributed collaboration and mobility in a product design team. In: Ackerman MS (ed) Proceedings of the 1996 ACM conference on computer supported cooperative work (CSCW '96). ACM, New York, pp 209–218

Brophy DR (2001) Comparing the attributes, activities, and performance of divergent, convergent, and combination thinkers. Creat Res J (1040–0419) 13(3–4):439

Casakin H, Kreitler S (2011) The cognitive profile of creativity in design. Think Skills Creat 6(3):159–168

Coughlan T, Johnson P (2008) Idea management in creative lives. In: CHI '08 extended abstracts on human factors in computing systems (CHI EA '08). ACM, New York, pp 3081–3086

Coughlan T, Johnson P (2009) Understanding productive, structural and longitudinal interactions in the design of tools for creative activities. In: Proceedings of the seventh ACM conference on creativity and cognition (C&C '09). ACM, New York, pp 155–164

Edmonds EA, Weakley A, Candy L, Fell M, Knott R, Pauletto S (2005) The studio as laboratory: combining creative practice and digital technology research. Int J Hum Comput Stud 63(4–5):452–481

Fernaeus Y, Tholander J (2006) Finding design qualities in a tangible programming space. In: Grinter R, Rodden T, Aoki P, Cutrell ED, Jeffries R, Olson G (eds) Proceedings of the SIGCHI conference on human factors in computing systems (CHI '06). ACM, New York, pp 447–456

Geyer F, Pfeil U, Höchtl A, Budzinski J, Reiterer H (2011) Designing reality based interfaces for creative group work. In: Proceedings of the 8th ACM conference on creativity and cognition (C&C '11). ACM, New York, pp 165–174

Guilford JP (1957) Creative abilities in the arts. Psychol Rev 64(2):110

Gumienny R, Gericke L, Wenzel M, Meinel C (2013) Supporting creative collaboration in globally distributed companies. In: Proceedings of the 2013 conference on computer supported cooperative work (CSCW '13). ACM, New York, pp 995–1007

Gutwin C, Greenberg S, Blum R, Dyck J, Tee K, McEwan G (2008) Supporting informal collaboration in shared-workspace groupware. J UCS 14(9):1411–1434

Hoeben A, Stappers PJ (2005) Direct talkback in computer supported tools for the conceptual stage of design. Knowl-Based Syst 18(8):407–413

Howard TJ, Culley SJ, Dekoninck E (2008) Describing the creative design process by the integration of engineering design and cognitive psychology literature. Des Stud 29(2):160–180

Rhodes M (1961) An analysis of creativity. The Phi Delta Kappan 42(7):305–310

Sedivy J, Johnson H (1999) Supporting creative work tasks: the potential of multimodal tools to support sketching. In: Proceedings of the 3rd conference on creativity\& cognition (C\&C '99). ACM, New York, pp 42–49

Treadaway C (2007) Using empathy to research creativity: collaborative investigations into distributed digital textile art and design practice. In: Proceedings of the 6th ACM SIGCHI conference on creativity\& cognition (C\&C '07). ACM, New York, pp 63–72

Zhu L (2011) Cultivating collaborative design: design for evolution. In: Procdings of the second conference on creativity and innovation in design (DESIRE '11). ACM, New York, pp 255–266

Using the *Bright Sparks* Software Tool During Creative Design Work

James Lockerbie and Neil Maiden

Abstract This chapter describes *Bright Sparks*, a web-based software tool that provides support for a codified version of the established *Hall of Fame* creativity technique. After summarising experiences with our manual use of the technique, the chapter reports codified knowledge, collected through practice, about more effective use of fictional personas in creative processes. This codified knowledge was embedded in the new web-based software tool to provide automated support to use the *Hall of Fame* technique. The tool has been developed to meet the needs of product and concept designers. The chapter ends with a report of the first evaluation of the application with designers, and future developments to extend the software tool.

Creativity Thinking in Early Design

Creative thinking is core to early design activities. The United Kingdom's Design Council defines design as shaping ideas to become practical and attractive propositions for users or customers, and it can be described as creativity deployed to a specific end (Design 2011). Design is both a creative and user-centred approach to problem solving that cuts across different professions, from art and design to engineering and architecture. As such, creativity is needed to generate new ideas that design can shape to become the practical and attractive propositions for users or customers.

To deliver more creative design processes over the last decade, design thinking has become accepted practice. Design thinking is a human-centred innovation process that involves observation, collaboration, fast learning, the visualization of ideas and rapid prototyping, all of which run concurrent to business analysis activities (Lockwood 2010). It has been successfully used in projects to design new workplaces, consumer products and even brands. However, one criticism that can be leveled at most design thinking processes is the lack of explicit use of creativity techniques from creative problem solving communities. Indeed, we observe an

J. Lockerbie (✉) • N. Maiden
Centre for Creativity in Professional Practice, City University, London, UK
e-mail: James.Lockerbie.1@city.ac.uk

© Springer International Publishing Switzerland 2016
P. Markopoulos et al. (eds.), *Collaboration in Creative Design*,
DOI 10.1007/978-3-319-29155-0_15

increasing disconnect between design thinking and creative problem solving, and believe that new techniques and tools that bridge the outputs of these communities are needed. In our previously reported work, we have explicitly and successfully used creativity techniques in complex requirements processes, for example (Maiden et al. 2004, 2007; Zachos and Maiden 2008). That said, most of these successes have been achieved during better-resourced, longer requirements processes.

A more recent challenge was to integrate creativity techniques effectively into shorter, more iterative agile development methods and projects. We have reported case studies (Hollis and Maiden 2013) that describe the use of selected creativity techniques in short workshops of less than 1-h duration in agile projects, and developed a battery of existing creativity techniques suited to such workshops. These techniques include constraint removal to open a solution space, creativity triggers that direct problem solvers to solutions with pre-defined qualities, and desktop walkthroughs to simulate a new design with physical models. Another one of these techniques was derived from Michael Michalko's *Hall of Fame* technique (Michalko 2006).

The *Hall of Fame* technique provides creative problem solvers with personas to guide the exploration of a space of creative ideas, for example guidance such as how might a well-understood persona such as *Margaret Thatcher* solve the problem. We successfully applied a variation of the *Hall of Fame* technique in one agile project with BBC Worldwide. This *Random Stars* technique was designed to provoke creative thinking by asking the right questions, at the right time, from different perspectives. It was adapted from *Hall of fame* (Michalko 2006) to the domain of television programmes to make it fun and relevant to the organisation and its staff. The facilitator chose 14 programmes, represented by their main characters such as *Vicky Pollard* from Little Britain, *Alan Sugar* from The Apprentice and the *Top Gear* presenters, then photographs of each one were printed out, cut-up and placed into a bag. Participants used the characters to generate new design ideas (Hollis and Maiden 2013).

That said, although successful, the wider uptake of the *Random Stars* technique beyond the one agile project at BBC Worldwide was not possible due to a lack of software support for the technique. Use of the technique depended on a human facilitator both to retrieve and select the relevant personas, then to guide designers and stakeholders to use these personas during the workshop. An opportunity arose to codify the technique and its use, then to make it available to design projects as a new web-based software tool.[1]

In this chapter, we will report our new *Bright Sparks* web-based software tool derived from Michalko's *Hall of Fame* technique and our earlier version of it in the *Random Stars* technique, the explicit codification of this technique, and the use of this codification to deliver the new creativity support tool. The next

---

[1]This new tool has been developed as part of the EU-funded FP7 COLLAGE project. Details about the project are available at: http://projectcollage.eu

sections describe the original *Hall of Fame* technique in more detail and its current derivations and uses, and the extended codification of this technique. The chapter then introduces *Bright Sparks*, our new, publicly available web-based tool that uses inspiration-based search techniques to deliver an enhanced version of the *Hall of Fame* technique to designers from different backgrounds. It ends with some first evidence from designer use of this software tool, and its on-going integration into design thinking processes.

The Original *Hall of Fame* Technique

The original *Hall of Fame* technique seeks to provide problem solvers with support from famous and relevant characters such as a Nobel Prize winner, former president or extraordinary artist. The technique is suitable for unstructured problem solving with open questions and answers. It is a simple and effective technique that allows problem solvers to explore how a group of famous people will help solve their challenge using their unique knowledge and practices. To do this, it uses quotes from these people to inform creative thinking by the problem solvers that can be sourced from books, newspapers, magazines, and other media. It also recommends the use of web searches, in which the problem solver uses the main verb of the problem they are trying to solve to find appropriate quotes, such as to minimise something or improve something. Vázquez (2013) suggests a sequence of actions for the approach, which we summarise as follows:

1. Define the challenge to be solved by the team and choose some famous people to act as advisors. Select some appropriate sentences from the Web. For example, how could *Rupert Murdoch* help solve your problem using his uncompromising style to running and acquiring businesses?
2. Seek to solve the challenge using the Internet-based advice from the famous person – in this case *Rupert Murdoch*. Use the results of web searches to inform your understanding of what Murdoch would do, then apply this understanding to solve the problem;
3. Repeat the exercise several times with new quotes by different people. Three famous people, two sentences each (six quotes overall) is a good number to ensure that new and creative ideas are generated.

As such, we treat the *Hall of Fame* technique as a technique that supports Margaret Boden's (1990) concept of exploratory creativity. It is a technique with which to search one or more predefined spaces of partial and complete possibilities that are to be discovered in these spaces. Each famous person delineates a subset of a search space through their predefined association with those ideas – some of the commonly understood reasons for their fame. Quotes and characteristics of these famous people provide simple-to-use rules for problem solvers to search these subsets of spaces to discover ideas.

As already reported, we successfully used the *Random Stars* version of the *Hall of Fame* technique in the early design processes during creativity workshops (Hollis and Maiden 2013). During these workshops, and each subsequent use of the technique, a human facilitator was required to direct the stakeholders in the use of the technique and to make a set of pre-selected personas available to the stakeholders. These personas were selected for the stakeholders based on the facilitator's domain knowledge and relevance of discovered personas to the design challenge. During the workshop, each persona was presented as a simple photomontage containing only one, often iconic, photograph of the persona. No names, words or other content were presented to the design stakeholders. To compensate for this lack of information, the facilitator would first encourage the stakeholders, as a group, to discuss all of the personas, recognise them, and start to explore their defining characteristics with respect to the design challenge.

The photomontage from the agile workshop at BBC Worldwide reported in (Hollis and Maiden 2013) depicted fictional characters from BBC television programmes, and were selected for use in the workshop as strong characters that all of the stakeholders would be familiar with. As such, each persona was designed to provide a degree of common ground between all stakeholders using that persona at that stage in the design process. During one short session of 20 min during the agile workshop, 10 new requirements were generated, for example capabilities to capture a user's emotional response to a programme or character, allow the users to indicate which actors they lusted for, and a tool to enable and encourage greater social inclusion.

A second photomontage that depicted famous people from Sweden was used to train a large group of consultants based in Stockholm to use the technique as part of a training workshop. Here, due to the distance between trainer and trainees, a more generic group of individuals was selected, to reduce the common ground between the facilitator and the workshop participants. This workshop revealed one challenge associated with the use of personas in creative design processes – none of the people present in the workshop recognised the Swedish actress Ingrid Bergman from her photograph in the photomontage, and the facilitator was required to step in and name the persona for the participants.

To conclude, these and our other experiences with the *Hall of Fame* technique are that it is an effective creative thinking technique in design processes, but only with human facilitation and expertise developed over time in the use of this technique with designers and stakeholders. Alas, these resources and expertise are not always available on demand in design projects. Therefore, we sought to develop a version of the *Hall of Fame* technique available to designers and stakeholders more widely, without the need for human facilitator involvement. Our knowledge of the technique and its uses was codified and implemented in a new, web-based software tool called *Bright Sparks*. The tool and its development is described in the next two sections.

Codified Knowledge for the *Bright Sparks* Software Tool

Bright Sparks implements an adapted version of the *Hall of Fame* technique that allows a problem solver to explore how well known personas (either fictional or real) and their more extreme characteristics and traits would solve a particular problem. Unlike the original *Hall of Fame* it does not use quotes, but creative clues applied to the personas for use during creative design tasks.

We sought to codify the technique to deliver a simple and effective web-based software tool that users could use with little or no preparation. Our aim was to remove the need for human involvement to prepare famous people personas or to facilitate the creative process.

Therefore, to develop the web-based software tool, we sought to externalise and codify our own knowledge about effective use of the *Hall of Fame* technique. We had applied *Hall of Fame* extensively and successfully in unreported Masters course teaching, in training tutorials, and in commercial projects with clients. However, this successful use had required us to extend the technique as published, to overcome common problems and misunderstandings that people had during the use of the technique. Therefore, we iteratively reflected, externalised and shared our knowledge of good practices with the *Hall of Fame* technique, so that the externalised knowledge could be codified and implemented in the new web-based software tool. The result was two types of codified knowledge – knowledge about the types of information about a persona to present during creative problem solving, and knowledge about the clues to use to prompt creative thinking about a persona. Each type is reported in turn.

Codified Knowledge About a Persona

The following types of information about each persona were externalised as important during a creative problem solving process:

1. The **name** and a **photograph** of the persona, e.g. *Jonathan Ive*;
2. The **role** of the persona, e.g. *product designer*;
3. Important **defining characteristics** of the persona, described in less than 100 words, e.g. *he believes in the importance of materials, and making them part of the design process*;
4. The wider persona **type**, to enable persona categorisation. Categories include *Design* and *Dutch personas*.

Our earlier experiences with the *Hall of Fame* technique revealed that many people cannot recognise famous people, both real and fictional, from their photograph alone, so both the name and photograph of each persona were codified. The aforementioned example of Ingrid Bergman was a good example of this. Furthermore, people often did not have sufficient knowledge of that persona's role or defining characteristics, so the human facilitator in a workshop would often have to supplement the name and photograph with simple verbal descriptions of each persona. In contrast, we chose not to include video content about personas, for two reasons. The first was the availability of suitable online video material – such material is not available for all personas. The second reason was a concern not to infringe copyright of materials available on the Internet. Finally, in preparing creative workshops, it was often useful to have predefined categories of personas, for example by role and by nationality, to ensure relevance and familiarity to the creative problem solvers.

Information about each persona that was applied in our earlier uses of the *Hall of Fame* technique was codified using the above structure. Moreover, we used these information types to provide a simple template to direct the search for and to manage information about new personas to be included in the tool.

Codified Knowledge About Creative Clues

Similarly, we externalised four different types of codified clue specified for use with personas during creative design tasks. One common problem with the *Hall of Fame* technique was that creative problem solvers needed some specific directions during the problem solving process, to direct them to use knowledge about the persona. Simply imagining the role of that persona in the process was insufficient. The resulting four types of codified clue were:

1. **The persona as stakeholder**: Imagine that the persona is a stakeholder in the design process, and explore what needs and knowledge that the stakeholder would contribute;
2. **The persona as inspiration or role model for the product or service**: Imagine that the persona has characteristics and/or qualities that the product or service should have;
3. **The persona as emotional inspiration**: Imagine that the persona is a source of emotional engagement in the design process, and evokes different emotions in the designers and other stakeholders;
4. **The personas are combined to generate new personas**: Imagine that the persona is combined with another persona to generate new combinations of characteristics and qualities that can then provide a new persona as a stakeholder or an inspiration or role model for the product or service.

From these types, we codified a larger number of concrete creativity clues that could be applied to most personas to direct creative thinking. These codified clues are specified in Table 1.

Returning to our earlier example, the codified knowledge includes not only the name, role and defining characteristics of Jonathan *Ive*, but also explicit instructions to creative problem solvers to imagine if *Ive* joins their project team, and consider

Table 1 The codified concrete creativity clues for use in the *Bright Sparks* web-based software tool

The personas as stakeholder codified clues direct the designers to ask the following questions:
Imagine the person is your client. What extra features or qualities would the product or service need?
What if the person joins your project team? What new ideas and concepts will the person come up with?
Put yourself in the position of the person. Can you envision their needs? What might these be?
What would the person do in your place? Push your thinking to the maximum.
Who else does the person make you think of? Someone similar or opposite? Imagine that person is your client. What would that person want from you?
Does the person have any friends or colleagues? What new ideas and concepts would you expect this person to come up with?
Do the opposite of what the person would want for your service
Imagine you interview the person for your project. What do you predict that the person would want?
The personas as inspiration or role model for the product or service codified clues direct the designers to ask the following questions:
What are the most notable capabilities of the person? Can you build these capabilities into the product or service?
What is the most important power of the person? Could you incorporate it into your product or service?
What is the person most well known for? Imagine your client is also well known for this. What would you do to keep such a client happy?
What is the most unappealing characteristic of the person? Imagine you have that characteristic. What needs would you have?
What strengths does the person have? Use these strengths when thinking of new ideas?
What weaknesses does the person have? Think of new ideas that could exploit these weaknesses in someone or something
How did the person become famous or infamous? Can your project follow the same route to fame or infamy? If so, what would you do?
List the 5 most important attributes of the person. Take each one in turn. Can it be modified to be an attribute of your new product or service?
Where does the person come from? What do people from there tend to want?
Dressed up as the person for a fancy dress party, would might you feel? How would it change you needs and desires?
Does the person have a catchphrase? Say it out loud. What new ideas come to mind?
In what environments would you encounter the person? What new ideas might being in that environment trigger?
Pretend to act like the person. Mimic them. How does it make you feel? What new ideas arise from acting like the person?
Who in your team has a real affinity to the person? Make them role play the person in your project
What values does the person have? Design your product or service to have these values
What emotions does the person make you feel? Use these emotions about the person to think about new ideas

(continued)

Table 1 (continued)

The personas as emotional codified clues direct the designers to:
Imagine that your new product or service is like the person. What qualities would it have?
The personas are combined to generate new personas codified clues direct the designers to ask the following questions:
Imagine the person and another person meet. What kinds of ideas might you expect the two of them to generate?
Combine the person and another person into a single character. What qualities would the product or service need?
What if the person and another person join your project together as design consultants? What new ideas and concepts will the person come up with?
Combine the person and another person into a single character. Can you envision their needs? What might these be?

what new ideas and concepts would he come up with, to incorporate the most important power of *Ive* into their product or service, and to imagine the qualities of the product or service if it was like *Ive*. Our purpose was to develop a software tool that removed the need for a human facilitator. Therefore, the codified knowledge was applied to develop *Bright Sparks*, a new web-based software tool for creative problem solving.

The *Bright Sparks* Web-Based Software Tool

One important requirement on the *Bright Sparks* software tool was to ensure creative problem solvers could use a form of the *Hall of Fame* creativity technique successfully without any external human intervention and guidance about the technique. Therefore, the codified knowledge was embedded into the software tool in order to seek to achieve this requirement. Another important requirement was to discover new personas and relevant knowledge about these personas without time-consuming external human intervention. Therefore, the tool was developed to exploit new forms of inspiration-based creative search that have been developed in the COLLAGE project.

Creative problem solving can be characterised as a process of information search and idea discovery in a large space of partial and complete solutions (Boden 1990). New computer-based search technologies have been adapted to support creative search of large information spaces made available by the Internet. For example, *Combinformation* is a web-based search tool that was designed to support creative manipulation of search queries and results (Kerne et al. 2008), *CRUISE* is a tool that was developed to support inspirational search of different internet and social-media sources, and we have developed computational term disambiguation and query expansion mechanisms to enable creative search in tasks such as the care of older people with dementia (Maiden et al. 2013). All of these tools have been

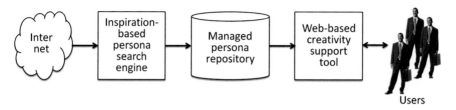

Fig. 1 The software architecture of the *Bright Sparks* web-based software tool

demonstrated to provide effective support for creative problem solving. Therefore, we designed *Bright Sparks* to exploit a form of internet-based search for new information about personas that can trigger creative thinking.

The software architecture of *Bright Sparks* is depicted graphically in Fig. 1. The architecture separates the inspiration-based search for information about personas from manipulation of this information during support for creative problem solving, for two reasons. The first is that personas and their information are independent of any concrete problem that people are seeking to solve creatively, for example the codified knowledge related to *Jonathan Ive* can be applied to a range of design problems. Therefore, the architecture includes a managed repository of codified knowledge about personas to guide creative problem solving. The second reason is that our experiences reveal that curated knowledge about personas is more likely to be effective than information retrieved from the Internet that is not curated, to guide creative problem solving tasks. Maintaining a repository of knowledge about personas that is searched, collected, curated then released to users for creative problem solving is predicted to improve creative problem solving. The current version of the repository has codified and stable knowledge about 213 different fictional and real personas. Moreover, if users need more dynamic forms of inspiration-based search, there are other software tools that are available for use alongside *Bright Sparks*.

Bright Sparks was designed and implemented to support the pain-free and playful exploration of information about personas that is a necessary pre-requisite in effective creative problem solving tools (Greene 2002). The design assumes that the users have already identified a problem that needs to be solved creatively, and therefore does not provide explicit support for problem identification. To use *Bright Sparks*, the problem solvers simply request the tool, at the press of a button, to retrieve codified knowledge about one random persona at a time, to select one of these personas to think creatively with, then to use the codified knowledge in the form of the creative clues to guide creative problem solving. Use of these codified clues continues until the problem solvers have exhausted the creative potential of the personas, then the user can request the tool to retrieve a new persona. New creative clues about each persona are presented in response to pressing a button.

The *Bright Sparks* creative problem solving web page, with the demonstration *Jonathan Ive* persona, is shown in Fig. 2. The web page presents the stored codified knowledge about each persona – the name, role, type, and characteristics, along

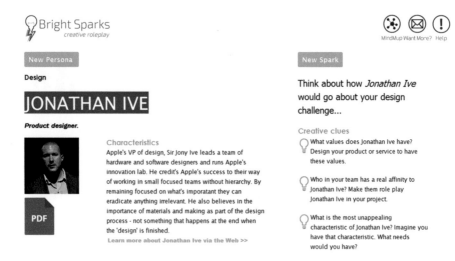

Fig. 2 The *Bright Sparks* web page for creative problem solving showing the *Jonathan Ive* persona

with an image. The page is divided into two parts. The left side presents information about each retrieved persona and features to manipulate personas, and the right side presents the codified creative clues instantiated for the current persona presented on the left side of the page.

The web page provides users with the following simple-to-use features, all of which are depicted in Fig. 2:

- **New persona**: to retrieve a new random persona. Because of this simple-to-use feature, personas can easily be skipped over if they are not suitable by repeatedly pressing the button;
- **New Spark**: to retrieve a new creative clue applied to the current persona. A retrieved new clue appears at the top of the list, causing the bottom clue to disappear from the list of clues. This feature allows the user to work through the clues until no more new clues are available;
- **Previous clue**: to walk backwards through the set of creative clues in the order of their generation, in order to be able to explore previously-generated clues;
- **Learn more about persona via the Web**: clicking this link opens a new Google search for this persona in a new browser window or tab with a pre-defined search query that includes the name and role of the selected persona. The persona role is included in the search criteria to remove results including unrelated namesakes from the search results;

- **PDF**: to generate a PDF version of the persona in a new browser window or tab. The web tool is read-only, but users will need to permanently record the outcomes of session, therefore a PDF outcome includes personas characteristics, image and creative clues. The user can then save and download the file through their browser.

Furthermore, Fig. 3 shows the presentation of other personas and creative clues that the *Bright Sparks* web page can present to creative problem solvers. The first two personas on the Figure – *Mata Hari* and *Alice Liddell* – are good examples of real and fictional personas that can be used in a wide range of creative problem solving processes. The other two personas – *Rem Koolhaus* and *Jakub Dvorsky* – represent personas that were retrieved and curated in the repository to provide explicit support for designers on collaborative design projects, in conjunction with design partners in our EU COLLAGE project. Examples of codified knowledge to guide creative problem solving with these two designer personas include imagining the five most important attributes of *Rem Koolhaus* and modifying each to be an attribute of your new product or service, and imaging that *Rem Koolhaus* and *Salvador Dali* meet in order to imagine what new design ideas that the two personas, together, might generate.

Figure 4 demonstrates how *Bright Sparks* supports users to explore more information about the current persona, and to generate a permanent record of the session about a persona. The top of the Figure shows the PDF print format of all codified knowledge about this persona that a user might want to record and take from the creative session with the tool, and the bottom of the Figure shows the typical Google search that is generated, in a new window, for *Jonathan Ive*.

The current version of *Bright Sparks* has been adapted to provide explicit support for designers, through curated knowledge management about a predefined set of 40 designer personas that were agreed with designer stakeholders. This current set of designer stakeholders are listed in Table 2, and reflect a current bias towards designer personas who would be familiar to designers in the Netherlands – many of the current designer stakeholders are based in this country.

Bright Sparks has been developed as a simple read-only web-based software tool, without user management or password protection, to maximise its take-up by designers and other creative problem solvers. We envisage a wide range of contexts of tool use, from individual use on a mobile device to collaborative use in large facilitated workshops, in which each persona is projected onto a large screen to guide creative thinking by larger stakeholder groups. As a consequence, new ideas that can be generated through use of the tool could be documented as audio notes on a mobile device, on paper post-it notes in creativity workshops, and as digital records on computing devices, and *Bright Sparks* has been developed to support all of these forms of idea recording. However, we recognise that some users will expect digital support for idea recording to be coupled with digital support for idea generation, and therefore *Bright Sparks* has been extended with an explicit link to *MindMup* (MindMup), an open source mind-mapping software tool.

New Persona

New Spark

Art

Daring exotic dancer executed as a spy.

PDF

Characteristics

Exotic dancer and courtesan she is perhaps one of the most famous female spies while known for daring outfits and her 'temple dance'. A free-spirited bohemian to some she was an icon of promiscuity with lovers in both Germany and France. During World War I she was found guilty of spying and 'causing the death of 50,000 Frenchmen' and executed by firing squad.

Learn more about Mata Hari via the Web >>

Think about how *Mata Hari* would go about your design challenge...

Creative clues

What weaknesses does Mata Hari have? Think of new ideas that could exploit these weaknesses in someone or something.

Imagine that your new product or service is like Mata Hari. What qualities would it have?

In what environments would you encounter Mata Hari? What new ideas might being in that environment trigger?

New Persona

New Spark

Books

ALICE LIDDELL

Heroine and dreamer from Alice's Adventures in Wonderland.

PDF

Characteristics

A daydreamer, she has an insatiable curiosity and values honesty and respect. Confident in her social position and education, she attempts to understand the strange fantasy world she finds herself in. Surrounded by Wonderland's nonsensical rules her fundamental believes are challenged at every turn.

Learn more about Alice Liddell via the Web >>

Think about how *Alice Liddell* would go about your design challenge...

Creative clues

Where does Alice Liddell come from? What do people from there tend to want?

What if Alice Liddell joins your project team? What new ideas and concepts will Alice Liddell come up with?

How did Alice Liddell become famous or infamous? Can your project follow the same route to fame or infamy? If so, what would you do?

Fig. 3 The *Bright Sparks* web pages for creative problem solving showing different personas and creative clues

MindMup Want More? Help

New Persona

New Spark

Design

REM KOOLHAAS

Dutch architect and architectural theorist.

PDF

Characteristics
Koolhaus has designed buildings all over the world, and written books on the development of urban life. His interest in urban living informs his work, and many of his buildings try to balance the needs of the building's primary users with the needs of the city and the people who live in it. He writes a lot about the speed of urbanisation and the impossibility of architecture to keep up with the pace of cutural change - by the time a building is finished it's already out of date.

Learn more about Rem Koolhaas via the Web >>

Think about how *Rem Koolhaas* would go about your design challenge...

Creative clues

List the 5 most important attributes of Rem Koolhaas. Take each one in turn. Can it be modified to be an attribute of your new product or service?

Imagine Rem Koolhaas and Salvador Dali meet. What kinds of ideas might you expect the two of them to generate?

Imagine that your new product or service is like Rem Koolhaas. What qualities would it have?

MindMup Want More? Help

New Persona

New Spark

Design

JAKUB DVORSKY

Czech Games designer and founder of Amanita Design studio.

PDF

Characteristics
Dvorsky's games are hand drawn and richly textured - they feel very atmospheric and 'handmade'. The goal is to be in the game - not to win it. As a designer he believes the most important thing is coherence of vision (story, visual and sound design) and therefore it's important to have a single auteur with a team of designers - not a collaboration.

Learn more about Jakub Dvorsky via the Web >>

Think about how *Jakub Dvorsky* would go about your design challenge...

Creative clues

What strengths does Jakub Dvorsky have? Use these strengths when thinking of new ideas?

What would Jakub Dvorsky do in your place? Push your thinking to the maximum.

Imagine that your new product or service is like Jakub Dvorsky. What qualities would it have?

Fig. 3 (continued)

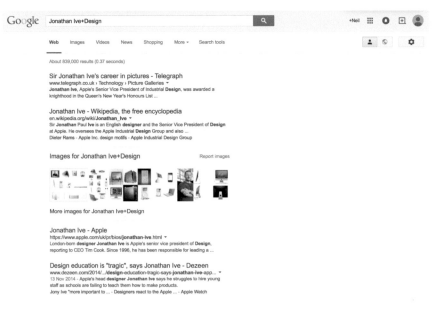

Fig. 4 The *Bright Sparks* web pages for storing permanent knowledge about a creative session with one persona and related Google Search

First Evaluations of the *Bright Sparks* Web-Based Software Tool

We undertook an early formative evaluation of the *Bright Sparks* web-based software tool in order to inform its future design and implementation. The aims of the evaluation were to identify any problems with the functioning of the tool and the implementation of the technique, and to discover the users' needs in the context of

Table 2 The current predefined designer stakeholders about whom knowledge is codified in the *Bright Sparks* software tool

Steven Jobs (entrepreneur and inventor)	Daan Roosegaarde (artist and innovator)	Philipe Starck (designer)	Jonathan Ive (product designer)
Marcel Wanders (designer)	Anton Corbijn (photographer and film director)	Jakub Dvorsky (games designer)	Gerrit Rietveld (designer and architect)
Tim Brown (speaker on design and design thinking)	Coen Brothers (film directors)	Christien Meindertsma (artist and designer)	Vincent van Gogh (artist)
Dali (artist)	Willem Gispen (furniture designer)	Antoni Gaudi (architect)	Rem Koolhaas (architect)
Dick Bruna (author and Illustrator)	Wassily Kandinsky (artist)	Damien Hirst (artist)	Olafur Eliasson (artist)
Leah Buechley (designer and engineer)	Christo (artist)	Richard Serra (sculptor)	Wim Crouwel (graphic designer)
Eric Carle (Graphic designer and illustrator)	Jeff Koons (artist)	Fiep Westendorp (illustrator)	Hella Jongerius (product designer)
Isabelle Beernaert (choreographer)	Mondrian (artist)	Maarten Baas (furniture designer)	Issey Miyake (fashion designer)
Viktor & Rolf (fashion designers)	Banksy (street artist)	Fred de la Bretoniere (fashion designer)	Franz Marc (artist)
Bernd & Hilla Becher (artist-photographers)	Arne Jacobsen (architect and designer)	Dieter Rams (product designer)	Ai Weiwei (artist)

their work and to capture any future revisions to the tool. A total of six participants with design experience provided feedback on the software tool.

All of the participants reported that the tool was easy to use, stating for example that it had a "*clear interface*", "*after one glance we got the idea and it is always the same*", "*the automation of it, ease of use*" and "*personas, simplicity, clues, flows quickly*". In contrast, some of the participants expressed concerns about some of the personas that were retrieved and presented by the tool, for example: "*Some personas weren't really famous to be considered of common knowledge*" as opposed to "*... even if a person doesn't know the persona, the blurb is helping and it could be a good trigger on persona technique led.*" That said, most of the participants acknowledged that they could simply skip unknown personas – as the tool was originally designed for. Some of the participants reported that the random retrieval of personas was not always what they wanted, for example: "*maybe category: social, physical so you can select the mind set, but random is fun too*".

One of our aims was to remove the need for human facilitation, and several of the participants provided feedback on the tool's automated guidance features. Some of the participants reported that the tool "*makes the technique easier to use i.e. don't have to think of own personas – fear of the blank page"* and *"it's a quick way to use the technique, means you don't have to think on your own"*. However, opinion was divided on the need for facilitation when using *Bright Sparks*. For those familiar with the *Hall of Fame* technique, facilitation wasn't deemed necessary. However, for those unfamiliar with the technique, some instruction would be required. In response to this feedback we have added a discrete pop-up window function with help information and user instructions to address this. In particular, the evaluation explored the effectiveness of the codified clues to guide use of each persona to generate creative ideas. Overall the clues were found to be useful, for example: *"yes, I found the questions to be helpful and stimulating. Avoid repetitions immediately"*, *"the clues. Makes it easier to get started"*, *"I think they were really good even though not all of them applied, we could always find some of them that worked"* and *"yes, although you don't use all it helps to start the thinking"*. That said, some requests for improvements were noted, in particular: *"some of the clues and repetition of some clues too"* and *"repetition of clues too soon. Better not to repeat clues right after they have just been displayed. Ok to repeat after more rounds of clues"*. Therefore, the design of *Bright Sparks* was adapted to avoid clue repetition for each retrieved persona.

To conclude, results from the first evaluation of *Bright Sparks* provided evidence for its effectiveness in creative thinking by designers, but also revealed some problems with the tool's presentation of codified knowledge to users. Therefore, we are currently extending the tool with features to, for example, enable a user to select from pre-defined categories of personas such as designers and superheroes, as well as add a simple search function so that the user can return to a specific persona.

Conclusions and Further Evaluation

This chapter has reported the design, implementation and first evaluation of a new web-based software tool called *Bright Sparks*, which can be used by designers and others to support divergent creative thinking. In particular, the software tool was developed to make the *Hall of Fame* creativity technique available to people without the need for human facilitation or other resources.

As such, we believe that the *Bright Sparks* software tool has the potential to make a novel and important contribution to design thinking. Design thinking has a focus on fast learning, divergent idea generation and rapid prototyping that takes place concurrently with business analysis, and we believe that the tool can be used effectively by both designers and lead stakeholders in these processes. In particular, the use of the tool can lead to the explicit use of one creativity technique in design thinking processes, thereby bridging the increasing divide between creativity research and design thinking processes that we reported at the start of the chapter.

Acknowledgments This work is supported by the EU-funded FP7 COLLAGE project number 318536.

References

Boden MA (1990) The creative mind. Abacus, Routledge, London/New York

Design Council (2011) Design for innovation: facts, figures and practical plans for growth. Available via http://www.designcouncil.org.uk/. Accessed 29 Oct 2015

Greene SL (2002) Characteristics of applications that support creativity. Commun ACM 45(10):100–104

Hollis B, Maiden NAM (2013) Extending agile processes with creativity techniques. IEEE Softw 30(5):78–84

Kerne A, Koh E, Smith SM, Webb A, Dworaczyk B (2008) CombinFormation: mixed-initiative composition of image and text surrogates promotes information discovery. ACM Trans Inf Sys (TOIS) 27(1):1–45

Lockwood T (2010) Design thinking: Integrating innovation, customer experience and brand value. Allworth Press, New York

Maiden N, Robertson S, Gizikis A (2004) Provoking creativity: imagine what your requirements could be like. IEEE Softw 21(5):68–75

Maiden NAM, Ncube C, Robertson S (2007) Can requirements be creative? Experiences with an enhanced air space management system. In: Proceedings of the 29th international conference on software engineering (ICSE 2007). IEEE Computer Society, Washington, DC, pp 632–641

Maiden NAM, D'Souza S, Jones S, Muller L, Panesse L, Pitts K, Prilla M, Pudney K, Rose M, Turner I, Zachos K (2013) Computing technologies for reflective and creative care for people with dementia. Commun ACM 56(11):60–67

Michalko M (2006) Thinkertoys: A handbook of creative-thinking techniques, 2nd edn. Ten Speed Press, Berkeley

MindMup. https://www.mindmup.com/#m:new. Accessed 29 Oct 2015

Vázquez F (2013) Hall of fame. http://hailtothecreativity.wordpress.com/2013/03/14/hall-of-fame/. Accessed 29 Oct 2015

Zachos K, Maiden NAM (2008) Inventing requirements from software: an empirical investigation with web services. In: Proceedings 16th IEEE international conference on requirements engineering, IEEE Computer Society Press, pp 145–154

From the Real to the Virtual: Developing Improved Software Using Design Thinking

Julian Malins and Fiona Maciver

Abstract We all have our favourite software applications that we are most familiar with, and our least favourite, which we find frustrating and difficult to use. This chapter addresses how applying a methodological approach based on design thinking can be used to identify opportunities for the development of effective software applications and their subsequent design and evaluation. It begins by reviewing the concept of design thinking in order to provide a framework for the subsequent discussion. Much of the current software that we take for granted has either evolved from a period when computers were not much more than sophisticated adding machines, or by attempting to provide a virtual analogue of the real world in a digital format. A number of the more established applications are now beginning to creak at the seams and don't meet our contemporary needs as the applications attempt to include more and more features. This chapter considers ways in which we can interrogate the real world in order to identify new opportunities and new approaches for developing applications and interfaces. It considers what criteria should be used to assess the effectiveness of a software application. The chapter reviews design thinking as an approach that can inform the development process.

Introduction

This chapter begins from the starting point that many of the software applications most commonly in use today suffer from significant design shortcomings, either because they have failed to sufficiently recognise the needs of end users or because they have become too complex and unwieldy. The chapter sets out to examine some of the ways in which everyday situations can provide essential clues to what new software could be developed and to examine some of the ways in which existing software could be improved. To address this issue we adopt a design thinking approach.

The term 'design thinking' is one which gained popularity approximately 10 years ago, thanks to the work of Tim Brown and others from the IDEO design

J. Malins (✉) • F. Maciver
Norwich University of the Arts, Francis House, 3-7 Redwell Street, NR2 4SN Norwich, UK
e-mail: j.malins@nua.ac.uk; f.maciver@nua.ac.uk

© Springer International Publishing Switzerland 2016   337
P. Markopoulos et al. (eds.), *Collaboration in Creative Design*,
DOI 10.1007/978-3-319-29155-0_16

consultancy company (e.g. Brown 2008, 2009; Kelley and Litman 2001). However, it has been around for considerably longer (Cross 1999; Johansson-Sköldberg et al. 2013; Kimbell 2011), and much has been written about it, both in support of it as a business management approach (e.g. Martin 2009; Wylant 2008) and criticising it as a form of 'smoke and mirrors' used to confuse clients (Badke-Schaub et al. 2010; Collopy 2009; Malins and Gulari 2013; Nussbaum 2011). Despite this controversy, there are a number of well-documented methods associated with the design thinking process that can be usefully applied for the identification of new opportunities and solutions which lend themselves to the development of new software.

Design thinking relies on a cognitive approach, which begins by requiring the reflective reframing of the initial starting point of a problem or question. It requires a holistic viewpoint that aims to integrate information and perspectives from as many sources as possible. It is characterised by its focus on human-centredness, employing experiential methodologies often based on visual methods. It is also generally collaborative and multidisciplinary in nature. The mind-set of the 'design thinker' has to allow for ambiguity, making it possible to tackle complex or paradoxical problems that may not have a fixed solution; this type of problem is sometimes referred to as a 'wicked problem' (Rittel and Webber 1973; Buchanan 1992). Design thinkers are required to be experimental and have a willingness to challenge existing assumptions, while at the same time retaining focus on the future needs of individuals. Figure 1 provides an overview of the design thinking process.

At each stage, a number of methods and techniques can be applied which, taken as a whole, provide a route map for software designers in identifying opportunities for new applications, and their subsequent development and implementation. It also provides a useful structure for this chapter.

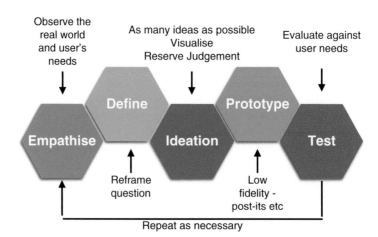

Fig. 1 Overview of the design thinking process (*Source*: Adapted from the Stanford d-school 'Virtual Crash Course in Design Thinking'(The d.school's Virtual Crash Couse in Design Thinking is available at this address: http://dschool.stanford.edu/dgift/))

Step #1 Empathise – Understanding User Requirements

Design thinking requires us to empathise with the end user in order to understand their particular needs either through a process of consultation or observation. There are a number of design thinking techniques that are commonly employed to gain insights into user requirements. These include the use of 'empathy maps' first credited to Scott Mathews and Dave Gray of the Dachis Group (Curedale 2012) and techniques such as customer journey mapping (Schneider and Stickdorn 2011). An empathy map is a template that can be drawn on a whiteboard and completed using Post-It notes. It tries to help the design team understand the context and motivations of the individuals for whom they are designing. Journey mapping can include real journeys during which touch points are noted as occasions at which particular events take place. At these moments in time or space, the dominant emotion is recorded: for example, in any journey or experience there may be points at which we get confused or anxious. By comparing a number of journey maps, and recording the impressions of the same journey made by different individuals, particular touch points can be highlighted and provide a focus for design improvements. Figure 2 shows a journey map illustrating a trip to the doctor. During the journey, the user's emotional reactions at particular touch points are noted, such as making the appointment, talking with the receptionist etc.

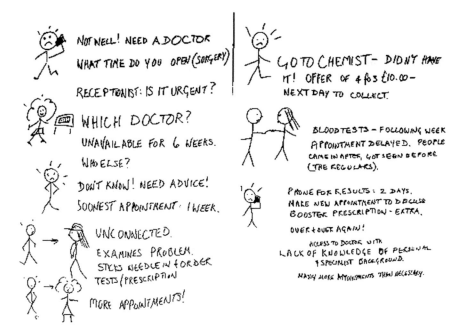

Fig. 2 Hand drawn journey map

The journey map successfully highlights points at which the user's anxieties could be alleviated by deploying design thinking to re-envisage and re-design the user experience.

A further approach termed 'empathic modeling', aims to directly simulate the needs of individuals with complex needs. This may entail for example, the wearing of blindfolds or mobility suits that artificially restrict movement to simulate old age or some other physical impairment (Malins and McDonagh 2008). Empathic methods allow us to develop an understanding of user requirements based on close observation of human behaviour. This involves finding ways in which you can not only observe what happens in a real world situation, but also go further by simulating the needs of a particular user.

Categorising users, for example on their level of experience in a given situation, allows the design team to address and understand the needs and issues specific to that group. One category of user includes the novice or naive user. This is an individual that has never been in the given situation previously and therefore approaches all problems by questioning the steps they have to take to achieve their goal. The developer uses this mind-set to undertake a task using the software as though they have never previously used anything similar. The first task, for example, might be to switch on a computer and interact with it. The familiar user will know how to switch the computer on and how to open the applications. By contrast, the naive user may struggle to find the 'on' switch, and immediately become confused when the device starts up. The naive user may be able to decipher how to use the *Start* menu on their PC or open an application using the Apple dock, however at first glance these tasks are not immediately clear. Cryptic and unhelpful error messages are common on our PCs. David Pogue, in his TED talk entitled 'Simplicity Sells',[1] cites an example of a confused user trying to respond to a computer problem by typing 'Error 11' repeatedly. When the helpdesk asked him why he was doing this, he responded that: 'the error message told me to "type Error 11"!' Amusing as this story might be, it highlights how easy it is to make assumptions about the comprehension of users.

As the naive user continues working with the computer, the shortcomings of current interfaces become apparent to the expert observer, alongside ideas and opportunities for design improvements. By using observation techniques, journey mapping, empathy maps and empathic modelling, the software developer can begin to identify opportunities for improvements in the way in which we approach every day situations. This could include every day tasks and the use of existing software. When used to observe everyday activities, the techniques begin to suggest areas for development. This could be around information sharing and communication, navigation or human-machine interfacing.

Design thinkers constantly look for situations in the everyday world, which throw up minor frustrations or problems. Users will generally have adapted naturally to these frustrations. To the design thinker however, they represent significant new

---

[1] See: https://www.ted.com/talks/david_pogue_says_simplicity_sells?language=en

opportunities for innovation. Every day tasks that we are familiar with suggest new opportunities for computer-supported applications. As an example, trying to find your luggage on the carousel at the airport might trigger the question "*is there a way it can be tracked and matched to its owner more easily?*" Like many ideas, which initally appear to be original thoughts, it turns out that others have already made a similar observation resulting in a new product based on the use of radio frequency identification chips.[2] Such chips can enable your luggage to communicate with your phone. Another common problem that many people may relate to is losing your keys just when you are about to leave the house. Perhaps in the future we will be using our smart watch to locate our keys - of course, keys themselves may become a thing of the past as biometrics embedded into technology become more commonplace. This concept relies on the ever-increasing phenomenon of ubiquitous computing or 'the internet of things' (Weiser 1991). The design thinker's mind-set aims to turn every problems into potential opportunities for innovation.

Related to this, the use of scenarios or storytelling is a further powerful design thinking method (Curedale 2012). There are many types of scenarios which can be based on actual experiences or entirely imaginary situations. By considering the every day tasks faced by an individual, we can start to identify possibilities for innovation. The following fictional scenario illustrates how this might work:

'A day in the life of Fred'
   Fred is woken up by his digital alarm. He gets up to use his bathroom. How healthy is Fred? The smart toilet tells Fred he needs more fibre in his diet and he needs to lose a few pounds. Fred leaves for work. His house locks up as his personal electronic identity moves with him via the subcutaneous microchip - no need for keys or cash. Too much information is now available to Fred. How is it tailored to his needs? Over time, his preferences have been stored in his personal cloud while case-based reasoning algorithms begin to tailor his specific data needs. Fred's virtual space extends around him, helping him interact with people he knows, or who may share common interests. As he walks, his ambient computing starts to tailor the advertising on nearby electronic billboards to reflect his profile. His subcutaneous chip allows him to interact with his environment. Doors open as he approaches and his office building welcomes him and switches on his computer system. His work environment is tailored to his specific requirements . . . [3]

Exploring this virtual world allows us to explore different possibilities and envision new applications. Just as some people react positively to this scenario, others may be horrified, viewing it as a gross invasion of privacy. The answer, then, might be another application that allows the user to take more control of when they are or are not connected to the virtual world, as demonstrated by the research conducted by Warwick et al. (2003). Extending this idea of storytelling as a source of inspiration has led to some significant software developments. Most famously, the cult science fiction novel *Snow Crash* anticipated the development of Google Earth and Second Life among other concepts (Stephenson 2003), and the

[2]For further information, see: http://www.trakdot.com/en

[3]These ideas have a basis in reality, see for example the Versatile Interactive Pan (VIP) in development by Twyford http://news.bbc.co.uk/1/hi/health/1433904.stm

earlier cyberpunk novel Neuromancer by Gibson (1984) is credited with bringing the phrase 'Cyberspace' into our mainstream language. It is also cited as having influence on the current form of the World Wide Web.

Extending this line of thought, we might ask 'what software applications are we currently missing?' Applications that extend our normal range of senses might be worth considering. This is of particular interest as our population ages. We look to software to supplement our existing senses, helping us to remember things we have forgotten, to anticipate our needs, or for example, to supplement our vision through the use of acoustics.

Many of the applications that we are most familiar with today began as an attempt to produce a digital model of a real world process. Word processors, for instance, were originally designed to replace the electronic typewriter, which in turn was based on the manual typewriter; spread sheets were developed to mimic book keeping practice; and computer aided design software was developed from the world of technical drawing. All of these applications use terminology and language directly borrowed from their real world predecessors. Word processors still refer to 'cutting and pasting', font size is measured in 'point' size, and the term 'leading', from the world of manual typesetting, is still used to describe the space between lines. Software developers often make use of these legacy terms and ideas to help users make the transition between older systems and newer technologies, or between the real and the virtual worlds. This is sometimes referred to as skeuomorphism. The design of interfaces may often include the use of skeuomorphs, which reference much older technologies to help give users a feeling of familiarity with the interface. These have long been a standard feature of many of our more popular applications and devices (Page 2014). Visual skeuomorphs are often similar in action, such as the use of dials and sliders on music software in direct imitation of a mixing desk. It is also possible to make use of auditory skeuomorphs such as the mechanical shutter sound on digital cameras. Skeuomorphism is not confined to software application development. Many of the technologies that we are most comfortable with begin by using terms that refer to earlier devices. Cars, for example, use 'horsepower' to describe engine capacity, digital radios sometimes refer to 'dials' for frequency adjustment and e-books go as far as providing animations of page turning. Books and e-books provide some clear examples of this phenomenon: it could be argued that the e-book would look entirely different had it been developed without reference to its real-world counterpart (Cope and Philips 2006). For example, speed-reading software (for example www.spreeder.com) presents words one at a time rather than in a conventional page layout. This software demonstrates that this is actually a more efficient way of presenting text-based information for screen reading, however this is not a standard option available on e-readers.

This initial phase, which we have termed Empathy, is mainly concerned with understanding users' requirements and identifying opportunities for development. The following section describes how this initial research can be used to begin to define a set of specifications for new applications.

Step #2 Define – Reflective Reframing

End users have always been an important source of inspiration for the development of new software solutions. This is referred to as user-centred design (Norman 2002; Norman and Draper 1986). However, there are limitations to this approach. End users can tell you what currently works well in an application, what they like and what they don't like, but may not be able to tell you about future possibilities, and anything about what does not yet exist. User-centred design is particularly helpful when it comes to suggesting possibilities for evolutionary development of software, but less helpful when it comes to radical innovation which may rely on something entirely new, or more commonly on the remixing of technologies from different domains.

An important step in defining the problem to be addressed is the reflective reframing of questions and design issues: this is perhaps one of the most crucial steps in the design thinking process. Many design problems are phrased in such a way as to predetermine the final outcome. A reminder of the importance of reframing the initial design objective is provided by the apocryphal story of NASA commissioning a 'pen' that could write in space and spending millions of dollars developing such an object. Meanwhile, the USSR solved the same problem by simply taking a pencil to write with in space. Rephrasing the problem avoiding the term 'pen', perhaps using the helpful term 'thingy' might have saved a lot of time and expense. This anecdote illustrates how important it is not to include terms that will predetermine the solution when considering any type of design problem.

The computer scientist Pranav Mistry provides an excellent example of the importance of framing questions that, at first sight, appear deceptively simple but allow the imagination free rein. In his TED talk entitled 'The thrilling potential of SixthSense technology',[4] he asks how we can interact more directly with the world around us. Mistry envisages writing a Post-It note and having it instantly appear on his computer, or taking a photograph by simply holding his hands to simulate a camera's viewfinder. Mistry's approach highlights a willingness to prototype with whatever comes to hand by cannibalising readily available pieces of technology in order to mock-up new devices. Such use of low fidelity prototyping is inherent in the design thinking approach. In software terms, this means using Post-It notes, cards and visual prototypes when working with end users to question some of the assumptions that may already be present in the way in which the design problem has been constructed.

This approach to software development also has limitations. For example, as software evolves and functions are added to the initial specifications, it becomes like a building that has been extended by multiple occupiers with no regard to design, architecture or planning regulations. The result can be unstable and chaotic.

---

[4]For further information see: http://www.ted.com/talks/pranav_mistry_the_thrilling_potential_of_sixthsense_technology?language=en

Fig. 3 Screenshot of Microsoft Word showing all tool bars constitutes 'visual clutter'

In developed software, menu structures can become attenuated, resulting in a more complicated and difficult to navigate solution. The screenshot of Microsoft Word in Fig. 3 illustrates the complexity and visual clutter when all toolbars are showing at once. Applications that have developed over a number of years, generations and iterations, end up requiring large increases in computer memory and processing power to run the application. Moreover, the package itself becomes less intuitive to the point where specialised courses are required to teach individuals how to use the software – for example, the European Computer Driving License Foundation provides courses for users wishing to become expert in the use of Microsoft Office.

The underlying metaphors that are used for the design of an application can have a profound effect on how the application is developed, presented and used. We can apply this idea to imagine new metaphors, and use these to explore interface ideas: for example, we might choose a symbol from nature such as a tree structure which might suggest information roots, main trunks and branches and fruits etc. Alternatively we might choose a man-made metaphor such as a building, suggesting rooms, floors and basements and services etc. This latter metaphor also begins to suggest a more three-dimensional view, in contrast to the two-dimensional views we are used to. Exploring these ideas can begin to suggest very different forms of interface.

Since the 'desktop' and 'window' metaphors have dominated interface design for such a long period, the commercial risk involved in introducing an alternative interface may be prohibitive: large corporations may be unwilling to develop radical alternatives that imply starting from scratch. However, this may still present opportunities for innovation based on the adoption of open innovation approaches which mean that software can be launched before being fully developed, provided it is free to download for individuals who are keen to provide on-going effort and feedback. The Google Chromium OS project operates on these principles.

The real world is a three dimensional space that we move through generally without effort, using all our senses for navigation. However, in the virtual world we are often restricted to a two-dimensional view. This two-dimensional view has a powerful influence on the design of new software, as does the computer's rectangular screen. The WIMP interface (window, icons, mouse, pointer) tends to predetermine a particular way of developing software, however new forms of interaction such as physical gestures and voice commands, are starting to open up opportunities for new applications that will allow us to go beyond the confines of conventional metaphors. Software and hardware manufacturer Oblong Industries is attempting to revolutionise the industry by developing gesture-based interfaces. Microsoft's Xbox Kinect[5] also provides opportunities for some promising interactive technologies.

One of the hardest ways to design is when there are no constraints. The more restrictions that a designer has to deal with, the clearer the design process becomes, since the solution can be assessed against well-defined criteria. For example, applications which are designed for a specific purpose, or that perform specific tasks, help provide focus for the design solution. Extreme users (that is, users with specific needs or requirements) provide the designer with a clear set of constraints. Although these constraints initially appear more challenging to the design process, the solution is likely to have a much wider application outside of the constraints imposed by that extreme user. Designing a navigation system for totally blind people, for example, imposes some very severe constraints. Once solved, however, the resulting navigation system can address essential problems that any user would find valuable: the initial specialised area of development may have a much broader application. The same principle applies when designing for extreme environments, such as space exploration, or oil extraction in the Antarctic. Solutions for these challenging situations could have a substantial application in every day life. Such extreme needs, requirements and situations can be powerful drivers for innovation.

The idea of the design thinking process requires the software developer to initially define what it is they are trying to develop, being careful to identify criteria that do not predetermine an inappropriate solution. Developing specifications based on the needs of users makes it easier to evaluate the software as it is being developed. Having defined the problem, the next stage is to generate as many solutions as possible.

Step #3 Ideation – Divergent Thinking

There are a considerable number of design methods devoted to expanding the number of ideas, sometimes referred to as 'divergent thinking' or as the 'divergent' phase of the design process (Design Council 2007). Standard methods include brainstorming, mind-mapping and other 'synectics' approaches (Curedale 2012).

Another very useful source of ideas is to draw lessons from natural systems, referred to as biomimetics. It is an approach that has been successfully used to innovate software applications. For example, the foraging behaviour of ants has been used in the development of new algorithms for processing very large data sets.

---

[5]For further information see: http://research.microsoft.com/en-us/projects/lightspace

The ant colony optimization algorithm (ACO) helps optimize the best route to a particular objective based on a probabilistic technique for solving computational problems that can be reduced to finding good paths through graphs. It is part of a family of algorithms referred to as metaheuristics, as proposed by Dorigo (1992).

Another source of ideas is based on direct observation from real life. For example, based on the observation that a large percentage of post-operative complications were a result of poor procedural practice and human error, a new checklist system was introduced in UK hospitals (Birkmeyer 2010; World Health Organisation 2009). The innovative check-list system is currently a paper-based solution, yet it has had the effect of dramatically reducing the number of post-operative complications. This example shows how, based on a set of observations, the problem was identified, a solution was then developed, evaluated and found to be effective. Developing the checklist into a software application requires a further step. A virtual version of this real world solution might provide an effective software application. The following section considers alternative forms of prototype in more detail.

Step #4 Prototype

There are three main approaches to prototyping: (1) rapid or low fidelity prototyping, mainly used in the early stages of the design process, (2) iterative prototyping which can make use of software development tools, and (3) evolutionary prototyping which builds on an existing application through a process of further development (Beaudouin-Lafon and Mackay 2003). While prototyping is shown in Fig. 1 as a separate step in the design thinking process, in practice it is integrated with the ideation phase, using low fidelity prototyping as a method for idea generation and clarification. In software design, low fidelity prototypes may deploy simple materials such as paper cut-outs, Post-It notes, or hand-drawn storyboards to illustrate the user's progress through an application or a website. The emphasis is on using either sketching techniques or software, such as Axure, to simulate the basic functions of an application, delaying the actual programming until an adequate picture of the problem and solution are conceived.

To ensure success of the finished article, the rule is to separate the ideas from their evaluation: in other words to reserve judgement. We now consider how judgement can be arrived at through a process of evaluation and testing.

Step #5 Test and Evaluate

So, what criteria should we be using to evaluate new software applications? One suggestion is that new applications should be sufficiently intuitive so that they do not require users to refer to a manual before use, let alone attend special training courses. This requires the software designer to provide metaphors that are meaningful to end users, and to use language and terminology familiar to the target user group. A more effective approach may be to allow the users themselves to customise elements of the interface, such as the terminology used in the menus, to fit their needs and match their own requirements of the interface.

Thanks to platforms that learn our preferences, such as iTunes and Amazon, users have become accustomed to software that provides options based on analysing user preferences. It is important to recognise that individuals have different cognitive thinking styles that directly affect the way in which they approach software applications (Kirton 2004). This might involve, for example, presenting information in a list structure for people who prefer a linear thinking style, or alternatively in the form of a map or web favouring a more holistic thinking style. Developing applications that are capable of learning our individual thinking styles, and thereby helping to filter and present information in a personalised manner, is a logical next step in software design.

Developing without being adaptable to end user requirements can often lead to a standard being set before the product has become mature. Once a convention becomes an industry standard, it becomes increasingly difficult to make design improvements. An example of this is the QWERTY keyboard layout that was originally designed for the manual typewriter, developed in such a way as to slow down typists so that the keys did not become tangled up (Noyes 1983). However, this layout is so well established that it has become almost impossible to replace (Noyes 1998). It is also important to avoid counter-intuitive actions such as dragging files to the trash icon on a Mac in order to eject a disc, or using the 'Start' menu to shut down your PC. The touch screen revolution illustrates that there are perhaps more intuitive ways to interact with software, and allows us to envisage alternatives to those options to which we have become habituated.

Another approach for software evaluation that is commonly used is referred to as the 'cognitive walkthrough' method (Rieman et al. 1995). Based on the systematic observation of a user undertaking a specific task, this can either be done using a checklist or in a more direct way by close observation. The user is set a specific task, such as starting a new project. The user then searches the interface for available actions, for example interactive elements or buttons, and then selects an action that seems likely to allow them to fulfill the task. Finally, the user performs the selected action and repeats these steps as many times as necessary to arrive at the desired goal. The walkthrough records the effectiveness of each of these steps. A variation on the walkthrough technique is to make use of eye-tracking technology,[6] which provides both qualitative and quantitative data on how a user has engaged with an application.

---

[6]For further information see: http://www.mirametrix.com/products

Conclusion

This chapter has introduced the idea of design thinking as an approach for identifying every day opportunities derived from real world observations. Such observations can spur low fidelity prototyping, which can in turn be used to innovate new platforms. When developing new applications, the importance of being able to critically reframe the initial design question becomes a crucial step when considering appropriate solutions, taking particular care to avoid terms that could unduly influence the possible outcomes. Exploring ideas by using alternative metaphors as part of the reframing process, as well as using scenarios, have also been described as alternative approaches for considering new starting points for the design of software applications. The incremental development of software has resulted in overly complex and difficult to use applications. Some solutions fall victim to their own success: as some software becomes industry standard, radical change becomes much harder to implement. Developing software that is both simple and effective still remains highly challenging. Observing the real world and being alert to opportunities presented by daily anomalies and minor frustrations provides us with new opportunities for innovation. The application of design thinking, with its emphasis on early prototyping and human centred design values, has a great deal to offer the software developer of the future. The challenge is to apply the design thinking approach to discover new opportunities for the development of effective applications which are capable of making a significant impact on our lives, to develop intuitive interfaces which can be navigated without the use of manuals or special training courses, and that allow end users to take control of the way in which information is presented to them.

References

Badke-Schaub P, Roozenburg N, Cardoso C (2010) Design thinking: a paradigm on its way from dilution to meaninglessness. In: Proceedings of the 8th design thinking research symposium. pp 39–49

Beaudouin-Lafon M, Mackay WE (2003) Chapter 52: prototyping tools and techniques. In: Sears A, Jacko JA (eds) Human computer interaction-development process. CRC Press, Boca Raton FL, pp 122–142

Birkmeyer JD (2010) Strategies for improving surgical quality—checklists and beyond. N Engl J Med 363:1963–1965

Brown T (2008) Design thinking. Harv Bus Rev 86(6):84–92

Brown T (2009) Change by design: how design thinking can transform organizations and inspire innovation. Harper Collins, New York

Buchanan R (1992) Wicked problems in design thinking. Des Issues 8(2):5–21

Collopy F (2009) Thinking about design thinking. FastCo. Design. http://www.fastcompany.com/1306636/thinking-about-design-thinking. Accessed 19 Feb 15

Cope B, Phillips A (2006) The future of the book in the digital age. Chandos Publishing, Oxford

Cross N (1999) Natural intelligence in design. Des Stud 20:25–39

Curedale R (2012) Design methods 1: 200 ways to apply design thinking. Design Community College Inc, California

Design Council (2007) Eleven lessons: managing design in eleven global companies. Design Council, London

Dorigo M (1992) Optimization, learning and natural algorithms (in Italian). PhD, thesis, Dipartimento di Elettronica, Politecnico di Milano, Milan, Italy

Gibson W (1984) Neuromancer. Berkeley Publications Group, New York

Johansson-Sköldberg U, Woodilla J, Çetinkaya M (2013) Design thinking: past, present and possible futures. Creat Innov Manag 22(2):121–146

Kelley T, Litman J (2001) The art of innovation: lessons in creativity from IDEO, America's leading design firm. HarperCollinsBusiness, London

Kimbell L (2011) Rethinking design thinking: Part I. Des Cult 3(3):285–306

Kirton MJ (2004) Adaption-innovation: in the context of diversity and change. Routledge, Hove

Malins JP, Gulari MN (2013) Effective approaches for innovation support for SMEs. Swed Des Res 2(13):32–39

Malins J, McDonagh D (2008) A grand day out: empathic approaches to design. In: Engineering and product design education conference. ETSEIB, Universitat Politècnica de Catalunya, Barcelona, Spain, 4–5 September

Martin RL (2009) The design of business: why design thinking is the next competitive advantage. Harvard Business Press, Boston

Norman DA (2002) The design of everyday things. Basic Books, New York

Norman DA, Draper SW (1986) User-centered system design. Laurence Erlbaum Associates Inc., Hillsdale NJ

Noyes J (1983) The QWERTY keyboard: a review. Int J Man Mach Stud 18(3):265–281

Noyes J (1998) QWERTY-the immortal keyboard. Comput Control Eng J 9(3):117–122

Nussbaum B (2011) Design thinking is a failed experiment: so, what's next?. FastCo Design. http://www.fastcodesign.com/1663558/design-thinking-is-a-failed-experiment-so-whats-next

Page T (2014) Skeuomorphism or flat design: future directions in mobile device user interface (UI) design education. Int J Mob Learn Organ 8(2):130–142

Rieman J, Franzke M, Redmiles D (1995) Usability evaluation with the cognitive walkthrough. In: Proceedings of the Conference companion on human factors in computing systems. ACM Press, Denver CO, pp 387–388

Rittel HWJ, Webber MM (1973) Dilemmas in a general theory of planning. Policy Sci 4:14

Schneider J, Stickdorn M (2011) This is service design thinking: basics, tools, cases. BIS Publishers, Amsterdam

Stephenson N (2003) Snow Crash. Bantam Books, New York

Warwick K, Gasson M, Hutt B, Goodhew I, Kyberd P, Andrews B, Shad A (2003) The application of implant technology for cybernetic systems. Arch Neurol 60(10):1369–1373

Weiser M (1991) The computer for the 21st century. Sci Am 265(3):94–104

World Health Organisation (2009) Surgical safety checklist. WHO, Geneva

Wylant B (2008) Design thinking and the experience of innovation. Des Issues 24:3–14

Design and Data: Strategies for Designing Information Products in Team Settings

Mathias Funk

Abstract This chapter aims at linking data and information to creative design, focusing on collaborative processes at early phases of the design with data. The chapter aims at providing clarity in a large space around design and data. Thus, it serves as a guide for design team's approach towards the challenges of data design. Consequently, design is one of the key disciplines involved in data and information visualization (Moere and Purchase 2011). This chapter starts with a short introduction of ideas and concepts in the intersection of data, information, and design. It looks at users and designers as the main stakeholders, and considered the purpose of designed information. Following this introduction, we first focus on design artifacts essential for collaborative data design practices. Secondly, we focus on what it means to integrate data with design and the potential roles of data in the data design process. The chapter outlines a general design process with methods and approaches towards early design challenges. Furthermore, this chapter concludes with an annotated bibliography to guide further reading. Along the chapter runs an example case of a real information product that helps for better understanding. It links the more theoretical elaborations to the application level of a concrete design case.

Introduction

Design often means to trust intuition and our senses about aesthetics, look and feel, or the emotions that our products stimulate. Design opportunities are sought especially at the early phases of designing. Thus, we might not only rely both on user research and ethnography, but also on intuition to find an issue to design for.

Consequently, what we need to achieve with design is (1) to effectively translate gathered, sensed, or observed data into information without losing important qualities such as meaning, connections, truthfulness; (2) to achieve a balance

M. Funk (✉)
Department of Industrial Design, Eindhoven University of Technology, Eindhoven, The Netherlands
e-mail: M.Funk@tue.nl

© Springer International Publishing Switzerland 2016
P. Markopoulos et al. (eds.), *Collaboration in Creative Design*,
DOI 10.1007/978-3-319-29155-0_17

351

between complexity, understandability, and ease of use by means of interaction; and (3) to allow a design process guided by a few core principles that reliably leads towards good designs.

With the data involved, the game changes (Bigelow et al. 2014). Thus, data is very special in the sense that it may appear as an ingredient of a design process. It may also be the subject of design, which is the primary focus of this chapter: information products such as interactive data visualizations, wearable health trackers, and mobile apps communicating data about everyday life. In times of ubiquitous "big data" with even more invading collection practices, we need to find ways to turn raw data into meaningful designs: "information products". Apart from that, designing with data is in itself a complex field. Hence, there are three main challenges addressed in this chapter:

1. Modality and Materiality of Data: Data is inherently intangible and ephemeral. It cannot be changed and modified. It expires easily, but can remain useful.
2. Meaning and Meta-data: Data gets meaning by contextualization, linking, and relating. These are the outcomes of processing and annotation.
3. Scope and Framing: Data available for design can be overwhelming in terms of quantity and quality. Nowadays especially, scoping and framing can be real problems in a data design.

In addressing these three challenges, the chapter provides guidance with clear-cut design artifacts and process phases. In this chapter, we will focus on *information products*, i.e., packaged design artifacts that represent and embed data into the context of the end-user. Thus, two main stakeholders are relevant in this context: the end-user and the designer in a team of potentially more technical or entrepreneurial stakeholders. We will focus on information products that reflect data and information meaningfully, and highlight the influence of data for the design team.

Example Case

As a general illustration for this chapter, an example case runs along the sections, and will develop with the chapter. After every section, the section context will be related to and explained with the design case: a personal wearable device visualizing different layers of (activity) data in a minimal way depends on the context the user is currently in. This is because our needs for information greatly differ with the context. For instance, when exercising, we are interested in the heart rate. However, when working, we might be more interested in nudges to relieve stress or that pushes us to take a break. This concept, *Qualica*, was designed for users in a changing world, in which visual media and communication activities increasingly dominate their reality and do so in a noisy way. Therefore, this concept leverages minimal visualization in context, instead of showing numbers, charts, or iconic visualizations.

Fig. 1 Qualica physical information product used in two different contexts: sport and work

Essentially, this is an information product combining data, contextual information with real-time data processing, interaction, and a minimal visualization in wearable design (see Fig. 1 for Qualica in different contexts). To better illustrate the design process, we assume that different roles are involved in the early design phases of this case. This separates the different concerns such as: a technical expert on bio-signal data and external APIs; an industrial designer who will not only design the device's form and interaction, but also the service around this; and a mobile application developer responsible for the development of the Qualica companion app. In the development of the chapter, we will see how the case evolves. The concept was taken from Pepijn Fens' Master thesis, including all illustrations relating to the example case (Fens 2014).

In the following section, data is introduced as material for design, and as material that potentially carries meaning. However, it is also challenging to design with, especially in team settings. The section introduces information products and elaborates on how different usage scenarios of information products influence the design (exploration and communication). The next section focuses on artifacts involved in the data design process, with special attention to separation of concerns and design information. This leads into a section detailing a design process that draws relations between artifacts and different phases of the design process.

From Data to Information Products

Data for design? What is[1] this strange ingredient that is found everywhere today, but hard to grasp and often harder to understand? Unlike other "materials", data is an abstract, dynamic, and less malleable resource which needs special attention. Data can be seen as granulate of meaning in space and time.

---

[1] In this chapter, the word "data" is used mainly in the singular form. For discussion: see Borgman (2015).

Throughout the design process, the needs for data and information change from exploration to clearer communication. In the later stages, the designer knows what to expect from the data and information sources. He or she has become familiar with the data, and has found something interesting and worthwhile for further designing. Thus, this aspect of the data now needs to be nurtured and emphasized for the final design of the information product.

Data as Material

The reality that surrounds us is increasingly captured in the form of data. Therefore, data has become an explicit resource that is known and potentially feared by the public. But what is this again?

According to Ackoff (1989), data, information, knowledge, and wisdom are linked in a hierarchical way. With data as the base layer, information can be derived. Furthermore, we derive knowledge from information when processing it, and potentially turn our knowledge to wisdom at some point. However, this view that suggests that data is the basis and also the most resourceful of the four was challenged (Tuomi 1999). Data is simply a form of information that is *optimized for machines* and that can appear in raw or processed forms. Data can be often empirical, that is, collected in a formal experiment or study, or coming from the field as field data. Raw data is often considered rich and truthful, but unusable for higher-level design activities. Thus, the central point is that data is an abstract concept and is often derived from the real world. Some authors have even suggested *capta* to be a more accurate description of what is commonly called data i.e., a likely biased interpretation of reality (Drucker 2011). Naturally, there are many views on data and information. Therefore, we will use the following interpretation in the remainder: data is processed by machines; information is processed by humans; and "information causes change. If it does not, it is not information." (Claude Shannon).

Material Meaning

The interesting future is not about data at all, it is about meaning.

Alan Kay

The challenge in designing with data is clearly in conveying meaning, to translate from the abstract to the semantically expression that touches us and that brings about a change. When designing with data as a material, we need to think about what is the meaning of data in the design process and how much we need to deal with it. At the same time, what would be the meaning for users? Would they need an explorative design, meant to extend the knowledge of a domain, a design subject, or simply reality? Would they need data presented for action, to inform, to influence, to engage, and to trigger? Consequently, designers are in charge of adding meaning to the material data. This is often in the form of contextual information, relational linking, or simply by explaining what is captured in the dataset. They need to consider the fact that transparency and honesty are related. Transparency links to the degree of access to the original dataset or data sources, whereas honesty links to the question on how a reduced image of the original can still truthfully convey a gist of what has been captured originally (Hullman and Diakopoulos 2011). In most data design projects, there comes the point when the original view of the raw data is lost. As a result, certain information takes its place, when a view more colored by the makers' influence is finally presented to the users. As we will see, it is advisable to step back from time to time and to think about why we need to introduce a certain piece of data into design. In addition, we also consider *why* the user needs to know something: are we solving a problem or are we exploring a (potential) dataset that might solve future problems directly or indirectly?

Information Products

Information products are ubiquitous nowadays, but seldom defined or seen as such: *information products are end-user products that package information in a user-friendly manner, can communicate easily, and possibly have a rich (interactive) experience.* Increasingly, we are able to capture these aspects from devices in the field, and the services connected to them. In the past, computers were not able to access more information than they would actually need to carry out their assigned tasks. With the introduction of smart phones and other smart devices, we learned to accept that embedded sensors capture more data than is actually needed. Thus, we arrived at the conclusion that every new sensor creates a new business opportunity, with design at the core of it. Increasingly, this means that we are also able to act upon ad-hoc collected data with our products.

Design for Exploration

End-user scenarios for information products are often leaning more towards communication than exploration. However, there are also explorative scenarios for end-users. One example are the map-based visualizations of radiation data (appearing after the Fukushima accident in March 2011[2]), which is fairly close to the raw data in real-time, simply extended or contextualized with spatial information. Still, a strong message can be understood by data explorers around the world. Explorative information products are useful when users (consumers and professionals) need to

---

[2] http://blog.safecast.org/maps/; last accessed: Dec 31, 2014.

access a vast amount of potentially diverse information through a comparatively limited interface. According to Heer and Shneiderman (2012), Common techniques adopted by designers include:

- Selection (for example, retrieving hidden contextual information of a precise point in time or location).
- Filtering (reducing the shown data by time or location; excluding or including data items).
- Zooming (decreasing the amount of space in view, increasing data points per location, and changing the level of abstraction).
- Navigation or browsing (showing a directory of available categories or search terms).
- Augmentation (dynamic linking of data items to external information or meta-data).

The shared expectation of such an explorative design is that due to the increased complexity, other benefits can be gained. Such benefits include: truthfulness and less potential bias, flexibility in how data can be viewed, and generally better insight into complex relationships between data items and between the data set and external information. An explorative view can mean also that meta-data and sources are disclosed and available in the user's interface. Consequently, this can provide another layer of information on how the data at hand was gathered, processed, and manipulated. Hence, it provides a lens to frame, understand, and interpret the data.

Design for Communication and Influence

What if data was used less in a explorative way, and more *to communicate and inspire action*? Good examples are personal health and activity tracking devices (Fens and Funk 2014), data visualizations in the public media, and other designs that access a data from a data source and translate it into a contextually enriched and visually lifted representation that appeals to the senses of the viewers. Such data designs focus often on the viewer. For various reasons, it excludes data about other users. Furthermore, an explorative view would include this, which leads to the idea that communicative designs can address a larger audience. This is possible because they disclose less potentially privacy-relevant information. Instead, they give an emphasis on collective and viewer-related aspects of data.

A central question is how can data lead whom to what action. This question introduces the notion of *actionable* knowledge. Actionable knowledge is a knowledge that is not tacit or bound to be forgotten soon, but is *useful*, and often leads to direct action. At the same time, the question extends towards persuasive designs that include a superficially reputable motivation or argumentation i.e., data from a credible source. Many applications of such a pattern however lacks credibility, as the chain from data collection to representation is broken and blurry in many places.

Example Case

Looking at the example case of an information *product*, the design is less explorative and more communicative. Therefore, the data that is sensed or drawn from external data sources, such as activity streams on the Internet, is not disclosed in full, but instead presented in an aggregated and highly condensed way through a simple physical display. This display is a multi-color LED bar encased in the wearable device as a casual accessory. While this can potentially lead to an explorative behavior on the user's side, the primary use case is informative, suggestive, and inspires behavioral change (towards the more healthy and aware).

There is also a companion app that links to the wearable device and allows for browsing collected data, exploring historical traces, or zooming into the specifics of a context. Several screens of this app are shown in Fig. 2. This gives the user a better understanding of the richness of the multi-layered data basis used in the Qualica design. When Qualica relates bodily activity to work statistics such as desktop application usage, the user might feel triggered to look into such a pattern. This app and its more explorative use scenario is not the focus of the remaining part of this chapter. It can be seen from Fens (2014).

After the introduction of information products, the following section focuses on the creation (process) of such products, starting with the artifacts of the data design process.

Artifacts for Designing with Data

The view that was presented in the previous section which places data in a pyramid with wisdom at the top, might work for a general reception of data and information from a user or consumer point of view. Nevertheless, this will not work for the data designer who might not be working alone. Therefore, a different understanding of data and information is necessary. We need to separate the design and underlying information to allow for collaboration between different data stakeholders and to support ideation in the early phases with a neutral representation of available information. Therefore, this is flexible enough for rapid prototyping and fast iterative cycles at the same time. This information in turn needs to be described in a way that clearly separates data sources and processing from the data workbench that the designer finally spends most time on. Finally, separation of concerns is a way to split a potentially monolithic entity into smaller parts that can be taken over by different experts in a collaborative setting.

These considerations lead to a new understanding of data and information in the context of design as a layered schema that stacks design infrastructure, design information, and the design vertically (cf. Fig. 3). What sounds rather technical and maybe even seem to be (over-)applying engineering principles to a design process, is an important forcing function. Architect Christopher Alexander (1964) noted earlier on how successful designs need to unfold and be the outcomes of experimentation,

Fig. 2 Qualica companion app, covering a more explorative usage scenario on personal health

iteration, mistakes, and failures. Design information is the breeding ground for such a process of repeated adaptations and "weathering" of a data design. Design information is a safe middle ground which helps to explore and iterate fast. Also, it constrains attempts to lose direction, momentum, and drive.

As shown in the figure on the left side, the *design infrastructure* encapsulates the means to acquire and process data and provides an interface to the data. *Design information* is the translation of what the infrastructure delivers into a view or representation that can be used during the design process as a full-fledged reference for what the data is. The *design*, finally, is the finished *packaging* of the data that allows for limited, but meaningful interactions with the underlying data sources.

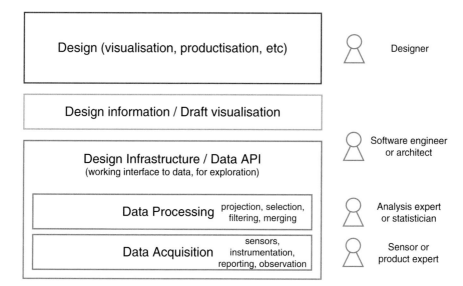

Fig. 3 Schematic overview of data design layers and possible experts

On the right side of the figure, a few expert roles were depicted (by all means are not exhaustive), which could work together in a collaborative design team. While all experts would be involved at all levels to some degree, the role next to the block shows their main expertise and responsibility. The figure shows clearly the need for good communication means between different professions and backgrounds. The described division into layers of design artifacts can facilitate this and shall be explained in detail in the following.

Data Infrastructure, from Source and Raw Data to Processed Data

Designing with data inherently means to gather, collect, and prepare the material of design, data, and related information, in a way that it is suited for the remainder of the design process. This is hard. Data material is different in that its collection often decides on its value. The amount can be important, the method of collecting contributes to its validity, and the analysis of data needs to be consistent, following a clear approach even at the very beginning. At the same time, the semantics and meaning of data, and their combination with other information and other factors, determines the value of data and its degree of degradation. Depending on its later use, data can loose or gain value over time. As an example, data that links to a competitive advantage looses its value quickly (cf. high-frequency stock trading or leaked secret cookie recipes). Nevertheless, if used as an evidence, suddenly it has a lot of value (cf. prior art for patents or evidence in a court trial). Treating data right involves a lot of intuitive aspects, which often leads to the pitfall of misunderstanding that "interesting" does not always equal to "valuable".

Data Sources and Data Collection

Regardless of whether the design finally incorporates a particular data set or not, data sources are essentially touch points with the users' reality. Data can be extracted from this reality by measuring or observing, for example, by applying technology to measure data from the environment (temperature, radiation, noise-level etc.) or by using human observation and interpretation of reality. These general approaches deliver different types of data. Therefore, they can be applied separately or together, as it is needed. Sometimes, it is more efficient or accurate to utilize technology for measurement. Sometimes also, the human mind is crucial to observe, report or interpret the desired data.

To frame the general challenge of data sources for more technical design: what is it that we can measure through products about users, their environment, their experience, and also their intentions, needs, and expectations? Increasingly, we are able to capture these aspects from devices in the field, and the services connected to them. Furthermore, two main areas of data attract the interest of designers: (1) data about humans, users, and their environment; and (2) data about the products they use and experience. Thus, we will focus on these two areas. The first one is interesting for designing *information products*, while the second one approaches data as a means to get better *insight* into a product's (potential) users and to adapt and tailor the design.

Consequently, there are many ways through which data can loose its value due to processing, and also due to the simple fact that reality can seldom be fully represented in data. We can generalize these aspects towards general quality criteria for data that should be useful for design. These criteria include:

- *Availability*: Data sources need to be available not only in the moment when data is captured, but also later on. This is required for reference or simply for updating a previously collected dataset. Readily available data is essential for an iterative design process and for establishing support infrastructure and design information.
- *Relevance:* Data sources need to be credible, valid, and honest for anything derived from them to be considered relevant and meaningful. Relevance over time is important in the sense that data can become easily outdated or "cold", resulting in less relevance for some applications.
- *Context:* Data sources belong to a particular context, which needs to be captured in some way for later reference and information enrichment. Thus, only with sufficient contextual information, a dataset can "land". Data that is deprived of its context becomes more abstract, and some contextual information needs to be added to re-create meaning.

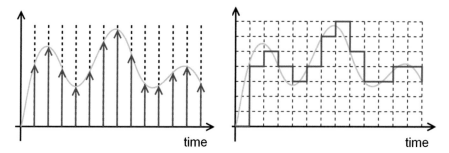

Fig. 4 Example of transferring a continuous (analog) signal into a discrete (digital) representation results in information loss in two dimensions: snapshots are taken in time (*left side*, horizontal raster) and values of these snapshots are quantized (*right side*, vertical raster) (http://en.wikipedia.org/wiki/Discrete-time_signal, last accessed Dec 31, 2014)

– *Accuracy:* Accuracy is combination of trueness and precision (according to ISO 5725-1) (Feinberg 1995). Trueness refers to how far the measurement or data point is from the ground truth, whereas precision essentially determines the distribution of samples. Low accuracy can result from close data points (high precision), and distance from the ground truth (trueness), and vice versa. Due to limitations in technology and storage, data is *sampled*, that means snapshots are taken (see Fig. 4). Thus, this automatically introduces lower accuracy.
– *Privacy:* Data can indeed invade a person's privacy, revealing or showing the potential to reveal personal and intimate information. Privacy problems are actually problems of contextualizing data, e.g. a person's identifiable details. As an example, without any context, "102,000" surely did not touch anyone's privacy. Nevertheless, it does only if it is brought together with the context "2013 taxable income in EUR of Joe Doe, living in ...". Context, true or false, can create a certain rooting in reality that lends great power to data. Privacy deserves special attention in the designing of data. Approaches to deal with privacy-sensitive information involve maintaining de-contextualization consistently. Such approaches need to be done right as dealing with data does not allow for even simple mistakes.

When talking about qualities, one assumes that such aspects should be maximized. This is not necessarily so. When designing with data, qualities call for balance and not for full maximization. Two examples: People do value privacy; and they are willing to compromise if they receive benefits to give-up privacy partly. Facebook and other social networks demonstrate this: users voluntarily reveal their private, personal, and even intimate information in exchange for social interaction, visibility, and perceived status. Another example is the accuracy of presented information. Modern sensors deliver extremely precise data that signal-processing applications can greatly benefit from. However, in contemporary data visualizations, one can observe that much less precise information is given. This is simply because it is not needed or is even considered harmful and distracting.

In addition, there are more aspects of data such as ownership, storage, security, and governance, which are relevant in differing degrees to enterprise data, "quantified self" data, and environmental data. However, these are beyond the scope of this chapter.

How to collect data from such data sources is more of a technical matter, but largely depends on the area of data in the design that we are interested in. It is advisable to identify and develop data sources that can deliver data steadily and with little effort. However, this is because "freshness" often determines relevance and meaning in this design space.

Sometimes, such fresh data which is generated continuously is called "online" data, and it often emphasizes its *connected* nature. However, "online" as a concept leads to "offline" data, which is about to leave a relevant time context and become *stale*. Both online and offline data can pose (technical) challenges. For instance, online data needs connectivity, real-time processing, and frequent updating of visual representations which need to be designed specifically for changing data. Offline data needs to be stored, might be more plentiful, and needs potentially stronger contextual enrichment as it lacks timeliness.

Often, it is beneficial if we can influence (online) data collection in terms of selection of sources, granularity, and semantics. Furthermore, changes in the data acquisition process lead to changes in the dataset and we might need to wait until we can process this new data.

Analyzing and Processing Data

Raw data, online or offline, to processed information is often a long winding road. Experts are needed to guide and facilitate understanding of what data source actually can reveal about our experienced reality. However, when analyzing data, different movements can be observed:

- *Down*: Trying to understand the rooting of data in reality, where it comes from, and what it means.
- *Up*: Trying to understand how abstraction can help generalize or connect to common knowledge and interaction.
- *Side-ways across the Dataset*: Trying to understand patterns and links between data items of the same source or reality (phenomenon).
- *Side-ways beyond the Dataset*: Trying to understand how data relates to other information beyond the dataset.

These movements are connected to skeptical analysis of what is there. Also, they help question our perception from time to time. A good example of a fallacy that might bias the analysis of a dataset is *Simpson's paradox*. Simpson's paradox is essentially a paradox about how aggregate statistics can mislead us. A pattern that can be found in distinct parts of the dataset does not appear when all data are combined together. For common sense protests, statistics knows it better (Blyth 1972).

All steps in a data processing chain determine its later value. Looking at the *collaborative* nature of early design processes, data collection and analysis starts as a manual effort and with the serious involvement of experts in sensors, APIs, signal processing, and data analysis. It is, however, desirable that this flow from raw data towards processed information should be highly automated for later stages of the design process (see the next main section). Thus, there is almost a direct link between the sources of data and the interfaces provided on the surface of the design infrastructure. If the sources of data are multiple users' subjective answers and are the sentiments of other qualitative contributions, it might be advisable to either work with automatically generated "mock" data or to begin with a larger body of historical data.

Automation is the key, but not for the price of architectural rigidity. What a designer wants is fluency and up-to-date data. Furthermore, the designer also considers the flexibility to change the data sources, to change the way the data is processed, and to interact directly with the data sources if needed. However, processing requires data. Thus, for an existing static dataset, this is easier. For dynamic data, especially in cases where the data is not captured yet, it is a bit more difficult. Furthermore, there is a distinction between offline and online data collection. Offline data is captured and will remain same regardless of what analysis that is needed from the data set.

Interfaces to Data

In the last step towards *design information*, processed information which is still in the engineering realm needs to be made accessible to designers. This can happen in files and folders for bounded datasets, or it can happen in technical interfaces that are potentially even online and continuously accessible, i.e., they deliver fresh data at any point in time. These interfaces are commonly called application programming interfaces (APIs). They are specifications of what an external party can provide or receive from the application. The external party might need to provide credentials to be authorized, but after this step, the door is open for data retrieval.

For data, such an interface is relatively common in the domain of web information systems. Nowadays, Internet giants open up their immense data stores towards developers who can make use of their knowledge and services. This has benefits for developers, but also costs. Especially when not paying for services or access, the developer locks herself into a specific eco-system of the API provider. Therefore, leaving the warm nest might not be so easy after a while.

Still, it is interesting to understand how large companies open their data caches. Certainly, the information is pre-processed and carefully "designed" to fit multiple use case scenarios. Exactly, this eco-system enables what we now know as the startup economy, a network of fast and highly versatile companies developed based on common technologies and information. Furthermore, we can translate this into the smaller context of a data design project. This is done by building interfaces to processed information from the desired data sources delivered on demand. At the same time, we can create a fertile environment for rapid prototyping and experimentation.

What is needed? Data needs to be opened, structured, consistent, and contextualized. Formats matter includes comma-separated value (CSV) files, spreadsheets (for instance, in Microsoft Excel format), databases (MySQL, SQLite, etc. together with their management interfaces), or specialized APIs to remote servers that provide information in readily consumable formats directly queried from internal databases.

These interfaces need not just to be there, they also need to be well documented. Source code, wikis, examples, tutorials, templates, and more formal code or API documentation help designers to create their main data design artifacts, i.e., *design information*.

Example Case

Looking at the example case, there are a number of *data sources* at different layers that contribute to the body of online and offline data that is visualized and communicated to the end-users. They include: bodily signals (cf. Fig. 5), work related log data (e.g., use of desktop applications), environmental information (weather, climate etc.), social data from social networks, and data about the current

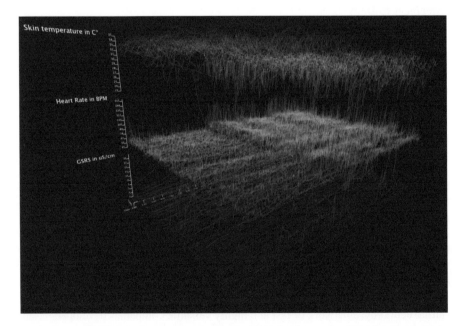

Fig. 5 Processed bodily activity data in a 3D plot. This visual overview of heart rate, galvanic skin response (*GSR*), and skin temperature is useful for looking at the "big picture" for spotting patterns, trends and correlations

Fig. 6 Layered approach to visualizing complementary data sources and contextual information

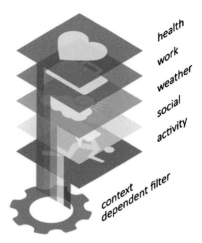

activity context (e.g., running, working, eating etc.). Some of these data sources are accessed actively at frequent moments, and their information is relevant for the current state of the visualization. Therefore, they are classified as "online". In Fig. 6, these online data sources can be gradually found at the bottom. There are also "offline" data sources that are queried less frequently and that do not contribute in real-time to the state of the visualization. Thus, their data is more stable and less volatile, and it provides a more general picture of the context and activity.

The data coming from these sources would be processed, for instance, to normalize the time or number format. For some data sources like weather or social activity streams, online data is available via APIs on the Internet. These data sources would need extra infrastructure to be accessible for the later stages. For wearable devices, data extraction can be difficult or at least cumbersome. At least, samples of relevant data needs to be taken and made available through the data infrastructure. The goal is to have all data programmatically accessible through various channels and to be available as "fresh" as possible.

Design Information

One of the designer's qualities is immediacy with material. To develop this for data, it means not just deeply understanding where data is originating from and what context it belongs to, but being able to work intuitively with specific data and information. Often times, the designer leans into craftsmanship (Megens et al. 2013). And, over time, we might even develop "data smell", which is an intuitive capability to sense the most interesting aspects of a dataset. As a (partial) craftsman, how many define a designer? Nowadays, building and shaping your own set of tools is an essential skill that can even precede the true expression of your craft.

Nevertheless, as a beginner in designing with data, it is tempting to pursue design directly from data sources or processed data. In addition, we might envision the final form already and are eager to proceed. Climbing Mt. Everest in a single attempt could be successful, but who would take the chances? Instead, a base camp is installed at a location from which the final attempts can be ventured. Thus, this is also convenient to retract to, in case of unforeseen events. The same is true for the attempt to design with data. The designer needs a base camp that allows for exploration, but always provides a good representation of the data at hand, with rough, but versatile visualizations and means to drill down. The designer also needs support to go back and fact-check their work, and to re-evaluate or calculate aggregates. Design information together with a documented approach helps in leaving breadcrumbs that let us backtrack from a cul-de-sac.

Another aspect is the communication between different stakeholders during the collaborative data design process. Thus, design information is an artifact that facilitates common ground among the team members and is the basis for creative decision processes (Kozlova 2011).

Coming back to the idea of offline and online data, the former is good to have a broad overview in relating time and space. Also, it enables the search for patterns and hidden links within the dataset. Such design information can be captured in tools, such as the commercial Microsoft Excel spreadsheet tool or specialized tools like Tableau, RAW, or even Matlab. The choice of tools depends on our familiarity with them, and on the degrees of freedom we need to fully understand dynamic data and to reveal the most interesting aspects of it with ease and fluidity. Consequently, several good visual overviews of tools can be found on the Internet.[3]

For *online* data, immediacy is often more interesting. What happens when certain values of different sensors comes in? What are the extreme cases? How can other information enrich the perspective of the given data? In these cases, design information might be a handcrafted visualization that fits our personal needs and which has grown over time. Many developers and designers have in the past ventured into tool making and maintain a personal set of helpers, resources, and craft support tools which allow them to work fast and intuitively. The goal is to establish this for data as well.

Behavioral Data for Design

As with almost every design process at the beginning, there is always a question asked: whom are we designing for and what are their needs and expectations? There are differences between nonprofessional and professional users in dealing with data visualizations (Quispel and Maes 2014). However, even the *makers* of visualizations are an interesting group to be taken into account in the context of design and data. While the above questions can help designers frame their product ideas, the inherent

---

[3]One example: http://keshif.me/demo/VisTools; last accessed on Sept 5, 2015.

natural tendency towards an idealized persona can also be misleading. There is another way: data about the user's intentions, needs, expectations, and also behavior can help form a more empirical view of who we are designing for, what are the needs they have, and how people might anticipate future designs (Sprague and Tory 2012; Brehmer et al. 2014). This is an explorative use of gathered data that over time provides an incrementally more accurate view on the users of a design.

Such data supporting the design process in exploration, design, and also in validation, requires appropriate data sources that will elicit properly contextualized data about the behavior of users within the target group and potentially their needs and expectations. The former data can be derived nowadays from the instrumentation of early prototypes, and later, products, whereas the latter information can be observed or derived from questionnaires, interviews, and other qualitative user's research methods. The combination of such different kinds of data leads to data design tools that optimally support the now data-driven design process (Funk 2011).

Subsequently, this is related to other data-driven design approaches such as evidence-based design in the healthcare domain (Evans 2010; Codinhoto 2013) and statistical hypothesis testing such as "A/B testing". Furthermore, it also involves the more general split testing approaches (Fogg et al. 2001; Kohavi et al. 2009), which are however out of the scope of this chapter.

Example Case

Looking again at the example case, design information is necessary to make better decisions in a design space filled with sensible options and a wealth of different data sources available through the data interface. The interface shown in Fig. 7 was developed to give an overview of all body related data sources. Thus, it allows for easy browsing and selection, and also for comparing and annotating points in time with contextual information that would be helpful in later stages of the design process. This interface could be used from the data perspective (towards the context given by annotations) and vice versa, and from annotations to the underlying "hard data". Based on the platform of the design information, different directions for the user interface design can be explored. For instance, a more communicative and information-limited product concept next to an explorative variant targeting different kinds of end-users, can be explored.

Generalization of Design Information into Design Tools

Design information is an encapsulation of data or information enriched by contextual information, which is tailored towards a clear user's base. When we turn from the end-users of the designed products to the designers and makers as users, the data is certainly useful in the design process for ideation, conceptualization, designing, and also validating a design. While design information and its user interface can be very specific to a design project, it can also be generalized towards a reusable

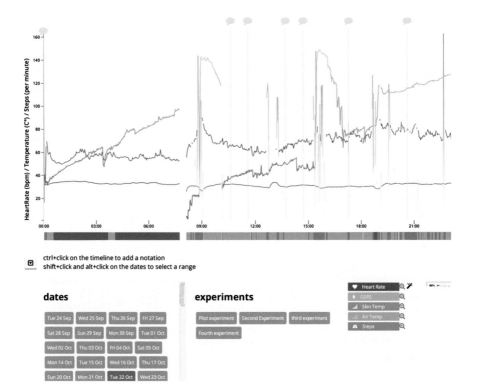

Fig. 7 Design information for Qualica, a rough data visualization providing versatile access to the processed data interactive control for selection, filtering, brushing and annotation

data "workbench" with easily accessible processing steps, data manipulation tools, and visualizations. Examples are interactive complex data visualizations such as process graphs and mapped data. Others include information services, professional dashboards, and other analytical business tools, i.e., domain-specific tools.

There is a tradition in crafts which states that craftsmen often need to create non-existent tools or adapt tools to their own practice or special use-cases. With data, designers need to embrace this thought as well. However, this though shows that data is complex and highly context-sensitive, which requires deep understanding and customized tools that can deal with such matter appropriately.

Design Product

Consequently, the *design product* that results from an elaborate design process is less, and at the same time, more than the design information. It is *less* in the sense that the final design usually reduces and condenses the given design information towards a clean and polished (visual) representation of the data that optimizes understandability, experience, and ease of use. It is *more* in the sense that the design contextualizes and roots the information in the user's and not the designer's reality. Semantic hints are taken into account in the presentation, and dynamic visual or physical presentations can adapt to the context of use. The designer might, for instance, use storytelling (Kosara and Mackinlay 2013) as a means to introduce the scope of the design, align the presented information with the reality of the users (Chuah and Roth 2003), and capture their attention and thoughts using a strong narrative.

Therefore, the designer capitalizes on the richness of the design information to optimize the design, and to iterate in cycles between the three layers as we will see in the next section.

Design Process

As we now know how to distinguish between the technical realm, *design infrastructure*, and the design space, the *design information* and the *design* itself, we turn towards the second view of the creation of information products. This view involves the actual process of designing with data. Two key points need attention. Firstly, having a layer of design information is essential for collaboration support, communication, fast progress, and stabilizing the design process at a later point. However, this needs to be established *fast*. Secondly, investing too much engineering efforts in an infrastructure or platform prematurely should be avoided because it will definitely change anyhow. Following this process is certainly a variant of an old engineering issue: achieving a good balance between generalization and structure, and flexibility and adaptability by improving a system without introducing too much technical debt.[4]

Consequently, the different phases of the data design process (as depicted in Fig. 8) will be explained in detail, starting with the bootstrapping phase.

Bootstrapping

Bootstrapping the design process can be as simple as assembling all people involved, sketching the challenge or brief, and getting a feeling for the different disciplines involved. A recommendation is prepared by letting everyone summarize their view on the project (why is the project relevant, interesting, and worthwhile?), their expertise (what brought them to the table?), and their input. Furthermore, everyone should be able to quickly grasp the challenges of the involved data. At the other end of the spectrum, they should consider what a potential design should look

---

[4]http://c2.com/cgi/wiki?TechnicalDebt, last accessed Dec 31, 2014.

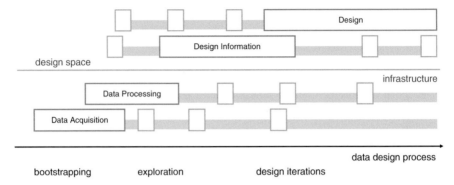

Fig. 8 Data design process with different layers of data in three phases: bootstrapping, exploration, and design iterations

like. On the data side, it is important to understand some initial aspects about the data such as: what data sources are available, how are the data generated and structured, which technology contains the data, and which tools could be of use for its analysis? And then again, what are our assumptions about the data? Are they factually accurate, potentially biased, or over-simplistic? It is important to be skeptical even if we do not consider all these questions as relevant, maybe our audience does?

In the process overview (cf. Fig. 8), the bootstrapping phase is dominated by data acquisition, analysis, and processing. However, this refers mainly to getting up to speed with the right tools and collaboratively developing a common communication channel – from data to design and back. It is the phase of divergence in unfolding all aspects and facets of data.

Example Case

Bootstrapping in the example case could mean that the bio-signal expert in the design team would prepare a short overview of possible sensors that could deliver data at a rate of one sample per 5 s without consuming too much power and being too bulky for a small wearable design. In parallel, the designer would find related research that connects sensor data to bodily phenomena such as arousal, relaxed state, or even sickness. In addition, the designer would search for examples and inspiration from related applications (cf. Fig. 9).

Exploration

When exploring data in the design, we often encounter a chicken-and-egg problem that unfolds. In a situation in which data is abundant, but not immediately accessible,

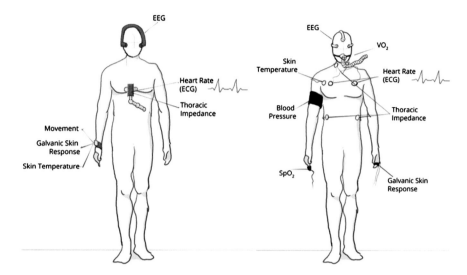

Fig. 9 Overview of data sources for Qualica

let alone, usable, we face the question, what can the data offer? At the same time, we need to ask, what does the design need? This problem unfolds further if more than one person is involved. Nevertheless, data and meta-data need to be shared, explained, critically evaluated, and discussed. In collaborative teams, such processes can easily stall if no means of sustained communication is in place.

Sampling: One strategy is to start with open exploration of the data itself. If it is not yet clear whether data can help us design or which aspects would be most relevant and useful, the process of starting with a data sample is a strategy to break out of this loop. Thus, this loop involves *seeding* a data exploration process. Such a seed is a small excerpt of data from a few (randomly chosen) data sources. It is so small that we can easily manage it, and at the same time, big enough that we can assess whether the chosen data sources are useful for further exploration. In a step-wise process, we can move through the available data sources and determine their value for the design challenge. This is not an easy task, but over time, a certain *smell for data* will develop and guide us towards more intuitive exploration of data. Once this step is done, we can move faster through the haystack.

Inspiration: Another strategy is to take the design challenge as a primer for exploring the data sources that could potentially inform the design process. An interesting approach is to make use of extensive inspiration material to identify similar or related designs that will in turn inform all design team members (especially the more technical ones) what the team might be looking for in the data. If you are stuck, change the visual paradigm or the visualization that might limit you.

Comparing these two exploration approaches, both have in common that they are used to break out of a potentially stalled process and aim at gaining momentum and moving forward with the design team. Paralysis by data is an unworthy thing to suffer from in design, but it happens all the time. As one moves along this process, it is worthwhile to document choices and decisions made. Therefore, the team can fall back on earlier thoughts and decisions when needed.

The next step is to formulate a short summary of what the data sources deliver, when, and which quality. Thus, it proceeds with constructing *design information* accordingly. Especially for information that is hard to get, privacy-relevant, protected, or rare, *design information* can mean in the first phases to collaboratively analyze the data sources and to build a mock-up of fake data that closely mimics the real data without revealing the truth.

Example Case

Exploration in the example case is based on the information about sensors within the device and a couple of open data APIs on the Internet. Different stakeholders such as business or domain experts could be consulted to get a better idea of a potential market fit. Together they document these starting points in a work book and conclude the session with a short sketch of a "landscape" of potential data sources, usages scenarios and connecting visualizations: the design space (see Fig. 10).

In the following exploration phase, the designer decided on the first approach to rely on the body signal data and link it to desktop application activity feeds from a desktop logging application. The sensor data is retrieved every 5 s. Therefore, by using the average value of a window of 40 s (~8 samples), a first good balance between reaction time and useful activity information was found. However, finding

Fig. 10 Sketch of the Qualica design space

patterns was difficult, as the sample size of the subjects generating the data was not large enough. Application usage data was derived and mapped to different types of applications and computer usage scenario such as work, entertainment, communication etc.

Based on this exploration phase, the design team decided to continue with a rough classification of activity as the reference context for visualization instead of pursuing unreliable correlations with body signal data.

Design Iterations

At the point of design iterations, once both infrastructure and design information have been established and explored, more traditional design processes can be weighed in an iterative process alternating between design information and the design. The remaining challenges are essentially about framing the design information in a way that best fits the target users and the desired user experience. Design information needs to be packaged and tested in the context of use by applying degrees of freedom for interaction with the presented information.

Example Case

Given the right interfaces to data and the insights from the exploration, the designer would start with a first functional prototype of the wearable interface. Consequently, a quick processing sketch was all the design team needed to get a better understanding of the visualizations. It is much faster than working with hardware and LEDs, although the design team could quickly move towards a prototype implementation with an Arduino board. Figure 11 shows the different

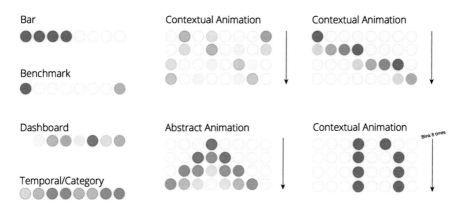

Fig. 11 Example case design iterations on the minimal visualizations of activity in context

visual patterns for a minimal visualization of application scenarios. The left column shows how a single row of multi-color LEDs could show static information such as a progress bar, a scale, and a categorical overview of activity. The middle and right columns show how animations could indicate a specific context or context absence.

The *design iterations* benefitted from quick access to design information, and if needed, changes to the data collection and processing functions. The process figure (cf. Fig. 8) shows how design information effectively separates the design from data collection and direct access to data sources, thereby preventing unnecessary complication in the overall process.

Conclusions

What we have seen in this chapter is not just an overview of the early steps when designing with data, there are stepping stones that we can use to guide our ways and not diverge too far from the original design goal. One of the most important lessons in this area is to trust in our gut feeling regarding data and the properties of a dataset, and at the same time, to question every single move. However, data and information are so abundant, but alien to us that we tend to forget the special positive traits and hidden pitfalls of the *materiality* of data. Another point is to communicate *meaning* and meta-data in the right way. Data gets meaning by contextualization, linking, and relating. This chapter emphasizes the utility of design information for *scoping* and *framing*, as data available for design can be overwhelming in quantity and quality. Especially nowadays, scoping and framing can be real problems in a data design. Thus, a clear set of design artifacts supported in the design process can help.

While this chapter focused mostly on *designing information* products as a category of more or less physicalized interactive designs that heavily rely on data and information, data design tools can be crafted in the same way with a stronger notion of reuse and generalization. The chapter was written also to balance the current emphasis on a "flat" graphical visualization of information. The potential design space around data is larger than that. It extends to physical products, apps, services, and systems that all carry the notion of communicating information to humans in an understandable, yet rich and expressive way that inspires action. Without design, this would be impossible.

Example Case

With the rich design information and some ideas in mind, the designer could start to dive into dynamic visualization of the incoming data. The design evolved naturally from simply signaling the different states in colors towards comparative views that would highlight differences between own state and the social network to the user. The design shown in Fig. 12 evolved from a few displayed values towards more

Fig. 12 Qualica; final
product form

contextual information. It adds text and icons from the social networks, and then goes back to reduced graphical "cues". The final design iteration was a minimal visualization consisting of several small dots, glowing and forming slow patterns over time. This concludes the early design phases, as the design was already at a stage where it could be evaluated with a few potential users.

Acknowledgements The example case that runs throughout this chapter was the final master project of Pepijn Fens (Fens 2014; Fens and Funk 2014), supervised in 2013/2014 by the author. Without this case, the chapter would have been much more difficult to read and understand. Thus, we are indeed very grateful for this contribution.

Further Reading

In the following, a few directions for further reading will be introduced briefly. There is a recent wealth of books on data visualization of which a few are presented here. First, pure data and information visualization is introduced well in books by Edward Tufte, an early advocate of presenting information to a reader effectively. You will find that he is (provokingly) outspoken against any kind of noise that masks data or information in their presentation. Also, he is quite close to design in his approach to understand and questioning needs to visualization from a data practitioner's point of view. A good starter is *Envisioning Information* (Tufte 1990), with *Visual Explanations: Images and Quantities*, and *Evidence and Narrative* (Tufte 1997) as a follow-up. More theoretical views on visualization are available from *Information Visualisation* (Spence 2001) and *Information Visualization: Perception for Design* (Ware 2012).

Second, there are recent books on *designing* visualizations and interaction with data, for instance, *The Functional Art: An introduction to information graphics and visualization* (Cairo 2012), *Now You See It* (Few 2009), or *Raw Data – Infographic Designers' Sketchbooks* (Heller and Landers 2014). The latter book does not only present finished designs, but looks behind the scenes and shows ways of working and translating data into masterful visualizations. Thus, a very practical guide is

Designing Data Visualizations (Iliinsky and Steele 2011). For an introduction into D3 as the currently most popular toolkit for visualization on the web, *Interactive Data Visualization for the Web* (Murray 2013) is highly recommended.

Other related disciplines, such as generative art can be inspiring as well. However, *Generative Gestaltung* (Groß et al. 2009) or *Design by Numbers* (Maeda 2001) are good starting points. A growing trend entails the use of visualization techniques in journalism and media. They are however, not the focus of this chapter. Unfortunately, little work is published so far for physical visualizations (Fens and Funk 2014). Also, multi-modal information products, which we also target in this chapter and an interesting list of physicalized visualization can be found here: http://dataphys.org/list/.

As for a bit more advanced reading on what is happening currently in the area of data visualization, there are three important conferences on visualization-related topics: IEEE VIS, ACM SIGGRAPH, and Visualized (non-academic) conference. There is also a relevant journal, ACM Transactions on Visualization and Computer Graphics, which publishes articles like *Mental Models, Visual Reasoning and Interaction in Information Visualization* (Liu and Stasko 2010), which are worth reading. For less academic and more practical data design resources, there is a lively community on Twitter and on different websites. Thus, be sure to check out http://www.datastori.es, www.visualization.org, and http://blog.visual.ly.

Collaboration in data visualization and interfaces mostly refers to the collaborative *use* of such interfaces and products, and not to their design or development. However, there are a few exceptions looking at what challenges research on collaborative visualization (design) has to tackle still. This is increasingly moving away from the collaborative *use* of visualization and visual analysis towards collaborative *design* (Heer et al. 2008; Isenberg et al. 2011).

A whole different area is demarked by literature on rationality, psychology, and statistics. *Everyday Irrationality* (Dawes 2001) is recommended for getting a general overview of how people experience information and interpret it (often in their favor or naively). To dive into statistics and behavioral psychology, there are again many sources to choose from e.g., *Becoming a Behavioral Science Researcher: A Guide to Producing Research That Matters* (Kline 2008) or *Straight Choices: The Psychology of Decision Making* (Newell et al. 2007).

References

Ackoff RL (1989) From data to wisdom. J Appl Syst Anal 16:3–9

Alexander C (1964) Notes on the synthesis of form. Harvard University Press, Cambridge

Bigelow A, Drucker S, Fisher D, Meyer M (2014) Reflections on how designers design with data. In: Proceedings of the 2014 international working conference on advanced visual interfaces – AVI'14. ACM Press, New York, pp 17–24

Blyth CR (1972) On simpson's paradox and the sure-thing principle. J Am Stat Assoc 67:364–366. doi:10.1080/01621459.1972.10482387

Borgman CL (2015) Big data, little data, no data: scholarship in the networked world. MIT Press, Cambridge, MA

Brehmer M, Carpendale S, Lee B, Tory M (2014) Pre-design empiricism for information visualization. In: Proceedings of the fifth workshop on beyond time errors novel evaluation methods for visualization – BELIV'14. ACM Press, New York, pp 147–151

Cairo A (2012) Functional art – infographics and visualization and exploration. Peachpit Press, Berkeley

Chuah MC, Roth SF (2003) Visualizing common ground, p 365. http://dl.acm.org/citation.cfm?id=939639

Codinhoto R (2013) Evidence and design: an investigation of the use of evidence in the design of healthcare environments. The University of Salford

Dawes RM (2001) Everyday irrationality: how pseudo-scientists, lunatics, and the rest of us systematically fail to think rationally. Westview Press, Boulder

Drucker J (2011) Humanities approaches to graphical display. Digit Humanit Q 5:1–23

Evans B (2010) Evidence-based design. Bringing world into cult. Comp Methodol Archit Art Des Sci 227–239

Feinberg M (1995) Basics of interlaboratory studies: the trends in the new ISO 5725 standard edition. Trends Anal Chem 14:450–457. doi:10.1016/0165-9936(95)93243-Z

Fens P (2014) Personal visualisation: a design research project on data visualisation and personal health. Eindhoven University of Technology

Fens P, Funk M (2014) Personal health data: visualization modalities and their perceived values. In: Skala V (ed) Proceedings of the 22nd international conference on centre European computer graphics visualization and computer visualization 2014. Plsen, Chech, pp 339–344

Few S (2009) Now you see it: simple visualization techniques for quantitative analysis. Analytics Press, Oakland

Fogg BJ, Marshall J, Kameda T et al (2001) Web credibility research: a method for online experiments and early study results. In: CHI'01 extended abstracts on human factors computing systems, pp 295–296. http://dl.acm.org/citation.cfm?id=634242

Funk M (2011) Model-driven design of self-observing products. Eindhoven University of Technology

Groß B, Laub J, Lazzeroni C, Bohnacker H (2009) Generative gestaltung. Schmidt Hermann Verlag, Mainz

Heer J, Shneiderman B (2012) Interactive dynamics for visual analysis. Queue 10:30. doi:10.1145/2133416.2146416

Heer J, Ham F, Carpendale S et al (2008) Information visualization. In: Creation and collaboration: engaging new audiences for information visualization, vol 4950, Lecture notes in computer science. Springer, Berlin/New York. doi:10.1007/978-3-540-70956-5, http://link.springer.com/chapter/10.1007/978-3-540-70956-5_5

Heller S, Landers R (2014) Raw data: infographic designers' sketchbooks. Thames & Hudson, Limited, London

Hullman J, Diakopoulos N (2011) Visualization rhetoric: framing effects in narrative visualization. IEEE Trans Vis Comput Graph 17:2231–2240. doi:10.1109/TVCG.2011.255

Iliinsky N, Steele J (2011) Designing data visualizations: representing informational relationships. O'Reilly Media, Inc, Sebastopol

Isenberg P, Elmqvist N, Scholtz J et al (2011) Collaborative visualization: definition, challenges, and research agenda. Inf Vis 10:310–326. doi:10.1177/1473871611412817

Kline RB (2008) Becoming a behavioral science researcher: a guide to producing research that matters. Guilford Press, New York

Kohavi R, Longbotham R, Sommerfield D, Henne RM (2009) Controlled experiments on the web: survey and practical guide. Data Min Knowl Discov 18:140–181. doi:10.1007/s10618-008-0114-1

Kosara R, Mackinlay J (2013) Storytelling: the next step for visualization. Comput (Long Beach Calif) 46:44–50. doi:10.1109/MC.2013.36

Kozlova K (2011) Visual histories of decision processes for creative collaboration. In: Proceedings of the 2011 annual conference on extended abstracts human factors in compututing systems – CHI EA'11. ACM Press, New York, p 1045

Liu Z, Stasko JT (2010) Mental models, visual reasoning and interaction in information visualization: a top-down perspective. IEEE Trans Vis Comput Graph 16:999–1008. doi:10.1109/TVCG.2010.177

Maeda J (2001) Design by numbers. MIT Press, Cambridge

Megens C, Peeters MMR, Funk M et al (2013) New craftsmanship in industrial design towards a transformation economy. European Academy of Design (EAD), Gothenburg

Moere AV, Purchase H (2011) On the role of design in information visualization. Inf Vis 10:356–371. doi:10.1177/1473871611415996

Murray S (2013) Interactive data visualization for the web. O'Reilly Media, Inc, Sebastopol

Newell BR, Lagnado DA, Shanks DR (2007) Straight choices: the psychology of decision making. Psychology Press, Hove

Quispel A, Maes A (2014) Would you prefer pie or cupcakes? Preferences for data visualization designs of professionals and laypeople in graphic design. J Vis Lang Comput 25:107–116. doi:10.1016/j.jvlc.2013.11.007

Spence R (2001) Information visualization, vol 1. Addison-Wesley, New York

Sprague D, Tory M (2012) Exploring how and why people use visualizations in casual contexts: modeling user goals and regulated motivations. Inf Vis 11:106–123. doi:10.1177/1473871611433710

Tufte ER (1990) Envisioning information, 4th edn. Graphics Press, Cheshire

Tufte ER (1997) Visual explanations: images and quantities, evidence and narrative. Graphics Press, Cheshire

Tuomi I (1999) Data is more than knowledge: implications of the reversed knowledge hierarchy for knowledge management and organizational memory. In: Proceedings of the 32nd annual Hawaii international conference on system sciences 1999. HICSS-32. Abstract CD-ROM full paper IEEE Computer society, p 12

Ware C (2012) Information visualization: perception for design. Morgan Kaufman, San Francisco

Decrypting the IT Needs of the Designer During the Creative Stages of the Design Process

Aggelos Liapis, Mieke Haesen, Julia Kantorovitch, and
Jesús Muñoz-Alcántara

Abstract Designers are frequently challenged by complex projects in which the problem space is unique, rapidly changing, and the information available is limited. In such cases, combining knowledge from different fields of expertise is required. Furthermore, collaboration during the design process is essential for achieving a meaningful and well-formed solution. Designers therefore regularly find themselves exchanging ideas and reflections in the form of emails, sketches, and images with a group of experts from different backgrounds, working altogether through the creation of a design, its development and proper implementation. This particular chapter focuses especially on issues of synchronous and asynchronous collaboration, team dynamics and the management and monitoring of the early stages of the design process. The overall aim is to identify the essential characteristics and needs of distributed teams when in remote collaboration during the early stages of the design process and to suggest a prototype environment based on the identified requirements and workflow.

A. Liapis (✉)
Intrasoft International, Markopoulou-Peania Avenue, Athens, Greece
e-mail: aggelos.liapis@intrasoft-intl.com; agliapis@gmail.com

M. Haesen
Hasselt University – tUL – iMinds, Expertise Centre for Digital Media, Wetenschapspark 2, 3590 Diepenbeek, Belgium
e-mail: mieke.haesen@uhasselt.be

J. Kantorovitch
VTT-Technical Research Center, Espoo, Finland
e-mail: julia.kantorovitch@vtt.fi

J. Muñoz-Alcántara
Department of Industrial Design, Eindhoven University of Technology, Den Dolech, The Netherlands
e-mail: J.munoz.alcantara@tue.nl

© Springer International Publishing Switzerland 2016   379
P. Markopoulos et al. (eds.), *Collaboration in Creative Design*,
DOI 10.1007/978-3-319-29155-0_18

Introduction

The working activities of design teams are supported by collaboration tools on many levels – communication, sharing documents, exchanging images, transferring media files, organising tasks, time-tracking, managing the progress of a project and "monitoring" work of team mates. The technology itself creates opportunities and imposes limitations and constraints while trying to mirror, mediate or augment the natural communication between team members. However, it can be the case that digital tools designed without attention to end-user requirements can even hinder designers rather than supporting the goal of the project at hand (Liapis 2008; Liapis et al. 2014; Malins et al. 2014). The biggest challenge for collaboration technology lies in the fact that design is a social process. It is crucial to establish a common base of understanding of the problem i.e. secure a mutual definition and acceptance of the context during the problem exploration and the paths to follow for solving this problem (Liapis 2014). Achieving a common understanding among all the team members would be expected to be easier when using video conference platforms, online shared repositories and other opportunities of digital communication (Ozcelik et al. 2011). Nevertheless, collaboration is not only about sharing information. To understand the significance of this specific technology, we should situate it in the context of the social field in which it is used (Exploring user requirements through mind mapping; Ozcelik et al. 2011). We should understand the social practices, the power relations and the points of tension between designers and other stakeholders (Exploring user requirements through mind mapping). A project might include interactions within the same organisation, between different organisations and even with mass communities of collaborators. Furthermore, these interactions could happen at geographical spaces of significantly different conditions – starting from the same room, the same floor, same building, same city, same or different country, and even across different time zones (Martens 2012).

This chapter explores the nature of early design work, and focuses especially on issues of collaboration and document sharing needs, aiming at identifying what are the essential characteristics of distributed creative collaboration within design teams. The reported research combines several research methods, such as: analysis of existing tools from a user perspective; surveys and interviews of designers regarding early design work, the tools and methods they use, with an emphasis on collaboration; case study analysis of eight (8) design projects conducted in the UK and Greece; an experimental evaluation of a tool to support early design.

Experiential Design Sessions

Here we describe some state of the art tools currently used by designers during the initial stage of the design process. During each experiential session, the evaluation of a specific tool was done with the goal to identify points of improvement. The findings from this section should go beyond usability evaluation and ultimately should be incorporated into the list of requirements for a prototype environment.

Mind Mapping and Brainstorming Tools

Mind mapping and brainstorming tools are used to create diagrams of relationships between concepts, ideas or other pieces of information. Their popular uses include project planning, collecting and organising thoughts; brainstorming and presentations – all in order to help solve problems, map out resources and uncover new ideas. Some of the properties of mind maps that have to be supported in order to attest the effectiveness of mind mapping include (Martens 2012):

- Keyword Orientation: the structural elements of mind maps are not sentences but keywords.
- Loose Syntax and Semantics: association is the only relationship between linked keywords.
- High-Level View: overview of a whole mind map in a glance.
- Evocative: a mind map evokes the context of the scene in which it was created.
- Semi-structured: a mind map can have a template structure but it can grow branches on demand to capture real-time verbal communication in semi-structured interview.

There are various tools available for providing mind mapping functionalities, most of which are web applications, making it even easier to use them anywhere from any web browser (Kung and Solvberg 1986). Based on a wide search on the available tools, it was concluded that the most prominent tools in this category to be potentially used by product designers within the prototype collaboration environment are FreeMind, Coggle and WiseMapping.

FreeMind: is a premier free mind-mapping software written in Java. The recent development has hopefully turned it into a high productivity tool. The operation and navigation of FreeMind is faster than that of commercial products such as MindManager because of one-click "fold/unfold" and "follow link" operations.

Coggle: introduces a new way of brainstorming and storing knowledge claiming that it provides users with a space for thoughts that works the way people do by avoiding the rigid ways of computers. The application is free, there are appropriate security mechanisms for data privacy allowing users to maintain control over the files they share with other collaborators.

WiseMapping: is a free web-based mind-mapping editor for individuals and businesses. The application is open source licensed under WiseMapping Public License Version 1.0 (WPL) and can be downloaded and installed on a local server. Apart from the editor the application provides users with a series of collaborative features and export/import functions allowing them to transfer their mind-maps to other commercial or free applications.

Sketching and Storyboarding Tools

With storyboarding, the designer turns ideas and goals of a projected user experience into something visual. In this way ideas are easier for other people to understand, and to give constructive feedback. Ideas are brought to life with storyboard shapes, text, animations, and all the other features that PowerPoint Storyboarding provides (Quevedo Fernandez et al. 2013). The following properties have to be supported by storyboarding tools:

- Editing and Re-use: the designer must often redraw features that have not changed. In order to avoid such repetitions, a manual translation to an electronic format is required;
- Design Memory: the sketches may be annotated, but a designer cannot easily search these annotations in the future to find out why a particular design decision was made. Practicing designers have found that the annotations of design sketches serve as a diary of the design process, which are often more valuable to the client than the sketches themselves.
- Interactivity: interaction between the concept and the user has to be supported. In order to actually see what the interaction would be like, a designer needs to "play with the computer" and manipulate several sketches in response to a user's verbalised actions. Designers need tools that give them the freedom to quickly sketch rough design ideas and to test the designs by interacting with them.

The old "Grid paper & Pen": Sometimes less is more and substituting grid paper and pen maybe harder than we software engineers thought. Sketching ideas on paper allows us to quickly visualize and play with different approaches to content structures, interface layout ideas and interactions

Sketch "The designers Toolbox": Sketch is an interface design app that is gaining popularity as a purpose-built alternative to tools like Photoshop and Illustrator. It's intended specifically for designing beautiful user interfaces for a range of devices and screen resolutions. Sketch is simple and quick to learn, and has a number of features that make it a joy to work with including built-in grid controls, linked styles and smart measuring guides.

Articulate Storyline: is an extremely user friendly tool to show the general structure of your projects, making it very easy to identify scenes, information flow and different relationships among pages (screens). In addition, it allows users to use annotated screenshots to communicate their ideas more effectively.

Storyboards 3D: is an extremely user friendly tool for quickly drafting and presenting your ideas. Users can position and rotate 3D characters and objects in all directions, include text blocks and speech bubbles, insert photos in every shot or scene, add notes and even record audio.

Conceptual Modelling Tools

The term conceptual model may be used to refer to models which are formed after a conceptualisation process in the mind. Conceptual models represent human intentions or semantics. Conceptualisation from observation of physical existence and conceptual modelling are the necessary means humans employ to think and solve problems. Concepts are used to convey semantics during various natural languages based communication (Martens 2012). Since a concept might map to multiple semantics by itself, an explicit formalisation is usually required for identifying and locating the intended semantic from several candidates to avoid misunderstandings and confusions in conceptual models. Within the prototype environment, conceptual modelling tools are envisaged to be used from various perspectives, serving a range of applications, from conceptual modelling of interactions within a process up to conceptual modelling of the designed ontologies.

Autodesk Inventor: enable users to create realistic representations of their designs. Clear visuals make it easier for stakeholders and customers without expert experience to understand engineering drawings and designs. The application allows users to quickly create photorealistic CAD renderings and animations that can convey ideas to managers, explain designs to manufacturers, and persuade customers that you have the best solution for their needs.

SOLIDWORKS: this platform delivers a new design experience focused on enabling users to create innovative products in a connected and truly collaborative environment. SOLIDWORKS® Conceptual Design and SOLIDWORKS Industrial Design solutions help users to easily develop, review, and select mechanical and stylized concepts before committing to detailed design and manufacturing. CAD and non-CAD users alike, including executives, design team leaders, and project managers can share information, participate in the design process, and easily aggregate data from any source to help make design decisions faster, from anywhere, on any device.

Web Survey: Collaboration Practice

In order to obtain general insights into design practices, designers' tool preferences and the settings in which they collaborate with other designers and team members, a web survey was conducted. This web survey consisted of 32 questions and was launched online in December 2013. Respondents were invited through different international channels (i.e. mailing lists, social platforms) for design practitioners, schools of design, human-computer interaction communities, and web design practitioners.

Results

The survey resulted in 82 responses; 32 female and 50 male respondents with an age ranging from 21 to 56 or older participated in the survey. The respondents were located in 16 different countries: 58 respondents currently reside in Europe, and the others reside in the United States, Canada, Australia and Asia. The respondents are involved in design activities and/or management in a wide range of design projects concerning visual/graphic design, industrial design, product design, vehicle design, user interface design, user experience design and interaction design.

When asking about the approaches to exchange information in design projects, several approaches were mentioned, including real time face-to-face meetings, the use of a war room/design studio, video conference, teleconference, chat / instant messaging / social media, e-mail, wiki / blog and cloud-based documentation sharing services. E-mail is the approach which is used most of the time or often for information exchange by the majority of the respondents. Next, real time face-to-face meetings and cloud based documentation are also frequently used by the majority of the respondents.

For some questions, three different situations were distinguished: (1) the individual creation of artefacts or documentation, (2) the collaborative creation of artefacts or documentation and (3) the use of artefacts or documentation to inform team members and other people involved in the design process. One of the questions asked what type of artefacts or documents they use in each of the three situations. The respondents use a plethora of documents and artefacts, including user/usability requirements, mind maps, scenarios, storyboards, sketches, presentations and reports. Surprisingly, we see only little difference between the use of artefacts and documents in each of the the three situations. A second question inquired the respondents about the use of media and devices in each of the three situations. Media and devices used the most, include pen and paper, PC and whiteboard/flipchart, while e.g. smartphone/tablet and camera are used less often by the respondents. Similarly to the use of documents and artefacts, there is only little difference between the three situations in the use of media and devices.

Furthermore, the survey inquired into the problems experienced in collaboration and communication. In the case of creating designs collaboratively, the problems that are reported about, concern communication problems and almost an equal number of technical problems. In the case of informing team members and other people involved in the design project, there are less technical problems while most of the problems that occur are communication problems. An open question to the respondents was regarding the specific problems they have to deal with in their projects. Communication problems that were mentioned include *"the status of progress"*, *"remote communication"* and *"misinterpreted design"*, while the technical problems specified include: *"versioning/tracking changes"*, *"difficult to 'create' and 'brainstorm' while in separate locations"* and *"compatibility"*.

Additional Insights

We interviewed two respondents during semi-structured interviews to obtain additional insights in their answers on the web survey. Both interviewees confirmed that they often use e-mail. They both explained that the use of the approaches depends on the type of tasks that are conducted at each moment in the project. One designer mentioned that if possible, he prefers face-to-face communication to discuss design ideas and decisions, while design elements are shared with team members by e-mail or cloud-based document sharing. The other designer explained that they often collaborate and communicate remotely. This infers that the people of his team often have to get used to other tools than the tools used in face-to-face communication.

Considering the difficulties that occur during collaborative design, one of the interviewees mentioned that technical problems are usually easier to solve than communication problems. Exactly the risk of having communication problems forces this designer to have as much face-to-face communication as possible.

At the end of the conversations with the interviewees, we inquired about their expectations of a collaborative design tool. One designer admitted that support for communication within design teams is very ambitious, while the other designer emphasizes that he is in favor of, adding support for collaboration and communication within design teams to existing design tools.

In conclusion, the biggest challenges supporting designers in an optimal and intelligent way, depends on the situation in which they are working at a specific moment (e.g. individual creation of designs vs. remote collaboration). Avoiding problems by supporting the communication within the design team seems to be an important requirement.

Second Web Survey: Collaboration Tools

A second web survey was conducted with the aim to explore any existing relationships between the different design activities that occur during the early stages of the design process, and the tools used to support those activities. The survey focused on examining what were the most frequently and popular tools that supported designers in their collaboration tasks, and the problems and challenges that a distributed collaborative design team faces while using these tools. The online survey was distributed through social networks (LinkedIn, Facebook) among 12 design-related groups in the Netherlands. 59 people (mostly indicating an age between 30 and 40 years) responded during the first two weeks of March, 2014. This response rate corresponds to 0.0015 % of the total number of members (40,632 people) that were reached through the online design groups. Out of the 59 respondents of the web survey, 42 are currently based in the Netherlands, 4 are based in the Czech Republic, 3 in Finland, 2 in Belgium and the rest reported to be based in Australia, Austria, China, Ireland, Italy, Mexico, Thailand and the United States. Having the possibility to select multiple options, most of the respondents indicated being currently active in the disciplines of user-experience design (32,2 %), interaction design (27,1 %), visual/graphic design (27,1 %), industrial design (27,1 %), web design (11,9 %), or interior design (11,9 %). The size of the latest team in which the respondents had cooperated (including themselves) was 3–4 people in 44,1 % cases, 5–9 people in 28,8 % cases, 2 people in 11,9 % cases, and a single person (i.e., only the respondent) in 10,2 % cases. The rest worked in teams with more than 9 members. Respondents indicated that they collaborate with the team members located in a different time zone in 15,4 % of the cases, whereas 32,7 % responded that they collaborate with the rest of the team within the same room, 17,3 % within the same floor, 9,6 % within the same building, 15,4 % within the same city, 28,9 % within the same country, and 7,7 % within the same time zone.

A first look at the results reveals that overall, 98,3 % respondents use the computer at work and 94,9 % use Internet for activities related to work, both on a daily basis. These numbers indicate that the Internet through its diverse online services actively supports the work of designers. A closer look on the topic of collaboration showed that 82,7 % of the respondents had used several online collaboration tools to support the work of their team in the past 6 months. We asked respondents of the survey to explain the reasons why they liked or disliked using online collaboration tools. The positive responses included mostly properties and functionality such as: easy to use, fast, simultaneous editing, multi-platform / multi device, accessible from anywhere, drag-n-drop functionality, chat, history tracking, sharing, real-time, everything at one place, comments, alerts / notifications.

On the other hand, negative responses mentioned the following annoying properties or insufficient functionality when using online collaboration tools: complicated setup of user account required, being unstable, slow, confusing, buggy, non-intuitive, creating conflicting copies, complicated user interface, mutual disturbance during simultaneous document editing. Additionally, we explored what was the relation between mobile services and team collaboration. The mobile adoption in our survey uncovered that 91,5 % respondents owns a smartphone (44,1 % iPhone, 40,7 % Android phone, 6,8 % Windows Phone), and 52,5 % respondents own a tablet (37,3 % iPad, 13,6 % Android tablet). 91,4 % respondents use a smart phone or a tablet for their work (35,5 % on a daily basis, 25,9 % a couple of times a week).

Interviews with Professional Designers

From the total number of participants of the survey, 25 respondents indicated that they wanted to participate further in the research, and provided their contact information. We selected 15 participants from the web survey, based on their design specialisation, trying to cover all the different expertise that was available. In the end only 9 designers were successfully interviewed covering the following types of specialisation: UI Designer, Industrial / Product Designer, student of Industrial Design, Visual and Graphic Designer, Interaction Designer, User Experience

Designer, Game Designer, Industrial and UX Designer. The results from the series of interviews indicate that brainstorming is done most often together with other team members or stakeholders, typically in one room, and on paper. Pinterest and/or a shared folder with interesting materials may serve as a source of inspiration. Usually, no special tools are used in the stages of creative design or idea generation; however, when tools are used, these include: UX Pin, Axure, Indigo Studio, Adobe Ideas, Mind mapping tool of Google Drive, and sketching tools such as Paper by 53.

Designers use most frequently Dropbox and Google Drive for sharing documents and various multimedia files. The Intranet shared repository is used frequently too, especially in case of confidential projects that cannot be shared on available public services. Pinterest is often used for sharing images, especially for inspiration (library of characters, scenes, UI elements, etc.), while code repositories like SVN, Bitbucket, or SourceTree are also used. Finally, files or URLs of the files are usually shared via email. Tasks are typically divided during group meetings, and one person (project manager / team leader) is usually responsible for the final decision at every process stage. Tools used for assigning tasks and tracking the process, include: Trello, Teambox, or Excel. Face-to-face meetings with other team members do not need to take place on a daily or weekly basis, sometimes they occur as rarely as on a monthly basis. However, organising time is done often individually, and is closely related to the task organisation. Some companies use Gantt charts (using online tools, such as Redmine), or shared calendars to keep track of the availability and the progress of the others. Companies use various tools for communication, mostly Skype, WhatsApp, and Facebook (groups) due to their advantage of real-time communication and immediate response. For non-real-time communication, email is a standard way of sending documents to other team members. Text messages or phone calls are preferred in cases of emergency. Communication with clients is done mostly via email and Skype, and often includes URLs of interactive prototypes or other deliverables that can be commented directly (in cases where the tool allows it – e.g., comments in Google Drive documents, etc.), or indirectly (taking a screenshot and writing comments on it, etc.). Clients typically do not communicate with designers on a daily basis – communication is usually mediated by the team leader or project manager, who filters, discusses, and/or prioritises the information before passing it to the rest of the team. Face to face contact is mostly preferred.

Problems are also often about tools that cannot be installed or accessed due to internal company network security policies, and tools that are not intuitive enough for the team members or clients to easily embrace. These cases are often solved in very inconvenient ways, such as printing out the documents, writing comments on them, scanning them again and sending them back to the other side, instead of using some direct, online commenting functionality. Another example of a problem is the involuntary creation of conflicting copies, when people work simultaneously on the same file. Keeping time-schedules, tasks, and project progress up-to-date is yet another common problem. Last but not least, using communication tools that include contacts that are not related to the work environment is often distracting.

User Requirements for a Collaborative Design Environment

This survey revealed some interesting insights regarding designers and the way they collaborate and communicate with team members and other people involved in design projects. At the end of the analysis of this study, we identified a list of opportunities that should be brought into discussion while formulating the list of user requirements regarding a collaboration environment:

1. Carefully consider which (collaborative) tasks are conducted by the design team and which stages in the design project can be potentially supported by features of technology.
2. There is the need for a tool that tracks all design decisions, the progress of the project etc.
3. It is important for a design platform to consider the difference between design tools and communication tools. It is likely that the best choice will be including a set of (existing) design tools balanced by supporting common communication channels.
4. The biggest challenges for a collaboration platform are not the technical issues, but the communication issues. By considering suitable and intelligent ways to support communication within the design team, the design environment can contribute to improved collaboration efficiency within design teams.

Surprisingly, the results from the survey clarified that social media sites were the most frequently referred tools for supporting collaboration. Tools that supported instant communication and file sharing were the second and third most referred collaboration technologies. Online editing of documents and integrated collaboration platforms appeared to be widely used. Finally, email and task tracking (project management) services were other recurrent tools. While each cluster of tools provides different core functionalities, (e.g. social media vs. instant communication), some of them have similar if not overlapping features. For instance, Facebook provides individual and group instant messaging. Another example is Google Docs, which permits the online editing of documents, and includes the functionality to chat with all the individuals that are editing the document at the same moment. A more detailed discussion about the specific tools should be centered in the context and the activities in which they are used. In extraordinary cases, users may even use one tool to accomplish a completely different task than the one the tool is intended for.

COnCEPT Collaboration Environment

Based on the results of the survey, COnCEPT, a platform to support collaboration in early design, is proposed (see Fig. 1). The COnCEPT framework identifies the high-level components of the system, and the relationships between them. Its purpose is

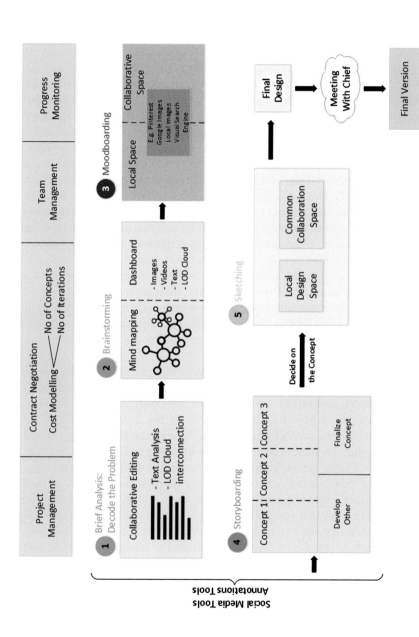

Fig 1 High level COnCEPT architecture

to direct attention at an appropriate abstraction of the system without delving into details (Liapis 2008; Liapis et al. 2014; Martens 2012; Exploring user requirements through mind mapping).

The design and deployment of the COnCEPT architecture is targeted at the support of a set of functionalities, based on the designers' needs identified from the analysis of the surveys and interviews. The phases for the deployment of design projects include the creation of new projects (describing essential project components such as the design brief, the client assigning the project, the cost model to be followed, details regarding the contract, project team members' selection, role and tasks assignment), as well as the management of existing ones, the Brief Analysis of the design combined with the initial brainstorming phase (realization of mindmaps as well as adding post-it-notes), the collection of a set of ideas and materials in a MoodBoard, the generation of a StoryBoard incorporating a set of design concepts, taking into account existing material and assisting the selection process of the desired concept, the Sketching of the concept, and finally the presentation of the final product to the client. Within all phases, proper logging of the activities of the team members and support for addition of annotations, when applicable are addressed. Upon finalization of each phase of the design process, the produced outcomes may be communicated to the client for receiving feedback, offering the option for comments as well as providing confirmation functionality for advancing to the next phase. While loosely coupled, tools are seamlessly integrated and unified by a dynamic semantic-based content management system, which allow the resulted outputs being propagated across different phases and can be utilised working on the project depends on the task at hand. This can be seen as a most compelling feature of the proposed platform.

In addition to the survey on the tools and interaction practices used by design teams, a research has been performed to understand better how professional designer approach to the management of design content. Designers have been asked to provide more insights on how they usually search for the content to facilitate the conceptualization of new product (e.g. using keywords, natural language, search by image), what are the sources of information used to generate new ideas (e.g. local company databases, personal collections, internet and particular internet sites) and the nature of the content to get the inspiration for their design activities used the most (e.g. images, videos, textual documents, etc.). A detailed analysis of the provided input revealed, that conceptual design is a knowledge-intensive process and designers often relay on resources available from local company's databases such as documents or sketches produced in the course of previous designs as well as on various external information sources. The external resources may include electronic books, images, music, online design journals and image collections such as getty, flickr, co.design, yatzer, designboom, designobserver, pinterest. Moreover general purpose search engines such as one provided by Google are mentioned as a daily source of information, independent on the nature of the product under design. From content management point of view, the understanding of vocabularies used by other team members is the technical limitation which is most experienced by the design team members.

Finding the relevant information from this large and distributed data space is usually both tedious and time consuming, and tends to become a challenge as the amount of information increases. This calls for advanced tools that are able to provide elegant mechanisms to organise various heterogeneous content while linking them with other related resources and concepts. Semantic technologies are proved to be beneficial enhancing the Information Retrieval or the ability to perform effective customised searches exploring the knowledge contained within annotations in order to access the heterogeneous content (Hollink 2006; Kobilarov et al. 2009; Carbone et al. 2010). Annotations with well-defined semantics provide the common vocabularies and ensure the interoperability of available information supporting knowledge sharing and collaboration across design teams.

Therefore such services as semantic annotation, search and recommendation which constitute the intelligence of the proposed collaboration environment are considered crucial to support the functionality derived from the user requirements. The annotation of content allows identifying relevant entities from content items. Semantic annotations enable the creation of semantically enriched metadata for the content produced by designers working on the new product concept or content selected doing web browsing activities in the discovery phase of the product design. Enhanced content metadata facilitates the customised task-specific search over available resources in the local project databases and in the Web. Once the entities are identified they can be automatically linked, for example, to open Linked Data (http://linkeddata.org/) sources on the web. As for the search functions, several approaches are considered, one is semantic search and another, in addition, benefits the intelligence of existing widely used search engines such as Google search as well as APIs provided by various popular internet sites used by designers (e.g. Getty, Flickr, museum, etc.). Semantic search, which is a search performed over RDF-JSON metadata, is used searching over material in the project space or local databases. In this case the objective is to understand the purpose of the search, so the keywords provided by user or derived from the analysing of the documents (e.g. design-brief, sketches, mind maps) are enriched with semantics (thanks to the information extraction tools) and compared with semantic metadata content descriptions. The results of both searches are aggregated in a single user interface for the designers' convenience. For the recommendation i.e. personalised search, the profiles of users are used as an additional parameter to rank search results (e.g. search from particular web sites/sources of inspiration).

Conclusion

This paper has reported on the results of the survey which uncovered some interesting insights regarding designers and the way they manage the design material and support tools, collaborate, and communicate with team members and other people involved in design projects. Then, based on the designers' needs identified from the analysis of the surveys and interviews, the COnCEPT architecture and functions to support the collaborative work of designers are defined. COnCEPT constructed as a standalone web-based platform, aims at integration of numerous examined set of collaborative tools for mind mapping and brainstorming, sketching and storyboarding tools as well as conceptual modeling tools which will allow professional designers to collaborate during the early stages of the design process. Such a large scale system contains many fine points that first need to be fully optimized in order to maximize its benefit to the creative industries, especially focusing to product designers. Specific care must also be given to the cultural and ethical issues arising when confidentiality, trust, security and IPR are involved, in order to assist professional designers to maintain awareness while promoting creativity.

While various commercial mostly stand-alone tools and open source software applications and platforms, such as the IBM- and Google-branded products, Live-scribe pen, eDrawings Professional, Matrix10, and others are available (The list of collaborative software), we still lack in collaboration tools, focused in serving the conceptual/creative stages and decision making techniques of the design process capable of providing both knowledge management and decision making techniques with the possibility of product design evaluation and the general management of various phases of design project.

Acknowledgements This work has been partially funded by the EC under the 7th Framework Programme, (ICT-2013.8.1: Technologies and scientific foundations in the field of creativity under grant agreement number FP7-ICT-2013-10 – 610725- COnCEPT COllaborative CrEative design PlaTform. The surveys have been implemented by the University of Hasselt and Eindhoven University of Technology, respectively.

References

Carbone F et al (2010) Enterprise 2.0 and semantic technologies for open innovation support. Trends in applied intelligent systems. Lect Notes Comput Sci 6097(2010):18–27

Exploring user requirements through mind mapping. http://www.change-vision.com/en/ExploringUserRequirementsThroughMindMapping_Letter.pdf

Hollink L (2006) Semantic annotation for retrieval of visual resources.PhD thesis, Vrije Universiteit Amsterdam

Kobilarov G et al (2009) Media meets semantic web – how the BBC uses DBpedia and linked data to make connections. The semantic web: research and applications. Lect Notes Comput Sci 5554(2009):723–737

Kung CH, Solvberg A (1986) Activity modeling and behavior modeling. In Ollie T, Sol H, Verrjin-Stuart A (eds) Proceedings of the IFIP WG 8.1 working conference on comparative review of information systems design methodologies: improving the practice. North-Holland, Amsterdam, pp 145–171

Liapis A (2008) Synergy: a prototype collaborative environment to support the conceptual stages of the design process. International conference on digital interactive media in entertainment and arts, submitted in DIMEA 2008, Athens, Greece, ACM Digital Library

Liapis A (2014) Computer mediated collaborative design environments: methods and frameworks to integrate creative tools to support the early stages of the design process. LAMBERT Academic Publishing, Germany. ISBN: 978-3-8465-0699-8, 2014.

Liapis A, Kantorovitch J, Malins J, Zafeiropoulos A, Haesen M, Gutierrez M, Funk M, Alcamtara J, Moore JP, Maciver F (2014) COnCEPT: developing intelligent information systems to support collaborative working across design teams. 9th international joint conference on software technologies, ICSOFT 2014, Vienna, Austria, 29–31 August, 2014

Malins J, Liapis A, Markopoulos P, Laing R, Coninx K, Kantorovitch J, Didaskalou A, Maciver F (2014) Supporting the early stages of the product design process: using an integrated collaborative environment. 6th international conference on engineering and product design education conference, EPDE 2014, University of Twente, Enschede, The Netherlands, 4–5 September 2014

Martens, JB (2012) Statistics from an HCI perspective: Illmo – Interactive Log Likelihood Modeling. In: Tortora G, Levialdi S, Tucci M (eds) Proceedings of the international working conference on advanced visual interfaces (AVI'12). ACM, New York, NY, USA, pp 382–385

Ozcelik D, Quevedo-Fernandez J, Thalen J, Terken J (2011) Engaging users in the early phases of the design process: attitudes, concerns and challenges from industrial practice. In: Proceedings of the DPPI'11

Quevedo Fernandez J, Ozcelik D, Martens JBOS (2013) A user-centered-design perspective on systems to support co-located design collaboration. In: Proceedings of the HCI'13

The list of collaborative software. https://en.wikipedia.org/wiki/List_of_collaborative_software. Accessed 8 Aug 2015